RECONSTRUCTION AND RESTORATION OF ARCHITECTURAL HERITAGE

PROCEEDINGS OF THE 2ND INTERNATIONAL CONFERENCE ON RECONSTRUCTION AND RENOVATION OF THE ARCHITECTURAL HERITAGE (RRAH 2020), 25-28 MARCH 2020, SAINT PETERSBURG, RUSSIA

Reconstruction and Restoration of Architectural Heritage

Editors

Sergey Sementsov

Saint Petersburg State University of Architecture and Civil Engineering, Saint Petersburg, Russia

Alexander Leontyev

Committee for the State Preservation of Historical and Cultural Monuments (KGIOP), Saint Petersburg State University of Architecture and Civil Engineering, Saint Petersburg, Russia

Santiago Huerta

Universidad Politécnica de Madrid, Madrid, Spain

Ignacio Menéndez Pidal de Navascués

Universidad Politécnica de Madrid, Madrid, Spain

CRC Press
Taylor & Francis Group
Boca Raton London New York

CRC Press is an imprint of the
Taylor & Francis Group, an **Informa** business

A BALKEMA BOOK

CRC Press/Balkema is an imprint of the Taylor & Francis Group, an informa business

© 2020 Taylor & Francis Group, London, UK

Typeset by Integra Software Services Pvt. Ltd., Pondicherry, India

Library of Congress Cataloging-in-Publication Data

Applied for

Published by: CRC Press/Balkema
    Schipholweg 107C, 2316XC Leiden, The Netherlands
    e-mail: Pub.NL@taylorandfrancis.com
    www.routledge.com – www.taylorandfrancis.com

ISBN: 978-0-367-65357-6 (Hbk)
ISBN: 978-1-003-12909-7 (eBook)
DOI: 10.1201/9781003129097
https://doi.org/10.1201/9781003129097

Reconstruction and Restoration of Architectural Heritage – Sementsov et al (eds)
© 2020 Taylor & Francis Group, London, ISBN 978-0-367-65357-6

Table of contents

Preface   ix

Organizers   xi

Organizing committee   xiii

Members of the organizing committee   xv

Sponsors & Co-organizers   xvii

Problems of historical and architectural, historical and urban planning research and restoration of monuments

An unnamed building of Bishop's house of Vologda according to newly revealed archival information   3
I.K. Beloyarskaya

Principles of embankment humanization in gray belt of Saint Petersburg   10
N.M. Drizhapolova, V.M. Supranovich & N.K. Yass

Laws of definition of subject of protection of houses and palaces of culture built in Leningrad in 1920-1930s   15
N.P. Dubrovina

Urban planning and compositional features of different territories of historical center of Saint Petersburg   20
G.O. Fedotova

Problems of adaptation of cultural heritage sites for people with limited mobility   24
T.S. Fridman

Integrated functional association — search for methods of cultural heritage sites adaptation (Russia)   30
L. Kaloshina

Specifics of forming a social and business complex in Kronstadt   37
O. Kokorina, D.A. Zinenkov, A.F. Eremeeva & S.A. Bolotin

Formation features of high nobility estates in neighboring districts of Saint Petersburg governorate on example of Peterhof district   42
E.A. Kozyreva

Architect J.-B. A. Le Blond and his role in construction of Grand Peterhof Palace   47
A.G. Leontyev

Fundamental differences of ensemble system of historical center of Saint Petersburg from European city centers 52
S.O. Markushev

Historic urban landscape in Saint Petersburg: To question of determining attributes of outstanding universal value of world heritage objects 56
N.V. Marushina

Saavedra's contribution to modern methodology of study of Roman Roads in Spain 61
I.F. Menendez Pidal de Navascues, M. Almagro-Gorbea, I. Moreno Gallo, I. Aguilar Civera, J. Mañas Martínez, J. Alonso Trigueros & A. Arcos Álvarez

Historical cemeteries as element of urban planning subject of protection of Saint Petersburg 67
A.V. Mikhailov

Territorial-planning constituents of urban planning components to be protected 72
A.V. Mikhaylov

Analysis of research on Bakhchisaray Palace 76
Z. Nagaeva & L. Budzhurova

City image in protection system of historical environment 81
A.Yu. Nazarova

Classification of ordinary development of Saint Petersburg in system of urban development regulation 85
I.L. Pasechnik

Renovation of coastal industrial areas. European experience of 2000s 91
O. Pastukh & A.G. Vaitens

Historical and architectural heritage of railway complexes in Belgorod region 95
M.V. Perkova, L.I. Kolesnikova & Y.A. Nemtseva

Principles of urban environment development based on identification features of historical city 99
F. Perov, O. Kokorina, U. Puharenko & Yu.N. Lobanov

"Garden cities" along railway lines in Saint Petersburg 104
N. Petukhova

Revitalization of architectural landscapes of Alexander and Babolovo parks of "Tsarskoe Selo" Museum-Reserve in 2014 - 2019 110
M.N. Ryadova

Real historical skyline of metropolitan Saint Petersburg by 1917 116
S.V. Sementsov

Saint Petersburg style of 21st century 122
M.A. Shapchenko

Use of cork coating in architectural monuments of Saint Petersburg 125
P.G. Shchedrin

Specific features of manors of nobility in remote uyezds of Saint Petersburg Governorate (through example of Luzhsky Uyezd) 130
E.Yu. Shuvaeva

Influence of architectural fantasies of Yakov Chernikhov on projects of Zaha Hadid   134
A.A. Smirnov & O.A. Kotlovaya

Decline of "one-storied USSR" politics in cities of Russia   138
A.V. Surovenkov, M.V. Skopina, G.V. Stukalov & Z. Tuhtareva

Organizational and technological industrial area renovation systems: Basics of functioning   143
D.V. Topchiy & V.S. Chernigov

Principles of forming functional landscape for wooden architecture museum in southern part
of Zaonezhye Peninsula   146
S. Zavarikhin, O. Kefala & T. Nesvitckaia

Specifics of rhythmic-and-metric as well as the architectural-and-spatial organization of
Kamennoostrovsky Prospekt   151
S.P. Zavarikhin, M.A. Granstrem & M.V. Zolotareva

Revitalization concept on historical area Near Zapskovye in Pskov   155
I.S. Zayats, E.S. Bakumenko & A.S. Perepech

Problems of urban and regional planning

Employment of collaborative information technologies in urban planning: Case study of
Alkut city, Iraq   163
A.M.H. Abokharima

New environmental consciousness   169
L. Baltovskij, V. Belous & V. Volkov

Urban planning model of ecologically balanced development of Siberia South   174
P.V. Skryabin

Problems of engineering reconstruction, performance of repair and reconstruction works on monuments

Linear multi-factor regression models in management system of Russian construction
industry   183
I.N. Geraskina & M.S. Egorova

Supply air jet simulation with machine learning   190
I.D. Kibort

Reconstruction of historical bridge using modeling and 3D printing   194
N.V. Kozak, D.A. Vabishchevich, A.V. Kvitko & M.P. Klekovkina

Preservation of historical buildings during the development of underground space in an
urban environment   199
R.A. Mangushev, A.I. Osokin, F.N. Kalach & S.A. Podgornova

Influence predicting of vibro-immersion and vibration removal of sheet piles on additional
deformations of new construction object   205
R.A. Mangushev, V.M. Polunin & N.S. Nikitina

Renovation of coastal industrial zones with possibility of using engineering geodesic dome
structures made of wood and polymer materials   209
O. Pastukh & D. Zhivotov

Buckling of shell roof structures under different loads 213
D.S. Petrov, A.A. Semenov & A.Yu Salnikov

System of automated quality control for operating materials of construction equipment 219
R.N. Safiullin, V.A. Treyal & R.E. Baruzdin

Environmental risk analysis in construction under uncertainty 222
E. Smirnova

Parameters of heat-shielding in prerevolutionary residential buildings 228
V.M Ulyasheva, A.Y. Martyanova & G.A. Ryabev

Assessment of quality of adhesive joints of amber mosaic 234
I.I. Verkhovskaia

Analysis of methods of nondestructive testing of heat pipelines 239
D. Zakharova & A. Potapov

Technological basis for use of composite materials when reinforcing wooden rafters in
heritage buildings of Saint Petersburg 244
D. Zhivotov, Yu.I. Tilinin & V.V. Latuta

Dust and environment: Saint Petersburg, Russian Federation, case study 248
A. Ziv & E.A. Solov'eva

Problems of training architects and restorers

Peculiarities of forming architecture of universal youth centers in Republic of Crimea 257
Z.S. Nagaeva & D.S. Mosyakin

Environmental values as component of professional ethics of student-architects 262
E.A. Solov'eva

Author index 267

Preface

The oldest Department of the Saint Petersburg State University of Architecture and Civil Engineering (SPbGASU) prepared Reconstruction and Restoration of Architectural Heritage 2020 together with leading scientists and experts in the field of cultural heritage preservation, conservation, restoration and renovation of monuments.

The contributions of **Reconstruction and Restoration of Architectural Heritage 2020** were presented at the eponymous conference (RRAH 2020, Saint Petersburg, Russia, 25-28 March 2020), and included key focus:

- Challenges in protection, conservation and restoration of cultural heritage objects in Saint Petersburg, regions of Russia and foreign countries;
- Methodical support of educational process and verification of study programs, including the format and requirements for the theses preparation on Bachelor, Master and PhD level;
- Benchmark events with the representatives of the Government of Saint Petersburg coordinating cultural heritage protection, the Russian Association of Restorers and leading design restoration workshops;
- Visiting sites related to the restoration and protection of the cultural heritage in Saint Petersburg.

We are well aware that the preservation of cultural heritage is the most important task of preserving the memory of civilizations. Through the efforts of the world community, UNESCO, ICOMOS, ICOM and other public organizations have been created to fight for the preservation of cultural heritage, including immovable architectural monuments.

The collection is devoted to aspects, stages and problems of preserving cultural heritage and includes an overview of professional work with monuments.

First, it provides a full-scale, broad and in-depth study of cultural heritage. Secondly, they allow us to implement the largest urban development projects on a regional scale, taking into account the specifics of design and urban planning and design and architectural works for the restoration of cultural and architectural monuments. Third, apply variable sparing methods of strengthening historical structures, identify the possibility of using non-destructive modern design and engineering solutions, and implement projects of modern "engineering reconstruction". Fourth, take into account the assessment of the economic components of the decisions made.

Doctor of architecture, Professor S.V. Sementsov

Reconstruction and Restoration of Architectural Heritage – Sementsov et al (eds)
© 2020 Taylor & Francis Group, London, ISBN 978-0-367-65357-6

Organizers

Saint Petersburg State University of Architecture and Civil Engineering (SpbGASU, Saint Petersburg)

Committee for the State Preservation of Historical and Cultural Monuments (Saint Petersburg)

Nonprofit Partnership "Russian Association of Restorers" (Saint Petersburg)

Charitable organization "Foundation for Preservation of Cultural Heritage" (CO "FSKN", Saint Petersburg)

Remmers GmbH

"NIiPI Special Representation»

Moscow Institute of Architecture (State Academy) (MARKHI, Moscow)

Voronezh State Technical University (VSTU, Voronezh)

Don State Technical University (DSTU, Rostov-on-Don)

Novosibirsk State University of Architecture and Civil Engineering (Sibstrin, Novosibirsk)

Tomsk State University of Architecture and Building (TSUAB, Tomsk)

Reconstruction and Restoration of Architectural Heritage – Sementsov et al (eds)
© 2020 Taylor & Francis Group, London, ISBN 978-0-367-65357-6

Organizing committee

Irina Drozdova, *DSc in Economics, Professor, Vice Rector for Science, Saint Petersburg State University of Architecture and Civil Engineering (Saint Petersburg)*

Sergey Sementsov, *DSc in Architecture, Professor, Head of Department of Architectural and Urban Planning Heritage, Saint Petersburg State University of Architecture and Civil Engineering (Saint Petersburg);*

Alexander Leontyev, *First Deputy Chairman, Committee for the State Preservation of Historical and Cultural Monuments Saint Petersburg (KGIOP), Professor, Department of Architectural and Urban Planning Heritage, Saint Petersburg State University of Architecture and Civil Engineering (Saint Petersburg)*

Reconstruction and Restoration of Architectural Heritage – Sementsov et al (eds)
© 2020 Taylor & Francis Group, London, ISBN 978-0-367-65357-6

Members of the organizing committee

Mikhail Golobordosky, *PhD in Architecture, Professor, Head, Department of Art History and Restoration, the Ural State University of Architecture and Arts (USUAA, Ekaterinburg);*

Alexey Gudkov, *PhD in Architecture, Professor, Dean, Institute of Architecture and Urban Planning, Novosibirsk State Unievrsity of Architecture and Civil Engineering (SIBISTRIN);*

Ekaterina Voznyak, *DSc in Architecture, Professor, Department of Architectural and Urban-Planning Heritage, Saint Petersburg State University of Architecture and Civil Engineering (Saint Petersburg);*

Valery Zalesov, *PhD in Architecture, Associate Professor, Head, Department of Theory and History of Architecture, Moscow Institute of Architecture (MARKHI);*

Zarema Nogaeva, *DSc in Architecture, Professor, Head, Department of Urban Planning, V.I. Vernadsky Crimean Federal University (CFU, Simferopol);*

Khanifa Nadyrova, *DSc in Architecture, Head, Department of Reconstruction, Restoration of Architectural Heritage and Basics of Architecture, Institute of Architecture and Design, Kazan State University of Architecture and Civil Engineering (KSUAE, Kazan);*

Olga Kormiltseva, *PhD in History, Associate Professor, Department of Architectural and Urban-Planning Heritage, Saint Petersburg State University of Architecture and Civil Engineering (Saint Petersburg);*

Viktoriya Pishchulina, *DSc in Architecture, Professor, Dean, Faculty "Schools if Architecture, Design and Arts", Acting Head, Department of Architectural Restoration, Reconstruction and History, Don State Technical University (DSTU, Rostov-on-Don);*

Larisa Romanova, *PhD in Architecture, Associate Professor, Head, Department of Restoration and Reconstruction of Architectural Heritage, Tomsk State University of Architecture and Civil Engineering (TSUAB, Tomsk);*

Alexander Slabukha, *PhD in Architecture, Professor, Department of Architectural Design, Institute of Architecture and Design, Siberian Federal University (SIBFU, Krasnoyarsk);*

Ekaterina Tribelskaya, *PhD in Architecture, Head, Department of Architecture, архиmекmуры Moscow State Academic Art Institute named after V.I. Surikov, the Russian Academy of Arts (MGAHI Surikov, Moscow);*

Tatiana Chernyaeva, *Acting Chairperson of Nonprofit Partnership "Russian Association of Restorers" (Saint Petersburg);*

Gennady Chesnokov, *PhD in Architecture, Professor, Head, Department of Composition and Conservation of Architectural and Urban Heritage, Voronezh State Technical University (VSTU, Voronezh);*

Sergey Fedorkin, *Director, Academy of Construction and Architecture, V.I. Vernadsky Crimean Federal University (CFU, Simferopol).*

Reconstruction and Restoration of Architectural Heritage – Sementsov et al (eds)
© 2020 Taylor & Francis Group, London, ISBN 978-0-367-65357-6

Sponsors & Co-organizers

1. Committee for the state preservation of historical and cultural monuments (Saint Petersburg)

2. Charitable organization "Foundation for Preservation of Cultural Heritage" (CO "FSKN", Saint Petersburg)

3. Remmers GmbH

4. Nonprofit Partnership "Russian Association of Restorers" (Saint Petersburg)

5. "NIiPI Special Representation»

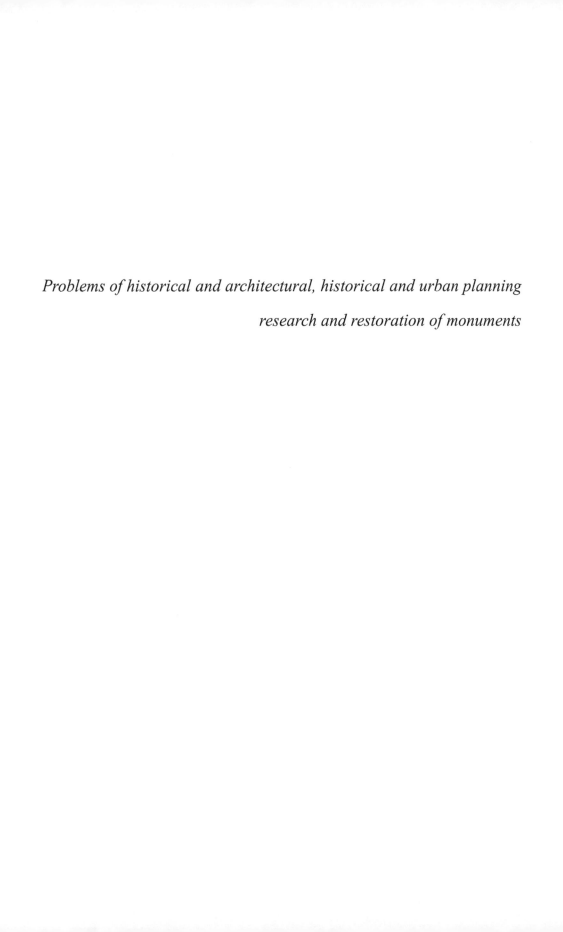

Problems of historical and architectural, historical and urban planning

research and restoration of monuments

Reconstruction and Restoration of Architectural Heritage – Sementsov et al (eds)
© 2020 Taylor & Francis Group, London, ISBN 978-0-367-65357-6

An unnamed building of Bishop's house of Vologda according to newly revealed archival information

I.K. Beloyarskaya
Vologda state University, Russia

ABSTRACT: Comprehensive scientific research was conducted based on reliable historical data, which resulted in determining the stages of construction, specifying the date of construction, and the original name of the object.

1 INTRODUCTION

The formation of scientific restoration in Russia began at the beginning of the XVIII century. At this time, there was a change in attitude to the heritage, expressed in the preservation, accompanied by repair work. In the era of Peter the Great that new attitude to historical heritage of Russia appeared. During this period, from the XVIII to the middle of the XIX century, restoration activities were mainly limited to renovation monuments. "The roots of restoration in Russia date back to the early XVIII century, the period of revaluation of spiritual and cultural values occurring under the influence of Peter's reforms... the Emergence of renovation as a method of restoration of destroyed monuments in their balances in the form in which they once existed, as a method that satisfies the requirements of science, came along with the aesthetic development of architectural tradition that goes back to antiquity and to the Renaissance". (Zverev, 2008, P. 4)

As a professional field of activity, restoration appeared in the middle of the XIX century, when "the Developing archaeological science of the beginning of the XIX century had a noticeable impact on the restoration practice only in the 30-40s of the XIX century" (Zverev, 2008, P. 7). At this time, a theoretical component appears in the restoration – the study of the monument, as a result, restoration takes its place in the system of Sciences.

In the late XIX – early XX century, in the epoch of scientific restoration in our country were identified theoretical problems of conservation, defined methods and the main aim of rehabilitation - preservation of maximum authenticity of the monument. Restoration methods and determining the authenticity of the monument were developed in the context of a certain level of development of culture and public consciousness in the historical period under consideration.

The beginning of the Soviet restoration is associated with the name of I. E. Grabar. He directed the Commission for the preservation and disclosure of monuments of ancient Russian painting, which was founded in 1918. Restorers in the post-revolutionary decades laid the foundation for modern methods of scientific restoration. The main provision of modern restoration methods is comprehensive scientific research, which determines the authenticity of the monument, the periods of construction, the time of construction, late layers, the history of the monument, the reliability of historical information, and the value characteristics of the object.

2 MATERIALS AND METHODS

The method of determining the value of the monument establishes the significance of scientific, aesthetic and historical criteria.

The purpose of the study: to Determine and justify the exact date of construction of the structure and the original name of the building. The methods based on historical-bibliographic and archival research, analysis of proportionality proportions, analysis of analogous objects, chronological analysis of the formation of the object - its reconstructions and extensions.

A comprehensive study of the building, including the study of historical, archival and bibliographic sources, the study of the history of the monument and its surroundings, systematization and analysis of analogous objects.

The official and generally accepted date of foundation of the city of Vologda is 1147, although the ancient Russian city Vologda is first mentioned in state acts from 1264.

From the first half of the XV century Vologda became part of the Moscow lands. By the beginning of the XVI century it was one of the richest cities. This was facilitated by the favorable position on the trade route from Moscow to Arkhangelsk, where the only sea port of the Russian state was located at that time. The city was mostly located on the "Lazy site", most of the territory of which was occupied by the fortified Detinets - the ancient Kremlin, the sprawling city stretched along the river bank downstream. Small villages were located on the other side of the river.

A radical change in the Central part of the city of Vologda was associated with the name of Ivan the Terrible and became a new stage in the development of the city plan. In 1566, Tsar Ivan the Terrible chose Vologda to build his residence there and planned to turn the Vologda Kremlin into one of the most powerful fortresses in the country. In 1567, on the day of Saints Jason and Sosipater, the construction of the walls of a new fortress begins, downstream of the river, which according to the Tsar's plan was to become one of the most powerful fortresses in the country.

In 1568, the English engineer Humphrey Locke was invited to Vologda to direct the construction of the Kremlin. The configuration of the plan of the Kremlin territory eventually received a diamond-shaped shape, surrounded on all four sides by water: from the East was the Vologda river, from the South the Zolotukha river, and from the north and west sides, artificial ditches were dug that connected with these rivers. Settlements are beginning to form around the Kremlin. Vologda city plan gets a four-part division: "City" (the territory of which was limited by the walls of the Kremlin), "Upper Posad", "Lower Posad"and "Zarechye". (Figure 1).

In the new city center – the Kremlin, in 1568, construction began on a new Cathedral in the image

Figure 2. Saint Sophia Cathedral in Vologda. Current state. Photo by the author.

and likeness of the Moscow Cathedral of The Dormition, built almost ten years earlier, and also dedicated to the feast of The Dormition of the Mother of God. The Cathedral was the first stone building in Vologda (Dunaev1914) (Suvorov 1863) (Figure 2). The Cathedral was located opposite the Archbishop's court, and to the South of the Cathedral was the Royal Palace.

Due to the changed political situation in the country in 1571, the Tsar urgently left Vologda and never returned there. With the departure of the Tsar, stone construction in the Kremlin stops. The construction of the Cathedral was also suspended. After the death of Ivan the Terrible, the walls of the Kremlin were not only completed, but also the digging of the canal on the southern and western sides of the fortress was completed. After the destruction of Vologda by the poles in 1612, construction activities developed in the city. The main city Cathedral is being renovated and the main throne is being consecrated in the name of Sophia the Wisdom of God. In 1622, the first bell tower of St. Sophia Cathedral was built. (Bocharov and Vygolov 1969).

Since the second half of the XVII century, stone construction begins in the Vologda Kremlin. Naturally, immediately after determining the location for the construction of the main city Cathedral, buildings of the New Bishop's residence begin to be erected next to it. The complex of the Bishop's court was built opposite the main city Cathedral in 1585-1587. At First, all the buildings of the complex were wooden, and the fence was also wooden. The building of the "State order" (the modern name of the Economy building) was the first to be built out of stone in the Bishop's courtyard. Then the first Episcopal chambers, The Church of The Nativity of Christ, the Gavrilovsky building, and The Church of The Ascension over The Three Holy gates (Trekhsvyatsky gates) are built in stone in an irregular quadrangle of stone walls of the "Bishop's court" with turrets at the corners. An unnamed building, originally called the Storeroom chamber on the cabbage cellar, was built between the Economy building and The Gate Church

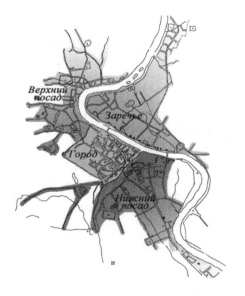

Figure 1. Plan-diagram of the city of Vologda in 1790. with the color of the city's suburbs ("posad"). Authors: S. A. Sharov and S. K. Regame (Regame et al., 1988).

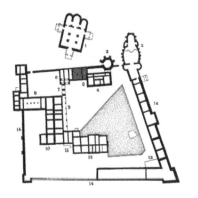

Figure 3. Plan of the Bishop's courtyard and cathedrals of the city of Vologda. (By V. S. Baniga, 1970).

of The Ascension, in one connection with these objects (Beloyarskaya 2011). (Figure 3).

1-St. Sophia Cathedral; 2-Resurrection Cathedral; 3-bell tower; 4-State order; 5-Nameless building; 6-The Gate Church of The Ascension; 7-transition to the gallery of the Cross chamber; 8-Consistory building; 9-cross chamber with the Church of the Nativity of Christ; 10-Ireneevsky building; 11-Gabriel building; 12-Joseph building; 13-service buildings; 14-walls and towers of the Bishop's house of their extensions.

3 RESULTS AND DISCUSSION

The unnamed building is located on the territory of the ensemble of the Bishop's house, to the right of the entrance through the Sophia gate (Water gate), in the same connection and between the volumes of the Economy building and The Ascension Church. This building is the middle part of a single complex of buildings located along the eastern wall, which is its northern part. The complex begins from the former Water gate of the ensemble, behind which the bell tower adjoins the wall from the outside and ends with The Three Holy gates, crowned by The Gate Church of The Ascension, and dividing the eastern wall into two almost identical in size parts. (Figure 4,a,b).

The complex of the first buildings of the Bishop's court appeared around 1585-1587, naturally in connection with the construction of a new city Cathedral in the central core of the city – the Kremlin. At the beginning of the XVII century, a wooden fence was built around the territory. A Church was built over the Holy Gates in 1653 in the name of three Permian Saints - Gerasim, Pitirim and Joseph. Inside the courtyard were residential and service buildings, as well as Bishop's cells, two wooden State cells and a hut of the Judgment Order. From 1658 to 1673, the Bishop's wooden mansions were replaced by stone Episcopal chambers, with a stone house Cross Church. By the end of the 1650s, in the south-eastern part of the territory, a stone building of the State Order was built, later called the "Economy building". The laying of the stone wall began in April 1673. And by the end of the summer of 1674, the last, eastern spinning wheel was completed. Immediately after that, between the Holy Gates and the northern wall of the Economy building, a storeroom building was built on the cabbage cellar, which in the middle of the XIX century, N. I. Suvorov will call it the "Nameless corps". (Suvorov, 1869) (Figure 5).

This object is still assigned not only the name "Nameless building", but also the date of construction - the turn of the XVII – XVIII centuries. However, research of archival and historical materials, technical condition, made dimensional drawings, carried out by the author's team of the Institute for the restoration of historical and cultural monuments "Spetsproektrestavratsiya", Moscow, in 2000, allowed us to conclude that the correct, historical name of this building is" Storeroom on the cabbage cellar" and the date of its construction 1675-76 years. (Perfileva, 2000).

As a result of the study of historical documents relating to the time of construction of the research object, it is determined that this building for its functional purpose and architectural solution (spatial structure, layout of interconnected rooms), is

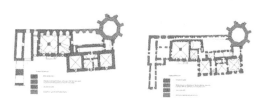

Figure 4. The complex of buildings of the church of the ascension, the storeroom of the chamber, the economy building and the bell tower. a – Plan of the 1-st floor, b- Plan of the 2-nd floor.

Figure 5. Unnamed building. Facade of the yard. Current state. Photo made by the author.

a continuation of the Economy building in the process of developing its functions and, in this regard, increasing the need for new premises.

Let's consider the historical facts of building construction. October 10, 1674. Vladyka from Moscow writes a letter to Vologda «...and on the winter way to tell you to buy rubble stone for the cabbage cellar... », November 22, 1674 again writes the order «...as this letter comes to you, tell to put rubble stone where more decent and closer to the place where The Three Holy gate is (Trekhsvyatsky gate) and in other places, where the thing to be in The Three Holy gates and other places where nicely the brick laying at discretion» (Perfileva, 2000).

In July 1674, the construction of the section of the eastern wall from the gate Church to the bell tower was completed. The laying of the Storeroom chamber and cellars dates back to 1675-in may, "the peasants from Fryazinova village dug a new cellar, the moat 21 fathoms without a half-arch under the wall and under the pillars, hammered piles and filled up the moat with rubble stone, and interior side in the cellar with rubble stone and pillars, all around smooth" (Expense book, 1674). By the end of the summer of 1675, the document testifies to the completion of work «in the Bishop's yard, all rubble was stripped and the yard was cleaned and the soil was spread... ». In the «Inventory of the property of Archbishop Simon» from November 10, 1676, the completion of the construction of the research object is recorded «in the Bishop's courtyard, the Bishop's State Order is a stone structure, opposite the Order of the Treasury Cell is a stone structure, between them are two porches with seeding, and above the canopies is a tent.

Under Order and under Treasury cell and under the porches are the cellars, to the Treasury Order and Treasury cell again attached residential tents, near the tent, that is close to Treasury Cell, the storage room is made and under it the cellar, and against residential tents to the storage chamber for the passageway the vestibule, and against the Storage chamber is the porch ...". (Perfileva, 2000).

Next, the windows are described, the details of the windows «...in two windows, two mica windows, the window shutters made of iron and wood", a list of items in the ward, utensils. This indicates that the alterations in the building of the Economy building were completed, the construction of the Storeroom chamber was completed, and both these buildings were united into a single administrative and economic complex by the beginning of 1676.

Thus, a more accurate date of construction of the object under study should be considered the date 1675.

The eastern wall of the Storeroom of the chamber on the cabbage cellar was built close to the eastern wall of the entire complex of the Bishop's court from the Holy gate to the Economy building, so the wall thickness doubled to 3 meters. The north wall at first half adjoined The Holy Gate, so that the other half of this wall had two windows to light the cabbage cellar. In 1691, a Gate Church was built over The Holy Gate, which was attached to the north wall of the building of the "Storeroom of the chamber on the cabbage cellar", in this connection, all the northern windows were laid, preserving the window niches inside the building on the upper and lower floors. When laying out the eastern wall of the cellars, the masons repeated in it the contours of large niches-embrasures, arranged in the wall of the fence and had in the depth of the external loopholes. In the result of cellar got on the east side apertures-airholes and a wide niches with embrasures. The appearance of the transverse wall, thickness 80 cm, in the lower floor «Pantry chamber on cabbage cellar», and specifically in the domed cruciform cabbage cellar, which has divided the cellar into two rooms, the cellar and the Salt Chamber, the researchers also refer to this 1680-90 years. On the western wall of the first floor of the building, there were two entrance openings and two windows between them. Windows are of the same size, and the front openings are different, this suggests that first there was one entrance to the two-column cabbage cellar, and when a cross-wall divided it two rooms – a Salt room and cellars, a second entrance appeared, which inputs the later is not possible to find out. The Salt chamber is covered by closed vaults with timbers that rested on a central column (1.5 x 1.5 m).

On the top floor there were three rooms, their layout depends on the layout of the lower floor, however, there are differences. Above the Salt chamber is a single-column Storage room, also square in plan, but its central column, also 1.5 m x 1.5 m in size, is located above the lower one, with an offset. The eastern wall is without openings, there are four niches measuring 50 x 50cm and one 75 x 75cm. Originally, the Storeroom had three windows on the north side, and three windows and a doorway on the west side. The north windows, as well as the north windows of the first floor were laid with the construction of The Gate Church. In the eastern part of the southern wall was a door connecting the Storage room with a Smaller Storage room (size 5, 8 x 4, 8m), covered with box vaults with two windows in the southern wall. From the small storeroom of the chamber through a narrow passageway people entered the vestibule common to the Storeroom of the chamber and the Economy building. Between the common hall and the Large storeroom was the Treasurer's living cell. The living Treasury cell was covered with closed vaults, lit by two windows on the western wall. The Treasury cell was heated by a Dutch oven. (Figure 6, a,b).

Initially, the two-pitched gable roof of the Storeroom building was covered with wood, but by the middle of the XIX century, as in all buildings of the

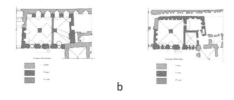

a   b

Figure 6. Diagram of construction periods: a– Plan of the 1st floor, b– Plan of the 2nd floor.

Bishop's court, the roof on the Storeroom was replaced with an iron roof with a change in height and the appearance of a third slope towards the Economy building.

The facades of the building are decorated with a modest decor-flat blades at the corners, belts of profiled cornices – ground, inter-floor and crowning, loosened above the pilasters, had lime whitewash. Picturesqueness of the main, western facade is given by asymmetrically located niches of door and window openings. The three western windows of the Big Storeroom and the two windows of the Small Storeroom had iron shutters; there is no mention of shutters on the other windows of the building. All the windows had wrought-iron bars.

The eastern wall of the Unnamed building on the second floor does not adjoin the outer fence of the monument of Federal significance «Vologda Kremlin Ensemble: Bishop's house and other buildings, XVII-XVIII centuries», but retreating from it, forms a corridor-passage that connected the Gate Church with the bell tower. The passage on the wall-fence of the Bishop's house complex was constructed simultaneously with the wall in 1674, in accordance with the rules of defense construction of the XVII century. In the corridor-passage, located along the outer side formed by the walls of the fence of the Bishop's house complex and the eastern wall of the Unnamed building, there are niches – embrasures, with narrow window openings, like fortress loopholes, part of the window openings are laid. At first, the loophole windows were closed with shutters. Window frames appeared, obviously, in the XIX century, they are in good condition. The window fittings are preserved and in the side walls of the niches remained the grooves for the crossbar fixing wooden shutters. The ceiling of the corridor is covered with a wooden ceiling, on which you can see forged nails. The wooden ceiling still has coils from early electrical wiring. The ceiling is flat, the same as in the continuation of the passage along the wall of the fence to the bell tower. Most likely, the ceiling has not been changed since the late XIX – early XX centuries. The passageway is bounded by two doors. The door from the Gate Church to the corridor appeared during the construction of The Gate Church. The door at the end of the corridor is

late. The exterior facades of the eastern wall-the fence of the Bishop's house complex, are subject to a single architectural solution of the exterior facades of the entire fence of the complex. The fence wall on the outside of the complex is divided horizontally by a narrow double belt. The lower level is a blank surface, the upper level is a composition of narrow windows-loopholes, located in a certain pattern along the wall plane.

The Unnamed building, in historical sources called the Storeroom chamber on the cabbage cellar, is a rectangular plan, elongated from North to South, two-story building with small original windows with a modest decor on the facades. There is an assumption that all the windows were scratched, although there is no documentary evidence of this. The dimensions of the building are 21 x 15 meters.

The research object is currently used by the Vologda State historical, architectural and art Museum-reserve. On the first floor there are two rooms, each with its own entrance. The salt chamber is almost square in plan, with a square column in the middle measuring 1.5 x 1.5 m, which is supported by closed vaults with decking. There is one entrance on the west side. Double doors are wooden. On the inner side of the doorways, grooves for the crossbeam supporting the door leaf have been preserved in the masonry. On the outside, the historic door hinges are preserved. On the same wall there are two windows in niches. On the north wall in the western part of the two windows laid during the construction of the Gate Church in niches. In niches in the brickwork, slots for latches and iron hinges for window shutters were preserved. On the east wall there are three niches-embrasures, in two of them there are laid windows. The floors are wooden. The walls and vaults are whitewashed. Interiors are devoid of decoration.

Cabbage cellar – dark chamber formed by separation of one two-columned chamber into two parts transverse stone wall (thickness 80 cm) adjacent to the northern side of the southern pillar size of 1.5 x 1.5 m, which is based torispherical vaults with face tray, takedowns and "oblique" arch is thrown from pillar in the southeast corner. The entrance from the west side. The doors are double wooden, the iron door hinges have been preserved on the outside. In the east wall there are two niches-embrasures with the windows blocked up. The floors are linoleum, inside the room at the entrance is equipped with a wooden porch. It is in poor condition. The walls and vaults are whitewashed. Interiors are devoid of decoration. The room is heavily filled with Museum inventory. (Figure 7).

On the second floor there are three rooms that can be accessed through the common hallways of the Economy building. The first room from the entrance is a small storeroom, covered with box vaults with two windows on the southern wall, facing the courtyard of

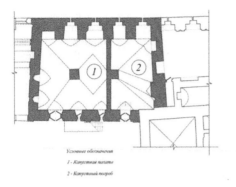

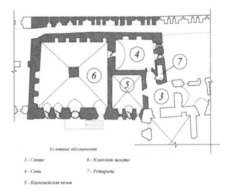

Figure 7. Explication of the storage rooms of the chamber in the cabbage cellar. 1st floor plan.

Figure 8. Explication of the storage rooms of the chamber in the cabbage cellar. 2-nd floor plan.

the Economy building. On the eastern wall there are three small niches. The walls and vaults are white-washed. Interiors are devoid of decoration. The floors are stone. A narrow arched doorway leads to the large Storage room. Traces of grouting cracks are in the arches. The room is not used.

Large Storage room - a square room in the plan, in the middle of a square column measuring 1.5 x 1.5 m, which is supported by closed vaults with additional vaulted parts embedded in the main vault. The vaults have a large number of plastered cracks. On the northern wall in the western part there are three niches-embrasures with laid windows, on the eastern wall there are four embrasures measuring 50 x 50cm and one 75 x 75cm, on the western wall there are four windows. The floors in the ward are made of stone. The walls and vaults are white-washed. Interiors are devoid of decoration. The room is currently unused.

Treasury of a living cell. You can get to the Treasury cell from the entrance hall of the Econ-omy building. A small bright room with a window and a stove located next to the window on the western wall. A box vaults ceiling, which also has a large number of plastered cracks. The floors are stone. The walls and vaults are whitewashed. The door is late and in poor condition, As well as all the rooms on the second floor, the living cell is currently not used. Currently, all the embrasures in the building of the "Storeroom of the chamber on the cabbage cellar" are lost. (Figure 8).

The last room on the second floor is a corridor that runs on the wall of the Bishop's house com-plex. The passage on the wall can be accessed from the altar part of the Gate Church through a wooden door. At first, the doorway was in the form of an arch, later the upper part of the archway was laid and the modern door is rectangular. The outer wall-fence of the complex has niches with window open-ings, most of which are laid. The window frames in the narrow window openings that imitate loopholes are well preserved, as are the window fittings. In the window niches of the preserved grooves for the

device shutter. The floors are wooden. The wall of the chamber's storeroom, which is also the wall of the corridor-passage, is blind. The ceiling is lined with boards, on which you can see the heads of forged nails. The second door, which leads to the open part of the passages on the wall-fence, looks more like a partition. On the western wall of the corridor, which is also the wall of the Storeroom building, there is a large vertical crack, approxi-mately the crack is located at the place where the closed arches rest on the wall. It is necessary to study the condition of foundations and floor struc-tures in this place. There is also a large crack in the upper part of the window opening niche that extends to the ceiling. This crack is likely the result of later strobing an indent for current needs in Soviet period. The crack also needs to be investi-gated. There are cracks in the outer wall of the fence. In general, the external wall-fence of the complex, in the described part, requires full research of its technical condition. The passageway is used as a storage room.

4 CONCLUSIONS

Based on the research conducted and studied arch-ival and bibliographic sources, as well as field sur-veys, we can conclude that the original name of the structure located between the volumes of the Econ-omy building and The Ascension Church in one con-nection with them is "Storeroom on the cabbage cellar" and the construction date is 1675.

REFERENCES

Zverev, V. V. 1999. From renovation to scientific restoration. http://www.art-con.ru.
Perfil'eva L. A. 2000. Monument of architecture of the XVII century. Economy and Unnamed building b. Bishop's house of Vologda, Historical-archival and bibliographic research 1 (2).

Regame, S. K. 1988. Combination of new and existing buildings in the reconstruction of cities, Stroizdat.

Dunaev B. I. 1914. Northern-Russian Civil and Church architecture. The City of Vologda.

Suvorov N. 1863. Description of the Vologda Cathedral of St. Sophia.

Bocharov, G. N. 1969. Vologda. Kirillov. Ferapontovo. Belozersk, Art.

Beloyarskaya I. K. 2011. Historical plan of the city of Vologda-an object of cultural heritage. Science, education and experimental design. MARKHI.

Banige V. S., Pertsev N. V. 1970. Vologda. Moscow: Art.

Suvorov N. I. 1869. Vologda Bishop's house. VEV 16: 646–654.

Michels, G. B. (1999). At war with the church: Religious dissent in seventeenth-century Russia. Stanford University Press.

Reconstruction and Restoration of Architectural Heritage – Sementsov et al (eds)
© 2020 Taylor & Francis Group, London, ISBN 978-0-367-65357-6

Principles of embankment humanization in gray belt of Saint Petersburg

N.M. Drizhapolova, V.M. Supranovich & N.K. Yass
Saint Petersburg State University of Architecture and Civil Engineering, Saint Petersburg, Russia

ABSTRACT: The relevance of this study is conditioned by the lack of access for Saint Petersburg residents and guests to the main panoramas and viewpoints from the embankments of the Neva River and other rivers and canals due to the modern new development of these territories (especially due to the renovation of the gray belt of the city). The purpose of the study is to find ways to restore the embankments' function of urban public space as well as determine the principles and specifics of further development of new embankments with account for their transformation into a pedestrian public space. In the course of the study, the following methods were used: analysis of the historical experience in embankment formation (as a part of the city structure) in Saint Petersburg, modern theoretical and practical materials on the revival of embankments in favor of pedestrians, as well as on-site surveys. Designs intended to return the historical Petersburg essence to embankments have been suggested, models to eliminate problematic nodes have been formed during experimental sketch design (by bachelor and master students majoring in architecture under the guidance of the authors of the article). The implementation of these suggestions in the design practice will restore the role of embankments in the city, which will increase its tourist potential. The materials of this article may be useful for urban planning authorities.

1 INTRODUCTION

"Water spaces of rivers, lakes and sea shores are one of the main city-forming elements of urban development, to which the planning and functional systems of cities themselves have largely adapted during their developing in space and time" (Sementsov et al. 2017). It is especially important for Saint Petersburg.

In Saint Petersburg, water spaces, embankments (especially, those of the Neva River) and the surrounding buildings were originally planned as a large parade public space.

The city was historically formed (since 1703) along the banks of the Neva River as a system of gradually expanding residential and public formations on separate gradually developed territories, starting with Zayachy Island (Gorodskoy), growing as a result of development along the Kronverksky Canal, Vasilyevsky Island, Admiralty, Petrograd, and Vyborg sides, and further on. Accordingly, the location of the public center changed, which was always tied to the embankments of water spaces. The water spaces themselves sometimes became huge public squares where social events took place every day (ships sailed, small boats scurried around, providing communication between the banks, temporary entertainment facilities were built on ice in winter, and entertainment events were held). By the beginning of the 20th century, a "multi-node urban center as a single spatial ensemble around the main space of the Neva" was formed "with the actual creation of the main water

square of the Imperial Saint Petersburg, the main compositional element of the entire spatially developed city center" (around the Neva, between Gorodskoy Island, the Spit of Vasilyevsky Island, Palace (Dvortsovaya) embankment, and embankment of the Vyborg side) (Sementsov et al. 2017). At the same time, a system of embankments and bridges was formed, and the outlines of the banks were put in order" (with a standard scale and height of development)" (Sementsov et al. 2017).

2 METHODS

The water areas of rivers and canals were gradually connected to the public spaces of the Neva River, developing a multi-node branched system of the center.

"Figuratively speaking, the system of open public spaces in the center of Saint Petersburg in the 1830s could be compared to a spreading tree where the Neva is the mighty trunk, and the other arteries are like branches. The relationship between the 'trunk' and 'branches' was provided by the coastal squares — Petrovskaya and St. Isaac's (Isaakievskaya), the Razvodnaya ground on the Admiralty side, and Rumyantsevskaya and Birzhevaya squares on Vasilyevsky island" (Lavrov & Speranskaya 2017).

"The role of natural and artificial waterways in the life of the city is significant... they not only became an ornament and a distinctive feature of Petersburg landscapes but also had a significant impact on the life of the city, the formation of its functional and planning structure" (Lavrov 2015). They were both transport arteries and public spaces, combining utilitarian and ceremonial functions. At the same time, Petersburg functioned as a "pedestrian city" mainly. "Researchers believe that the 'pedestrian city' can have an area of up to 12 square kilometers, and the transition beyond this boundary requires a fundamental change in intra-city communications with a view to the active use of transport" (Lavrov 2015).

The ensemble character of the water space development with its connection to the general composition of the urban multi-node center "continued until the 1990s" (Sementsov et al. 2017).

As the city grew and technologies developed, new vehicles appeared, from small ships and carts to horse-drawn railways, the first electric trams on the Neva ice (Lavrov 2015), and all the way up to modern high-speed transport and metro. On the outskirts (now areas of the gray belt of the city), industrial enterprises were growing, which were maintained, among others, by water transport, therefore, they were closer to the waterways (Obvodny Canal, the upper reaches of the Neva, etc.). The city gradually lost its "pedestrian character", and with it its charm, its scale of perception of space and the environment. First of all, the embankments were affected, where uninterrupted-flow highways were located in the "undeveloped" space in order to please the winning "technicism". It is difficult for pedestrians to walk along these highways. The possibility of perceiving typical scenery spots from the embankments has disappeared. For example, a front view of Smolny Cathedral from the opposite side of the Neva River is only possible from an uninterrupted-flow highway.

The purpose of the study is to find ways to restore the embankments' function of urban public space as well as determine the principles and specifics of further development of new embankments with account for their transformation into a pedestrian public space on the modern stage of Saint Petersburg development.

Certainly, it is not about all the embankments without exception. Priority nodes of public centers and spaces (Gelfond 2013), including embankments, should be determined. Ways to solve this problem should also be different in order to ensure the diversity of the environment of embankments and take into account different urban conditions. It is also necessary to determine the formation pattern of an expressive silhouette of the development in order to include new embankments in the system of historical panoramas organically.

Water and pedestrian areas are the main charisma of Saint Petersburg. This system (with difficulty) has been preserved in the historical center of the city. In new residential areas, it is almost absent (with rare exceptions). Currently, the "gray belt" of the city is being actively developed, which provides for a radical change in the purpose of territories: multifunctional public-and-business as well as residential functions instead of industrial ones (Drizhapolova 2017). However, embankments in these zones are usually occupied by highways. Development on coastal areas (Drizhapolova 2015) is carried out without a unified concept for the development of the entire embankment, linking it with the historical part and the silhouette of the water space. But these territories are adjacent to the historical part of the city. The historical code of the perception of Saint Petersburg water spaces is violated. Modern "new" embankments resemble the counter of an antique store, where "unique masterpieces" of modern architecture are located in one row, each of which claims a special accent role, ignoring the general decision, according to which something should serve as a background. The sections are delimitated in accordance with a pre-existing function (most often, a posteriori), without account for their new function and affiliation with the coastal part of water spaces.

It is required to save pedestrian embankments at the legislative level urgently, at least on individual nodal sections, recreating the multi-node system of the center and polycenters. It is necessary to create a conceptually defined architectural and urban-planning panorama and a diverse and expressive silhouette of new embankments as a natural continuation of their historical counterparts (Drizhapolova 2017).

Is it possible in the current situation, or is it already late, as many experts think?

To solve this problem, the authors conducted the following studies.

1. On-site surveys. These surveys have shown that even without major changes in the development, we can turn an embankment into a public space just freeing it from cars. Figure 1 (a, b) shows one of the typical examples. Two photos are taken at the same time from the same point in different directions. In Figure 1a, there are almost no pedestrians, although the embankment is surrounded by historical buildings (one can see the Buddhist datsan, historical buildings, attractive environment), but along the water the Primorskoe highway runs. Figure 1b shows the area where

Figure 1. View of the embankment at the entrance to the central park of culture and Leisure (a, b).

the highway turns deep into the development and the embankment is free of cars. There are a lot of walking and resting people on the shore, although the adjacent buildings are less attractive.

2. Determining territory delimitation principles. In particular, we studied the historical delimitation of the city (Drizhapolova 2017). The study showed that historical land delimitation (industrial and civil) has different scales: plots intended for public and business functions should be re-delimitated more finely, forming human-scale residential blocks. But the delimitation along the embankment is performed according to a different principle: not in 2–3 rows, as inside a block, but in one row. As a result, an embankment in two "lines" (double embankment) is formed. At the same time, both embankment lines 1 and 2 have their own individual features and characteristics. This makes it possible to perceive the changing development of the embankment from different angles with a variety of overlays of objects from lines 1 and 2 on each other. Figure 2 shows an example of such a double embankment.

Research of domestic and foreign experience in developing embankments in historical cities, including their architectural, image composition, as well as the formation pattern of the silhouette, panoramas, and viewpoints (Drizhapolova 2017, Kurbatov 2017).

According to the results of this research, described in the article "Principles of the transformation of urban industrial territories adjacent to water bodies", it is necessary to follow the general principles of forming the silhouette and image-bearing components of the embankment development:

– complete and comprehensive development of a site with the introduction of a whole range of functions (Supranovich & Vabishchevich 2019);
– compliance with the "scale of the object and context", based on the historical traditions of the

embankment design and modern trends in architecture (Supranovich & Vabishchevich 2019);
– silhouette formation of buildings that directly depends on the arrangement of the environment and its scale; new buildings should form the image of the city perceived from the water (Supranovich & Vabishchevich 2019);
– preservation of the "memory of the place" based on a detailed analysis of the history, stages of territory development formation and identification of lost elements in the urban planning structure (Baranov 1964, Supranovich & Vabishchevich 2019);
– compliance with the "protected status" — compliance with the requirements for adapting protection or cultural heritage sites to a new functional use (Supranovich & Vabishchevich 2019, Supranovich et al. 2017);
– adaptations to modern functional needs — conversion of preserved buildings (Supranovich & Vabishchevich 2019, Supranovich et al. 2017);
– environmental friendliness and sustainable architecture (using high-quality materials, taking into account the environmental factors of the urban environment and a high percentage of territory landscaping) (Supranovich & Vabishchevich 2019).

3 RESULTS

These studies allowed us to confirm the possibility of organizing pedestrian public spaces on the embankments in the gray belt of Saint Petersburg and formulate some basic requirements for their formation.

1. Pedestrian public spaces should be organized in strictly defined places, with account for the presence of significant urban-planning nodes and expressive viewpoints and panoramas.
2. A wide variety of techniques specific to each case should correspond to many urban-planning situations related to embankments in the gray belt of the city.
3. The priority should be given to the comprehensive transformation of large industrial areas adjacent to the embankments (Drizhapolova 2017).
4. Land plots intended for industrial construction should be re-delimitated with account for the scale of new blocks and in accordance with the new civil purpose.
5. When re-delimitating the territories of former industrial enterprises on the banks of water bodies, along which a highway runs, an additional pedestrian embankment can be provided — an alternative route, which can be formed by adjacent buildings in the following ways:
– be laid parallel to the main embankment with the arrangement of open spaces facing the water;

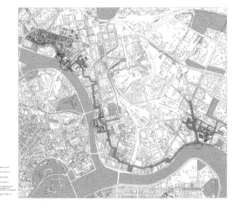

Figure 2. Diagram for the concept of creating a duplicate pedestrian embankment – a green public space.

- be zigzag-shaped, periodically going out to the main embankment (with vehicles going to an underground tunnel in this place), with the arrangement of viewpoints or coastal areas with the possibility of descending to the water or overhanging the water;
- periodically change places, leave the embankment and return to it alternately in the form of a highway or pedestrian space;
- be located in "dashes" in places where vehicles go in a tunnel.

6. If there is no territorial resource for the organization of a duplicate pedestrian embankment, it is possible to provide passable public spaces as part of objects placed along the main (transport) embankment.

4 DISCUSSION

The above requirements to the architectural development of the coastal areas in the gray belt of Saint Petersburg with the arrangement of public pedestrian spaces were tested in experimental design of integrated development of the coastal area on the Vyborg side, and areas along the Obvodny canal in the Admiralty District by master students of the Department of Architecture.

Figure 2 shows a diagram for the concept of creating a duplicate pedestrian embankment — a green public space that demonstrates all of the above principles. The concept was developed by second-year master students as part of their graduation theses in 2020.

Figure 3 shows an example of the organization of a public polycenter along the Obvodny Canal at the intersection with Moskovsky Prospekt, developed as part of an experimental integrated concept for the development of the territory along the Obvodny Canal in the Admiralteysky District by master students in 2017. There are duplicate

Figure 3. Example of the organization of a public polycenter along the Obvodny Canal at the intersection with Moskovsky Prospekt.

pedestrian embankments, with one of them being a part of a creative center on the northern side of the canal.

5 CONCLUSIONS

It is possible and necessary to save the embankments in the gray belt of Saint Petersburg for the purposes of the organization of public pedestrian spaces. It should be done immediately, until this territorial resource is completely exhausted. To achieve this goal, it is required to use administrative resources. Integrated concepts for the development of the main embankments should be developed, linked to each other and to the existing historical environment. Based on these concepts, changes and clarifications should be made to the main documents of territorial planning and urban zoning. In particular, the rules of land use and development can identify subzones in places of embankments with special regulations that specify the placement of accents and background objects (updated height parameters) in order to achieve the specified silhouette of panoramas. Also, it is possible to provide for an easement on coastal areas — a requirement to ensure the unhindered passage of pedestrians, if a duplicate pedestrian embankment cannot be allocated with red lines that save it from development. These measures will help to save embankments from haphazard mixed-character development and ensure the coexistence of pedestrians and transport at a legislative level.

REFERENCES

Baranov, N.V. 1964 *Composition of the city center.* Moscow: Stroyizdat.

Drizhapolova, N.M. 2015. Specifics of new housing construction in the central districts of historical city (on example of Admiralty District of Saint Petersburg). In Ye.B. Smirnov et al. (eds), *Architecture –Construction – Transport. Proceedings of the 71st Scientific Conference of professors, teachers, scientists, engineers and post-graduates of the university, Saint Petersburg, October 7–9, 2015*: 162–167. Saint Petersburg: Saint Petersburg State University of Architecture and Civil Engineering.

Drizhapolova, N.M. 2017. Complex transformation of the territory adjoining the Obvodny Canal in Admiralteisky District Saint Petersburg. In Ye.B. Smirnov et al. (eds), *Masters' hearings. Proceedings of the Inter-Regional Scientific and Practical conference in the framework of the 7th Inter-Regional Creative Forum "Architectural Seasons at Saint Petersburg State University of Architecture and Civil Engineering", Saint Petersburg April 18–21, 2017*: 24–27. Saint Petersburg: Saint Petersburg State University of Architecture and Civil Engineering.

Gelfond, A.L. 2013. *Architectural design of public space: teaching aid.* Nizhny Novgorod: Nizhny Novgorod State University of Architecture and Civil Engineering.

Kurbatov, Yu.I. 2017. Conditions of forming a full-fledged configuration of the architectural form. *Bulletin of Civil Engineers* 4 (63): 23–25.

Lavrov, L.P. 2015. Saint-Petersburg: the destiny of a pedestrian in a regular city. *Vestnik of Saint Petersburg University. Arts* 5 (1): 184–203.

Lavrov, L.P. & Speranskaya V.S. 2017. And now when the trees became big… About restoration of the open spaces of historical value in the center of Petersburg). *Vestnik of Saint Petersburg University. Arts* 7 (2): 232–248.

Sementsov, S.V. et al. 2017. Rivers and canals as the main public spaces of town-planning compositions and functional systems of the largest cities. *Water and Ecology* 4 (72): 86–94.

Supranovich, V.M. & Vabishchevich, D.A. 2019. Principles of the transformation of urban industrial territories adjacent to water bodies. In *Challenging Issues of Architecture and Design: Proceedings of the All-Russian Scientific Conference of Students and Young Scientists*: 158–160.

Supranovich, V.M. et al. 2017. The Admiralty District territory renovation, of Staro-Petergofsky, Narvsky Prospect, Bumazhnaya Street and Obvodniy Chanel embankment area. In Ye.B. Smirnov et al. (eds), *Masters' hearings. Proceedings of the Inter-Regional Scientific and Practical conference in the framework of the 7th Inter-Regional Creative Forum "Architectural Seasons at Saint Petersburg State University of Architecture and Civil Engineering", Saint Petersburg April 18–21, 2017*: 42–43. Saint Petersburg: Saint Petersburg State University of Architecture and Civil Engineering.

Reconstruction and Restoration of Architectural Heritage – Sementsov et al (eds)
© 2020 Taylor & Francis Group, London, ISBN 978-0-367-65357-6

Laws of definition of subject of protection of houses and palaces of culture built in Leningrad in 1920-1930s

N.P. Dubrovina

Saint Petersburg State University of Architecture and Civil Engineering, Saint Petersburg, Russia

ABSTRACT: The article is the result of a comprehensive study of a special type of buildings – Houses and Palaces of Culture in Leningrad of the 1920-1930s, including the identification of the main problems of their use and maintenance. The experience of restoration and adaptation of Soviet architecture of the first third of the 20th century totals several decades, and the results of the first restoration work on such buildings cannot be called satisfactory. It becomes obvious that the "new heritage" requires the formation of a special research and restoration methods. "The Subject of Protection" is currently the main document regulating the permissible degree of intervention in the restoration and adaptation of the architectural heritage. The article is based on the results of the detailed author's study of the Houses and Palaces of Culture in Leningrad and the analysis of the existing subjects of protection for such buildings. The main features of the Houses and Palaces of Culture in Leningrad are determined, methods for supplementing and refining the subjects of protection for such objects is proposed.

1 INTRODUCTION

The relevance of the study is due, on the one hand, to the growing interest in Soviet architecture of the 1920-1930s, and on the other hand, to the unsatisfactory state of many particularly unique monuments of "constructivism". The bibliography of works by Russian and foreign scientists on architectural trends in the first third of the 20th century is quite extensive. The works by A.G. Vaitens (1995), R. Dayanov (2018), B. M. Kirikov (Kirikov & Stieglitz 2008) S. V. Sementsov (2012), T.A. Slavina, M.S. Stieglitz are devoted to urban planning and architecture of Leningrad of the studied period, including some monuments of the architectural avant-garde.

The works of T.A. Slavina, S.V. Sementsov, A.V. Mikhailov (2017) and others are aimed at the development of scientific and methodological support for the protection of cultural heritage objects, including methods for determining the subject of protection and conducting historical and cultural expertise.

However, it is worth noting the insufficient knowledge of the architecture of the "constructivism" period in Leningrad. Briefly, without details, the general lists of monuments of the architectural avant-garde are compiled. Only some features of individual buildings have been studied in detail, the studies of this kind are mainly associated with the refinement of the dates and construction history. The palaces of culture, which are a new and unique type of buildings, are studied fragmentarily.

For the study, the largest separately standing newly built (with the exception of Palace of Culture of Communications Workers) Palaces of Culture of the Central, Petrograd, Vasileostrovsky, Moscow-Narva, Volodarsky districts of Saint Petersburg in the constructivism style were selected: Lensovet's Palace of Culture (the cultural heritage object of regional significance), Palace of Culture named after S.M. Kirov (the cultural heritage object of federal significance), Palace of Culture of Communication Workers (a newly identified cultural heritage object), Palace of Culture named after Maxim Gorky (the cultural heritage object of federal significance), Palace of Culture named after Gaza (a newly identified cultural heritage object), Ilyich House of Culture (a newly identified cultural heritage object), Kapranov House of Culture (destroyed), Palace of Culture named after N.K. Krupskaya (a newly identified cultural heritage object), Palace of Culture named after V.I. Lenin (a newly identified cultural heritage object). For each Palace of Culture, an analysis system has been adopted, including the urban planning situation, architectural and artistic description of the building and the building's construction history. The purpose of this research is to propose the methods for supplementing and elaborating the subjects of protection of the Houses and Palaces of Culture in Saint Petersburg of 1920-1930s to ensure the most complete preservation of the most important characteristics of this type of buildings.

2 METHODS

The methodological basis of the study is a comprehensive approach to the study of the architectural heritage of the 1920-1930s, Houses and Palaces of Culture of Leningrad (now Saint Petersburg), namely: the study of archival and published scientific, bibliographic and iconographic sources on the research topic; analysis of existing subjects of protection of the Houses and Palaces of Culture in Leningrad (now Saint Petersburg); on-site examination; cameral processing of the completed studies with the preparation of detailed graphic models. In order to formulate the methods for elaborating and supplementing the subjects of protection for Leningrad's Houses and Palaces of Culture, the following tasks have to be successively solved:

a) identifying the main problems of the content and use of the "constructivist" Houses and Palaces of Culture in Leningrad;
b) revealing the unique architectural and artistic features of the studied objects, the loss of which will cause irreparable damage to the cultural heritage objects;
c) analyzing the existing subjects of protection compiled for the Houses and Palaces of Culture in Leningrad (now Saint Petersburg) of the 1920-1930s;
d) giving suggestions for supplementing and elaborating the existing subjects of protection.

3 RESULTS

A) The study of the current state of Houses and Palaces of Culture in Leningrad (now Saint Petersburg) made it possible to formulate the main problems (Dubrovina 2019) of their maintenance and use, among which are the unsettled areas of the state heritage protection system, including activities aimed at preserving, maintaining and successful functioning of the buildings of the Palaces of Culture (including the definition of the subject of protection).

B) Based on the detailed study of all Houses and Palaces of Culture in Leningrad of the "constructivism" style, a number of features that need to be included in the subject of protection, can be distinguished.

All studied Palaces of Culture were designed as part of the urban planning, cultural and educational center of the All-Union (Palace of Culture named after Maxim Gorky, Palace of Culture named after S.M. Kirov (Dubrovina 2020)), urban (Lensovet's Palace of Culture, Ilyich House of Culture, Palace of Culture of Communications Workers) or of regional significance (Kapranov House of Culture, Palace of Culture of Textile Workers, Palace of Culture named after V.I. Lenin, Palace of Culture named after I.I. Gaza).

Stylistic attraction to classic forms can be considered a distinctive feature of the Houses and Palaces of Culture in Leningrad. The very first Palaces of Culture, built before 1930, do not yet acquire radical constructivist forms, although they already belong to this architectural direction (for example, Palace of Culture named after

Maxim Gorky, Palace of Culture of Textile Workers). Three Palaces of Culture were sustained in pure forms of constructivism - Ilyich House of Culture of, Palace of Culture named after V.I. Lenin, Kapranov Palace of Culture (destroyed).

All Palaces and Houses of Culture were designed with functional division into theatrical and club parts (Figure 1). The first Palaces of Culture have a compact planning scheme: the club sector is located around the central core of the theater sector (Palace of Culture named after Maxim Gorky, Palace of Culture of Textile Workers). Also, the club sector is not isolated in Palace of Culture of Communication Workers, as this is an example of the reconstruction of an existing building, carried out in extremely cramped conditions of dense historical development of the center of Saint Petersburg. In buildings built after 1927, the club sector is more developed; it is a separate part of the building, capable of functioning separately from the theater sector.

In all the studied objects, the theater sector implies a certain set of ceremonial rooms - the entrance hall and the foyer system, built on the principle of flowing space. The exception here is the Palace of Culture of Communication Workers, as this is not a newly constructed facility, but the result of the reconstruction of the existing church. Club sectors (in case of their expressiveness and isolation) are also equipped with a developed system of entrance halls and foyers, which are built on the principle of flowing space along with other rooms: main staircases, dining rooms, lounges, libraries, front rooms, etc. This structure of rooms was first used for the construction of the houses of culture and is the most

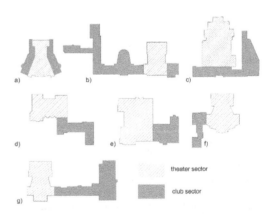

Figure 1. The schemes of the functional zoning of the houses and palaces of culture in Leningrad into the club sector and theater sector: a) Palace of culture named after maxim Gorky; b) Palace of culture named after S.M. Kirov; c) Lensovet's palace of culture; d) Ilyich#ulture; e) Palace of Culture named after V.I. Lenin; f) House of culture of textile workers; g) Palace of culture named after I.I. Gaza. Author's scheme.

important feature of the monuments, the basis of their space-planning system and architectural and artistic solutions to the interiors.

C) An analysis of the existing subjects of protection compiled for the Houses and Palaces of Culture in Leningrad of 1920–1930s was made. Comparing the results obtained with the identified features of the Houses and Palaces of Culture, one can note a number of problems of state protection of the "constructivism" style houses and palaces of culture. The subject of protection is only the preserved historical elements of buildings relating to the construction period. The completeness of the subject of protection depends on the safety of the building. A purely fixative approach to identifying the subject of protection is observed. In all investigated and analyzed documents, the subject of protection is detected pointwise, the building is "divided" into individual elements and zones. The subject of protection in most cases is not the most important features and characteristics of the Palaces of Culture.

Urban importance of the Palace of Culture. One of the most important features of this type of building is the Palace as part of an architectural and urban planning ensemble or complex of buildings, the urban planning dominant, and the center of a new district of Leningrad.

Historical spatial planning solution of the buildings built on the principle of flowing and transformable space. The subject of protection is only the spatial planning solution of each Palace of Culture within the boundaries of the bearing walls. Such an approach to determining the subject of protection does not ensure the preservation of the historical spatial planning structure of buildings with a developed system of vestibules, foyers, main staircases, halls, lounges, libraries and other rooms.

Later changes and unrealized design decisions. In some cases, it is necessary to include in the subject of protection unrealized design decisions based on historical and cultural expertise (an example is the corner tower of the Lensovet's Palace of Culture (Figure 2). Having traced the full construction history of the Palaces of Culture, the reasons for the changes, authorship and dating, we can conclude that it is necessary to include later changes of some objects in the subject of protection.

The historical functional purpose of the building as a whole and of its individual rooms. The need to preserve the general functional purpose of buildings (the Palace of Culture), the general functional division into theatrical and club sectors (compact or with a developed club sector), as well as the functional purpose of separate rooms (entrance hall, foyer, auditorium, spectator sector premises, corridors, libraries, sports complexes, ceremonial halls, etc.), since a radical change in the functional purpose of the building as a whole

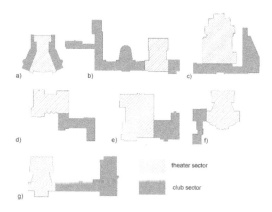

Figure 2. Lensovet's Palace of culture: a) project by E.A. Levinson and V.O. Munts, 1931, Central Saint Petersburg state archive of literature and art; b) current situation, photograph of the author.

or in the main rooms in all the objects under study entails serious changes in the spatial planning structure and decorative and artistic solutions to the interiors.

Historical decorative and artistic solution to interiors. Only preserved elements of decorative and artistic decoration of interiors are included in the subject of protection. In some cases, it is necessary to include historical interior solutions in "constructivism" style in the subject of protection.

Elements of memorial value. To determine the elements, objects, characteristics that make up the historical and cultural potential of the building, a thorough study of the history of the building, connection with outstanding personalities, important historical events is necessary. The subject of protection may be the historical name of the Palace of Culture, commemorative signs, late layers, which have important memorial significance.

D) For a more successful conservation of the most important characteristics of the Houses and Palaces of Culture in Leningrad (now Saint Petersburg), supplementing and elaborating the subjects of protection is required. The subjects of protection should be supplemented by graphic schemes and include the following sections:

Urban planning significance of the House of Culture as part of a city, district

Spatial solution of the building

Spatial planning solution of the building

Architectural and artistic solution of facades

Structural systems of the building

Interiors:

Historical decorative and artistic decoration and three-dimensional solution (including preserved decoration details) - zone I

Historical spatial solution (including preserved decoration details) - Zone II

All preserved historical elements

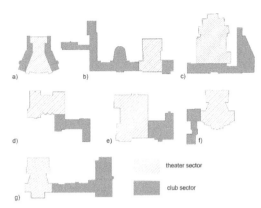

Figure 3. Graphic diagrams of existing and proposed subjects of protection based on the example of: a) Palace of culture named after S.M. Kirov; b) Lensovet's palace of culture; c) Palace of culture named after maxim Gorky; d) Palace of culture named after I.I. Gaza; e) Ilyich house of culture. Author's scheme.

The historical functional purpose of the premises
Items of memorial value
When drawing up graphic diagrams of existing and proposed subjects of protection, the transition from the fixative character of this document to the fixative-analytical one (Figure 3) is most clearly visible.

4 DISCUSSION

In this article, the "constructivist" houses and palaces of culture in Leningrad are considered for the first time as a special type of building with unique architectural and artistic characteristics and features. The main principles of the protection of such buildings are introduced into scientific circulation, taking into account such features to the fullest extent.

5 CONCLUSION

The analysis of the existing subjects of protection of the Houses and Palaces of Culture in Leningrad of 1920-1930s showed that the existing subjects of protection do not reflect the most important characteristics of the phenomenon of the Palace of Culture, which form the basis of stylistic, architectural, spatial solutions: urban planning significance, spatial planning solution according to the flowing space principle, architectural and artistic solution of interiors, functional zoning. This indicates the insufficient study of the phenomenon of the Palace of Culture, the need to form a special methodology for the state protection system of such objects. The subjects of protection for constructivist Palaces of Culture as the main tool for state

regulation of all construction works on the monument require additions and clarifications.

6 RECOMMENDATIONS

The results of the study can be recommended to architects in developing projects for the restoration and reconstruction of Houses and Palaces of Culture in Saint Petersburg and other cities, experts and art historians in the formation or updating of the subject of protection of objects of cultural heritage of the "avant-garde" era, as well as in the educational process of universities in the education of bachelors, masters, postgraduate students.

ACKNOWLEDGEMENTS

The author wants to express gratitude to her supervisor S.V. Sementsov who for many years patiently helps in the work aimed at studying and preserving the architectural heritage of the 20th century. The author expresses recognition to researchers whose work is aimed at ensuring the protection of cultural heritage sites, including T.A. Slavina, S.V. Sementsov, M.S. Stieglitz, B.M. Kirikov, etc.

REFERENCES

Allan, J. 2007. Points of balance: Patterns of practice in the conservation of modern architecture. *Journal of Architectural Conservation* 13 (2): 13–46.

Dayanov, R. M. 2018. LOSPK Sporting Palace and Palace of Culture: the authorship, construction time and continuance in compositional techniques. *Bulletin of Civil Engineers* 1 (66): 10–16.

Dubrovina, N.P. 2019. Palaces of Culture in Leningrad. Problems of maintenance and preservation. *Urbanistika* 3: 7–17.

Dubrovina, N.P. 2020. Urban role of the Leningrad palaces of culture of the first third of the XX century on the example of Petrogradsky and Vasileorovsky districts. *Bulletin of BSTU named after V.G. Shukhov* 3: 76–81.

Hatherley, O., 2016. *Landscapes of communism: a history through buildings.* New York: The New Press.

Kirikov, B.M. & Stieglitz, M.S. 2008. *Leningrad avant-garde architecture: A guide.* Saint Petersburg: Kolo.

Metspalu, P. & Hess, D.B. 2018. Revisiting the role of architects in planning large-scale housing in the USSR: the birth of socialist residential districts in Tallinn, Estonia, 1957–1979. *Planning Perspectives* 33 (3): 335–361.

Mikhailov, A.V. 2017. The main directions of the evolution of demand for cultural heritage sites preservation. the safeguarding of the intangible features. In V. P. Solomin et al. (eds), *Natural and cultural heritage: interdisciplinary research, conservation and development. Multi-authored monograph based on the Proceedings of the 6th International Scientific and Practical Conference, October 25 –26,2017:* 69–72. Saint Petersburg: Publishing House of Herzen State Pedagogical University.

Sementsov, S.V. 2012. Urban-planning of Petrograd—Leningrad: revolutionary defeat of 1917–1918 to the renaissance of 1935. *Vestnik of Saint Petersburg University. Arts* 2 (1): 130–143.

Saint Petersburg Central State Archive of Literature and Art. Fund 345, List 3, Case 19, Sheet 2. Letter of January 19, 1961, No. 5/438.

Vaitens, A.G. 1995. The architecture of constructivism in Leningrad: ideas and results. In Zh. M. Verzhbitskii & Iu.R. Savel'ev (eds), *One Hundred Years of the Study of Russian Architecture: collection of scientific articles*. Saint Petersburg: Saint Petersburg Repin State Academic Institute of Painting, Sculpture and Architecture.

Urban planning and compositional features of different territories of historical center of Saint Petersburg

G.O. Fedotova
Saint Petersburg State University of Architecture and Civil Engineering, Saint Petersburg, Russia

ABSTRACT: A system of parameters is determined for each of the studied areas of the historical center of Saint Petersburg (a fragment of the core of the historical center of the 18th century in the area of Morskaya Streets, a fragment of the historical center of the 19th century in the area of Troitskaya Street, a fragment of the periphery of the historical center of the second half of the 18th - early 20th centuries in the area of Malaya Kolomna) and their comparative analysis was carried out, the identity of the architectural and planning organization of the zones was revealed.

1 INTRODUCTION

This work is based on basic research in the theory and history of architecture and urban planning. N.V. Bagrova, M.A. Kushchenkova, A.V. Makhrovskaya (Makhrovskaya 1986); E.A. Planck, S.K. Regame's works are devoted to the issues of the integrity of the architectural environment, the transformation and the continuous development of urban environment with a valuable historical and cultural heritage, and the problems of the historical environment reconstruction. Theoretical and practical aspects of the perception of the urban environment are investigated in the works of K. Lynch (Lynch 1982); L.P. Panova, S.V. Sementsov, L.I. Sokolov, S.A. Khasieva, Z.N. Yargina (Yargina et al. 1986). The principles and methods of a comprehensive analysis of the urban historical environment, associated with the preservation of significant characteristics of the urban environment and the achievement of a combination of new buildings with the historical ones are considered in the works of L.S. Martyshova, Yu.S. Popkov, M.V. Posokhin, A.E. Gutnov (Gutnov 1984), B.L. Shumilin, S.K. Regame (Regame et al. 1989), D.V. Bruns, G.B. Omelyanenko, S.V. Sementsov, A.S. Schenkov (Schenkov 2003), Z.N. Yargina. The formation of the urban planning system of Saint Petersburg and its architecture is described in the works of A.G. Vaitens, V.S. Goryunov, S.P. Zavarikhin, B.M. Kirikov, E.I. Kirichenko, L.P. Lavrov (Lavrov 2015), V.G. Lisovsky, V.I. Pilyavsky, S.V. Sementsov (Sementsov 2007, 2013), T.A. Slavina, K.A. Sharlygina (Sharlygina 2019) and others.

However, earlier studies have not disclosed a comprehensive systematic approach to studying the development of individual fragments of the historical center of Saint Petersburg. In the majority of works on this subject, only some urban or architectural issues are considered.

This article formulates the conclusions of studies of three different areas of the historical center of Saint Petersburg, which were formed during the 18th - early 20th centuries:

- a fragment of the historic core in the area of Morskaya Streets (Nevsky Avenue, Admiralteysky Avenue, Voznesensky Avenue, Moika Embankment);
- a fragment of the historic core in the area of Troitskaya Street (now Rubinstein Street) (Fontanka Embankment, Lomonosova Street, Zagorodny Avenue, Vladimirsky Avenue and Nevsky Avenue);
- a fragment of the periphery of the historical center in the area of Malaya Kolomna (Angliysky Avenue, Dekabristov Street, Pryazhka Embankment, Lotsmanskaya Street, Rimsky-Korsakov Avenue).

2 METHODS

The methodology and research methods are based on an integrated approach, including:

- study, analysis and generalization of material (theoretical, methodological, historical and archival, on-site, etc.), reflecting the features and sequence of formation of the studied areas;
- an urban planning analysis of the formation of the architectural and planning system of the territory;
- a comparative analysis of the prevailing characteristics of the historical environment.

The source base of the study was:

- written (archival, literary), graphic (historical plans, historical art images, fixation and design archival drawings, historical photographs) sources. The main sources on the history of the

development of buildings in Saint Petersburg are stored in the funds of the Russian Academy of Sciences, Russian State Archive of Ancient Acts, the Russian State Historical Archive, the Central State Historical Archive of Saint Petersburg, the National Library of Russia, the Central City Public Library named after V.V. Mayakovsky, the Scientific and Technical Library of Saint Petersburg State University of Architecture and Civil Engineering;
– materials of an on-site inspection of the territory (measurements, photo fixation, results of an on-site visual examination).

3 RESULTS

The main stages of formation and morphotype of development of the studied fragments of the urban environment of Saint Petersburg have been determined.

The planning system of Morskaya Streets area, as well as the space-planning solution for the development, were mainly formed by the end of the 18th century. The framework of the territory is formed by a radial-ring structure with five beams at the base and a system of front squares. The historical land surveying of the sites has been completely preserved. The development is super-dense, of a mesh type, with perimeter and firewalls. The morphotype of the area's environment can be attributed to the development of the "core" of the center of Saint Petersburg in the second half of the 18th century.

The transitional nature of the Troitskaya Street area in the city structure (from the suburban territory to the city center) determined the features of the formation of the framework and fabric. The foundations of the planning structure laid in the 18th century significantly influenced the formation and parameters of the urban fabric, which appeared mainly in the 19th century. The development of the area can be attributed to the morphotype of the historical center of Saint Petersburg of the 19th century with the inclusion of elements of manor buildings in the first half of the 18th century (Fedotova 2019).

The architectural and planning organization of the development of Malaya Kolomna area is determined by its peripheral location. The framework of the territory was formed in the 18th century at a time according to the general plan. Historical land surveying is partially lost. Development according to the spatial planning solution - cellular, with the perimeter and firewalls, with a single facade along the red line - basically developed only by the beginning of the 20th century. Thus, the morphotype of the environment of the studied area can be attributed to the development of the periphery of the center of Saint Petersburg in the second half of the 18th - early 20th centuries.

The compositional role of the studied areas in the urban planning system of the historical center of Saint Petersburg is presented in Figure 1.

The Morskaya Streets area is part of the ensemble of the central Neva, the central squares of Saint Petersburg; it includes an ensemble of three beams with the center at the Main Admiralty, in particular, the ensemble of Nevsky Avenue. The ensemble of the central squares includes dominants of regional and city significance: the central tower with the spire of the Admiralty, 74.0 m high, and St. Isaac's Cathedral, 102.0 m high. This ensemble was formed in stages not based on the project master plans for the settlement of this zone by sequentially clarifying the framework, mainly formed by the middle of the 18th century, the development of a system of urban planning dominants and the formation of a great super dense cellular perimeter development by means of successive "point" reconstruction of the development of plots.

The Troitskaya Street area, which was formed gradually through the evolutionary development of the territory from the suburban to the historical center, is not included in urban planning ensembles of the city level, with the exception of the part adjacent to Nevsky Avenue (the section from the Fontanka River to Vosstaniya Square). The area borders on Vladimirskaya Square ensemble, including the dominant of urban importance – the Vladimir Cathedral, with a height of 68.0 m. The area includes the ensemble of Rubinstein Street, starting from Nevsky Avenue and ending with the ensemble of "Five Corners" with the dominant of local significance – a tenement house with a tower located on the corner of Rubinstein Street and Zagorodny Avenue. Also, on Rubinstein Street there are two buildings-ensembles built in the early 20th century, organizing the space of the street - the former tenement building of Count M.P. Tolstoy, with a system of courtyards, and ex-residential complex of the Saint Petersburg Merchant Council with court of honor.

Malaya Kolomna area is part of a single ensemble of Malaya and Bolshaya Kolomna, with a developed system of hierarchically organized areas and dominants. The study area borders on the ensemble of the Repin Square (with the dominant of local importance – an assembly house ending in a tower); and also includes the ensemble of the Kulibin Square (formerly the Voskresenskaya Square), which previously housed the dominant of urban significance - the Resurrection Church, 55.8 m high (now lost).

3.1 The elements of architectural compositional structure

The elements of the architectural and compositional structure of this level are determined by the organization system of the facade line; a system of alternating dominants of local significance, emphasis and background development; the number of hierarchical levels of tectonic and decorative-plastic design of facades.

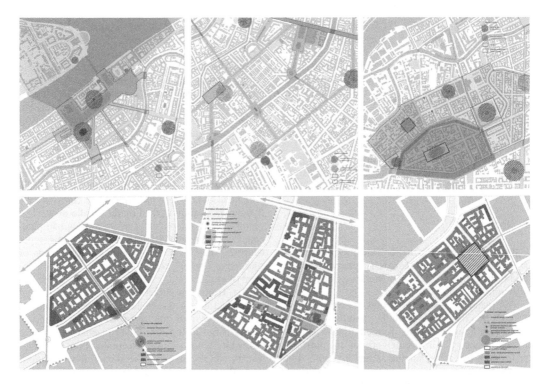

Figure 1. Urban planning and compositional role of the studied areas in the structure of the historical center of Saint Petersburg: on the left — the core of the historical center of the 18th century (Morskaya Streets area), in the middle — the historical center of the 19th century (Troitskaya Street area), on the right — the periphery of the historical center of the 18th - early 20th centuries (Malaya Kolomna area).

The development in the Morskaya Streets area has a formed front with front facades along the red line under a single cornice with small height differences of 1-3 floors, a developed system of dominants and emphasis of local importance and a large number of hierarchical levels of organization of tectonic and decorative-plastic design of facades.

The development in the Troitskaya Street area has a formed front of building with front facades along the red line, individual sections have compositionally organized gaps, and there are also sections with an unformed red line. The system of dominants and local emphasis is developed, but has a greater pitch and lower density than in the center core zone.

Buildings in Malaya Kolomna area have a formed front of the development with front facades along the red line. However, there are a lot of plots with unformed buildings or buildings that violate the historical structure in the area. Dominants and emphasis are located rarely and irregularly and do not form a single system.

4 DISCUSSION

In Saint Petersburg, a special principle of differentiation by type of development (center of the city – periphery - suburbs) in accordance with the "centrality" (according to S. Sementsov), which was maintained (and specified) in the process of formation and development of the city, has developed. As the border of the city gradually shifted, and the suburban territories were originally included in the center's development, the modern historical center of Saint Petersburg has a heterogeneous structure: individual zones have special planning and landscape-compositional system parameters that formed during the evolutionary development of the urban environment and determine its uniqueness. The study of these parameters will reveal the characteristics that determine the "centrality" of the development of the historical center of Saint Petersburg, and work out a strategy for ensuring its successive development.

5 CONCLUSIONS

1. The studied areas are included in the ensemble of the historical center of Saint Petersburg, as well as individual systems of ensembles of regional, city and local significance. The closer the area is to the core of the historical center, the more developed the system of ensembles in which it is included is.

2. The urban planning framework and the quarterly land surveying sites laid at the beginning of the 18th century significantly influenced the formation of the fabric of the studied areas.

3. The closer the area is to the core of the historical center and the stronger its role in the structure of the historical center of the city is, the more complete the structure of the fabric it has, with a higher building density and a more developed system of emphasis and dominants.

REFERENCES

Fedotova, G.O. 2019. Urban planning system and architectural treatment of the former Troitskaya street area in Saint Petersburg. *Architecton. Proceedings of Higher Education*: 4 (68).

Gutnov, A.E. 1984. *The evolution of urban development*. Moscow: Stroyizdat.

Lavrov, L.P. 2015. Reconstruction of the historical core of Saint Petersburg landscapes in 1870–1910. *Vestnik of Saint Petersburg University. Arts* 5(3): 78–96.

Lynch, K. 1982. *The image of the city*. Translated by V. L. Glazychev. Moscow: Stroyizdat.

Makhrovskaya, A.V. 1986. *Reconstruction of old residential areas of large cities: case study of Leningrad*. Leningrad: Stroyizdat.

Regame, S.K. et al. 1989. *The combination of new and existing buildings during the reconstruction of cities*. Moscow: Stroyizdat.

Sementsov, S.V. 2007. Town planning aspect of Saint Petersburg residential area and Saint Petersburg agglomeration. 1703–2006. *Vestnik of Saint Petersburg University. History* 2 (3): 63–70.

Sementsov, S.V. 2013. Introducing the principles of preserving the architectural-urban development heritage of Saint-Petersburg on the base of regulations of its three-century town planning development. *Vestnik of Saint Petersburg University. Arts* 3 (2): 190–211.

Sementsov, S.V. & Voznyak, E.R. 2017. Compositional structure of the facades of buildings in the XVIII century and its projection in the architectural and urban environment of Saint Petersburg. *Bulletin of Civil Engineers* 4 (63): 55–60.

Sharlygina, K.A. 2019. *Experience in the reconstruction of historic residential buildings in Saint Petersburg*: Saint Petersburg: Petropolis.

Schenkov, A.S. 2003. *Manual on historical and architectural pre-design studies of historical settlements*. Moscow: Moscow Institute of Architecture (MARKHI), Scientific Research Institute of Theory and History of Architecture and Urban Planning (NIITIAG).

Yargina, Z.N. et al. 1986. *Fundamentals of the theory of urban planning*. Moscow: Stroyizdat.

Problems of adaptation of cultural heritage sites for people with limited mobility

T.S. Fridman
Saint Petersburg State University of Architecture and Civil Engineering, Saint Petersburg, Russia

ABSTRACT: Today, the problem of creating an accessible environment is one of the most urgent ones that modern architects and restorers face. When adapting objects of cultural heritage, a number of difficulties arise, associated with the uniqueness of the architectural appearance of the building being adapted, the value characteristics of cultural heritage objects, the complex spatial and structural planning of buildings, the insufficient development of the legislative framework for working with heritage in this area, the inability to introduce most modern design achievements to adapt the environment into the historical space of the building. In this regard, at present, most of the historical buildings of Saint Petersburg and other cities of Russia continue to be inaccessible to people with disabilities, which is unacceptable in modern society. It is necessary to find a balance between the preservation of the historical heritage of cities and the development of a modern, accessible and comfortable living environment for all.

1 INTRODUCTION

In Europe and the USA, the problem of adapting buildings for people with limited mobility arose and was actively developed in the early 1990s. In the USA, the impetus for this was the movement for the rights of disabled veterans of the Vietnam War, in Europe – discrimination against people with disabilities.

The emergence and spread of disability rights in Europe happened incredibly fast. Twenty years ago, only two states, Sweden and Germany, included disability in their Constitution as a specially listed basis deserving of the protection of equality, and no European state had national legislation setting out measures to combat discrimination on the basis of disability. Today, more than thirty European countries have ample opportunity to protect the equality of persons with disabilities (SIMC 2017).

The international way to disseminate norms and formulate legislation in the field of protecting the rights of people with disabilities and, as a result, ensure an accessible environment for them, began to be actively developed with the entry into force and ratification of the UN Convention on the Rights of Persons with Disabilities in a number of countries around the world (SIMC 2006).

Prior to this, people with disabilities were not considered as legal and equal recipients of services and had to be excluded and isolated from the rest of society and provided with separate institutions necessary for their full-fledged life activity, since they cannot function in a "normal society" (*Federal Law 2012*).

Disability policy and its research around the world revealed that disability consists of personal, social factors and the impact of environmental accessibility. Given all these factors and the fact that disability is associated with a number of such disorders, it can be argued that the accessibility of the environment affects the disruption and functioning of people with disabilities as full members of society (*Construction Rules 2016*).

Now the movement in the field of architecture and design is beginning to take shape, approving a new set of design principles known as universal design. Universal design is based on the notion that today's hard-to-reach environment is mainly the result of inattention to the needs of users with special needs, as well as the fact that a well-planned environment can be more cost-effective.

Today, one of the most important tasks in state policy is the coordination and settlement of interests in the preservation of the unique cultural heritage of Saint Petersburg and the development of the city as a large modern megalopolis, requiring significant changes and interventions in the historical environment (*Decree of the Government 2015*).

Reasonable new inclusions can not only serve as the development of the urban planning and historical fabric of the city, but also become a great investment project, which will serve as a good impetus for the development of the city on the world stage. Adapted public buildings, higher education institutions, museums, theaters, hotels can positively affect the social and economic situation in the city, which will be expressed in an increase in the flow of people to the target areas of adaptable objects due to their easy accessibility and created comfortable environment.

The state cultural policy of Saint Petersburg is aimed at the rational use and development of the cultural heritage of the city. Priority is given to the problems of society and various groups of the state as a whole. When updating monuments, one should not forget that they are of value both for the state and society, which is the key to a successful state policy in the field of development of historical heritage (*Decree of the Government 2015*).

By means of cultural heritage and later developments, contact can be established between generations, on the educational and utilitarian side (*Federal Law 2007*).

When developing a methodology for working with monuments while creating an accessible environment and introducing devices for people with limited mobility, the main problem is the struggle between preserving a unique architectural appearance and at the same time meeting the requirements of people with disabilities. In addition, it is worth noting that the difficulties encountered in the design, coordination and implementation of projects negatively affect the investment climate of the city and make it not attractive enough for investment, and as a consequence for development.

Considering historical architecture and its changes due to the large number of restructuring activities, it can be compared with a multi-level architectural "template". The "template" in this context, as a rule, was not a complete architectural image, but only represented an external shell ready for further development and transformation. Many historical buildings that have acquired the status of monuments over the years, for the most part, upon closer examination, have the status of a "template" and, in essence, should be fully adapted to the needs of the population for their modern efficient use (Davydova, E. M., Radchenko, V. Yu. 2015).

The most striking example in connection with the growing interest and demand for education in order to create an accessible environment is the adaptation of cultural heritage objects of buildings of higher educational institutions. Considering and analyzing the spatial planning structure of buildings for educational purposes on the example of the Saint Petersburg State University of Architecture and Civil Engineering, S.M. Kirov Saint Petersburg State Forestry University, Ilya Repin Saint Petersburg State Academic Institute of Painting, Sculpture and Architecture at the Russian Academy of Arts, the author identified the main problem areas that require special measures to create an accessible environment for all groups of the population.

Saint Petersburg State University of Architecture and Civil Engineering is located in the Admiralteysky district of Saint Petersburg and is a complex of buildings located in two blocks that does not have a separate territory. Moving between the buildings is difficult for people with disabilities due to the large number of pedestrian crossings through the carriageway, 2nd and 3rd Krasnoarmeyskaya streets and Egorova street.

The inner space of the main building, an object of cultural heritage, has a complex spatial planning structure and pronounced value characteristics, which makes it difficult to introduce existing modern devices for people with disabilities. A large number of stairs, ways across the floors with a height difference on the same level, elevators that are not suitable for use by disabled people in wheelchairs, as well as narrow aisles along corridors with doors opening to the outside, require special measures for their adaptation.

The building of S.M. Kirov Saint Petersburg State Forestry University, located in the Vyborgsky district of Saint Petersburg, unlike the complex of buildings of Saint Petersburg State University of Architecture and Civil Engineering, has a separate park territory.

The main entrance to the university is located on Novorossiyskaya street and the central alley of the university park. Along the central alley there are pedestrian zones available for moving a disabled person in a wheelchair. Moving disabled people with other mobility impairments will be difficult due to the lack of information on the territory of the university and the park, tactile and contrasting markings, sound and light beacons.

Before entering the main historical educational building there is a parking for people with disabilities. There is a large pedestrian zone in front of the university entrance.

The entrance area to the university is difficult for people with disabilities. Historic doors are not equipped with closers and have thresholds that do not meet the requirements for building accessibility for people with limited mobility.

At the entrance to the building there is an entrance staircase leading from the main entrance to the target areas of the object, classrooms, workshops, an assembly hall and the administrative part of the building.

The buildings of the university complex are located throughout the park. Along the main alley, in addition to the main educational and administrative building, there are residential historical buildings of the teaching staff of the former academy. Some university art workshops are located in the western part of the park and are difficult to access for people with limited mobility, as the path to them passes directly through the park and is not fully adapted for people with limited mobility.

Ilya Repin Saint Petersburg State Academic Institute of Painting, Sculpture and Architecture is located in one building and for its adaptation it is not necessary to carry out work to adapt the whole complex.

The Academy building is located on Vasilievsky Island in a quarter limited by 4-5 lines, 2-3 lines of Vasilievsky Island, Universitetskaya Embankment and the Academic Garden adjacent to the quarter, facing Bolshoy Avenue of Vasilievsky Island.

The building of the cultural heritage site where the Academy is located has not been reconstructed since its foundation, and therefore it has preserved a historical layout that was not intended for the needs of people with disabilities.

2 METHODS

The analysis of the spatial planning structure of the academy showed that the building is difficult to access from the point of view of transport routes. There are no metro stations near the building. The nearest public transport stop is 150 meters from the building on Universitetskaya Embankment.

The historical territory adjacent to the Academy building is sufficiently adapted for people with disabilities, sidewalks have comfortable ramps for the disabled on a wheelchair. Despite this, the historical territory needs adaptation and implementation of devices for other groups of people with limited mobility, tactile and contrast markings, light and sound beacons.

The entrance zone to the building is located on the waterfront and needs adaptation for all groups of the population. Historical entrance doors, entrance stairs with the threshold in the entrance area leading to the academy's special purpose area and the museum, historical floor coverings of steps and the entire floor surface, which are the subject of protection, numerous elevation changes from the entrance lobby to functional areas of the university make it difficult to move.

Based on the analysis of the spatial planning structure of a number of cultural heritage sites, the main ways of moving people in buildings, evaluating their value characteristics, subjects of protection, as well as the possibility of using modern means to ensure the accessibility of the environment in the historical environment, the following problem areas have been singled out:

1. the entrance area
2. the public area
3. the administrative area
4. the leisure area
5. the target areas
6. the territory of the building.

3 RESULTS

As recommendations for measures to create an accessible environment in the buildings of higher educational institutions, the following principles were developed: the compromise principle, which is to create an accessible environment in compromise with the value characteristics of the adapted building, to a greater extent relying on the needs of society in ensuring the accessibility of the building than on the value of the adapted zones, subjects of protection, the principle of integration, the organization of additional entrances and traffic routes in the building by adding new volumes to the historical volume of the building, and the individualization principle (Figure 1).

The individualization principle implies the creation of an accessible environment through the use of devices for low-mobility groups of the population, developed taking into account the architectural and artistic characteristics of the historic building and having an individual design.

Consideration of devices for people with limited health abilities used to create an accessible environment for a monument of architecture as an architectural element can minimize the damage caused by similar reconstructive measures to the historical environment and the value characteristics of the monument.

The introduction of new architectural elements can be developed in accordance with the choice of architectural prototype in the reconstructed environment of the monument, addressed to the techniques and motives of the historical style. In such cases, there should be, first and foremost, an integral connection between the function and the style solution of the new architectural element.

The individualization principle implies the creation of an accessible environment through the use of devices for low-mobility groups of the population, developed taking into account the architectural and artistic characteristics of the historic building and having an individual design.

Consideration of devices for people with limited health abilities used to create an accessible environment for a monument of architecture as an architectural element can minimize the damage caused by similar reconstructive measures to the historical environment and the value characteristics of the monument.

The introduction of new architectural elements can be developed in accordance with the choice of architectural prototype in the reconstructed environment of the monument, addressed to the techniques and motives of the historical style. In such cases, there should be, first and foremost, an integral connection between the function and the style solution of the new architectural element.

Quite often, historical buildings already have great potential for their competent inclusion in the accessible environment of the city. Some historically designed details can be reinterpreted and interpreted as a device for people with disabilities with a slight change and, accordingly, minimal introduction into the historical environment.

Interpretation in the heritage sector has many uses, and a number of definitions are used to describe it. The basis, however, is the transfer of information from the institution to the visitor. Interpretation is the means by which we transmit this information. Often this is associated with an object, artifact, sample, or work of art. The best interpretation is to attract as many visitors as possible to the historical object and convey information about it in

The integration principle

The main provisions of the principle:

Organization of additional travel routes to create an accessible environment for people with disabilities:

1. atrium spaces

2. additional elevators from +0.000

3. additional public areas for all population groups

On the example of the building of the Academy of Arts	*An example of organizing an accessible environment using the integration principle*

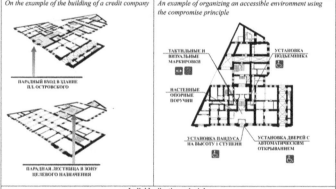

The compromise principle

The main provisions of the principle:

The compromise principle is the use of modern methods of adapting the historical environment, based more on a rational balance of the needs of society and the value of the preserved historical environment.

On the example of the building of a credit company	*An example of organizing an accessible environment using the compromise principle*

Individualization principle

The main provisions of the principle:

The individualization principle is the use of means for people with limited mobility in the context of the architectural and artistic features of the historical environment of each particular building and the development of devices for people with disabilities according to individual constructive and design solutions.

On the example of Tovstonogov Bolshoi Drama Theater	*An example of organizing an accessible environment using the individualization principle*

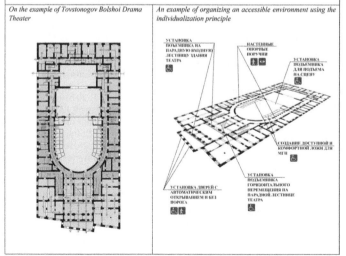

Figure 1. The principles.

a meaningful and relevant way, and not just information on the walls.

Applying the proposed principles for creating an accessible environment, two concepts for creating an accessible environment were developed on the example of the building of Saint Petersburg State University of Architecture and Civil Engineering and the building of Ilya Repin Saint Petersburg State Academic Institute of Painting, Sculpture and Architecture. It is proposed to create atrium spaces in the courtyards of the buildings in order to provide people with disabilities with accessible entrance, leisure and training areas, including an increase in exhibition and training areas for all users.

It should be noted that with this in mind, it can be argued that the optimal appearance of the architectural monument and the historical environment as a whole is a synthesis of its original appearance and later implementations, justified by the needs of society and improving the quality of the environment (Fridman, T.S. 2017, 2018).

Today, adaptation of monuments for people with disabilities is an important social problem in society. To create an accessible environment for all groups of the population, it is necessary to make significant changes in the planning structure and in the appearance of most monuments.

Most of the historical buildings of Saint Petersburg underwent significant changes and numerous introductions in their structure.

Often, monuments have a constructive problem; their appearance and style are completely different from the inner saturation of the building. Later changes made it possible to use the cultural heritage object for some specific goals and the space of the building was used effectively in accordance with public needs.

Later developments in most cases pose a problem for the movement of limited mobility groups: additional stairs and, as a result, elevations, become an insurmountable obstacle for them.

Bearing in mind the cramped conditions of the internal space of the buildings for the installation of devices for people with disabilities, it is necessary to analyze them and provide limited mobility groups with alternative ways for efficient and full use of the entire space of the monument, without violating their rights, in accordance with applicable law (Fridman, T.S. 2018).

In view of the problem of the complex structure of monuments of architecture that has been developing over the years, it is difficult to call events for creating an accessible environment an idiom.

Today, it is obvious that the construction of historical cities and the internal space of architectural monuments is multi-layered. This also depends on the modern interpretation of classical styles in modern architecture and on the development of structures, building materials and the ideals of society. For the most part, the dominant array of historical buildings in the city dictates the direction of

movement in the development of further architectural solutions and new inclusions in the current environment.

New architectural inclusions and styles in historical cities are always ambiguously accepted by society and denied by conservatives, who advocate the maximum preservation of the authenticity of architectural monuments. Often, these monuments require serious transformations for their effective use, adaptation to new needs, and involvement of all groups of the population in the process of use (Castell, L).

4 CONCLUSIONS

Thus, in spite of the active work and support of the state in the field of creating and developing an accessible environment, the development of legislation, regulatory framework, technologies and the socialization of low-mobility groups (a lot of work has already been done in this area), when adapting architectural monuments, a number of problems arise that need to be solved by applying individual principles for working with the heritage and the historical environment as a whole, based on a balance between the preserved heritage of historic cities and outside activities to create an accessible environment for all population groups (Blattner A. et al. 2015).

REFERENCES

Convention for the Protection of Human Rights and Fundamental Freedoms. URL: https://www.dostupnigorod.ru (Date of access: February 20, 2017).

Convention on the Rights of Persons with Disabilities adopted by Resolution of the UN General Assembly of December 13, 2006 No. 61/106. URL: https://www.dostupnigorod.ru.

Federal Law of the Russian Federation of May 3, 2012 N 46-FZ "On ratification of Convention on the Rights of Persons with Disabilities". URL: https://www.dostupnigorod.ru (Date of access: January 18, 2019).

Construction Rules 59.13330. 2016 "Accessibility of buildings and structures for people with limited mobility. Updated version of SNiP 35- 01-2001" approved by Order of the Ministry of Construction and Housing and Communal Services of the Russian Federation of November 14, 2016 no. 798/pr "On approval of Construction Rules 59.13330 "SNiP 35-01-2001 "Accessibility of buildings and structures for people with limited mobility". URL: https://www.dostupnigorod.ru (Date of access: May 3, 2019).

Decree of the Government of the Russian Federation of August 6, 2015 no. 805 "On amendments to the rules for recognizing a person as a disabled person". URL: https://www.dostupnigorod.ru.

Decree of the Government of the Russian Federation of June 17, 2015 no. 599 "On the procedure and terms for the development by the federal executive authorities, executive authorities of the constituent entities of the Russian Federation, self-government bodies of measures to increase the values of accessibility indicators for facilities and services for disabled people". URL:

https://www.dostupnigorod.ru (Date of access: February 20, 2017).

Federal Law of November 24, 1995 no. 181-FZ "On the Social Protection of Persons with Disabilities in the Russian Federation". URL: https://www.dostupnigorod.ru (Date of access: January 18, 2019).

Davydova, E. M., Radchenko, V. Yu. 2015. Social topics in design engineering as a means of forming an active life position of students. *Privolzhsky Scientific Journal* 11 (51): 107–111.

Fridman, T.S. 2017. The influence of legislation on creating an accessible environment for groups of population with limited mobility. *Science and business: Development Paths* 11 (77): 41–44.

Fridman, T.S. 2017. Modern trends in the adaptation of architectural monuments for people with disabilities. *Science Prospects* 5 (9): 29–31.

Fridman, T.S. 2018. The problem of accessibility of historical buildings for people with limited mobility. "Reasonable adaptation", a question of terminology. *Science and Business: Development Paths* 4 (82): 47–50.

Fridman, T.S. 2018. A device for adapting historic buildings for people with limited mobility as an architectural element. *Science Prospects* 3 (102): 55–58.

Fridman, T.S. 2018. Creating an adaptive environment for people with limited mobility in historical buildings of higher education. *74th scientific conference of professors, teachers, scientists, engineers and graduate students of the university. Saint Petersburg.* SPbGASU: 146–148.

Castell, L. *Adapting building design to access by individuals with intellectual disability.* Department of Construction Management, Curtin University of Technology, Perth, Western Australia.

Blattner A. et al. 2015. Mobile indoor navigation assistance for mobility impaired people. *Procedia Manufacturing* 3: 51–58.

Reconstruction and Restoration of Architectural Heritage – Sementsov et al (eds)
© 2020 Taylor & Francis Group, London, ISBN 978-0-367-65357-6

Integrated functional association — search for methods of cultural heritage sites adaptation (Russia)

L. Kaloshina
Saint Petersburg State University of Architecture and Civil Engineering, Saint Petersburg, Russia

ABSTRACT: Modern research in the field of the adaptation of cultural heritage sites implies local work with a particular monument or complex. In the new business environment, investment valuation of sites is becoming important. Half of the manors of the nobility have low commercial appeal. Based on the master's thesis on the integrated adaptation of Izhora Plateau manors, it is suggested to consider a method of integrated functional association, with the creation of a hierarchical functional system (work – accommodation — recreation). This system can be used for typologically similar and different sites. The most significant and valuable site of cultural heritage serves as a key to integrated association. Additional functions can be introduced in less commercially appealing heritage sites. This system makes it possible to create multi-functional complexes that will be independent and viable.

1 INTRODUCTION

The article is based on the experience of teaching candidates for a master's degree. Master's theses were composed of different subjects, but they were united by one topic — the adaptation of cultural heritage sites. When selecting a function, the entire range of conditions was taken into account: significance, value of the site, degree of preservation, availability of protection measures, location, etc. When a specific situation was considered in detail, it turned out that it was not enough to offer an adaptation option, but it was necessary to support the work of the function itself, its development, duplication, and the inclusion of related functions. Therefore, we had to search for, identify, and apply new techniques of a comprehensive approach to the adaptation of architectural monuments. The advantage of active creative search was that the project was an academic one, and we were not limited by legal issues or budget. The identification and analysis of new architectural techniques have not only scientific but also practical interest for the adaptation of cultural heritage sites.

The article addresses a master's thesis on the integrated adaptation of manors on Izhora Plateau in Leningrad region.

In modern scientific literature, the following studies can be distinguished: preservation of local historical complexes (Ageev 2005); the role of the historical and cultural landscape (museum preserves, national parks) (Vedenin 2018); formation of a system of historical and cultural territories (Shulgin 2004); principles of forming a regional tourist and recreational environment (Azizova-Poluektova 2015); economic development of regions (Malinina

2007, Skuridina 2012); manor and park complexes (Guseva 2008); current functions of manors' use, national and international adaptation experience (Krasnobaev 2009); issues of preservation and utilization of manors in different regions (Aksenova & Klavir 2014, Krasnobaev 2009, Molodykh 2019, Molodykh & Yenin 2010); analysis of the functional and planning composition of manors (Belyankina 2008).

These papers consider the issues of adaptation in terms of a particular site (complex), its territory, landscape, and connections with the environment. The studies are based mainly on examples of implemented works on the adaptation of cultural heritage sites.

Molodykh & Yenin (2010) indicate that in the practice of preserving and adapting provincial manors, "...there is a number of possibilities for developing new functions — church centers (Patriarchal Metochion in Sviblovo, parish center in Altufyevo), elite suburban institutions (Serednikovo, Lermontov–Stolypin manor), museums (Ostafyevo, Vyazemsky manor) and recreation centers (Sukhanovo, Vvedenskoye), farms (Artemyevo, Veres); ... projects developing estate tourism include the Wreath of Russian Estates program, Pleshanov's Manor hotel, and volunteerism (Voronino, Pryamukhino)."

According to Shulgin (2004), today we have an understanding that "...the protection and use of separate (point) sites cannot be effective outside of the surrounding historical and natural environment. It is necessary not only in terms of the perception of the monument but also its viability..."

The purpose of this article is to show the possibility of an integrated, economically mutually beneficial option of adapting several sites (in this case, manors) into one functional system.

2 MATERIALS AND METHODS

"For centuries, the Russian manor — boyar, princely, peasant, noble, merchant — was the main form of the spatial organization of private life. Along with the city and village, it belonged to the basic types of settlements in Russia. According to Guseva (2008), manors of the nobility began to appear in the 15th century due to the distribution of land to military people — the nobles". After the victory in the Great Northern war, the Ingrian Finnish lands were passed to Russia, extensive lands were passed by Peter the Great to his favorite Menshikov, and after his fall became the possession of different owners. Among them are Razumovskys, Cherkasskys, and Golitsyns. The largest of the lands were considered not as a place of residence but as a source of additional income. By the 19th century, many manors changed their owners several times. Only at the beginning of the century, these land plots began to acquire the classic features of a rural noble manor with permanent residence of the owners. A noble manor included: a manor house with household buildings made of brick and local limestone, a garden and park complex, and agricultural land. The main residential complex was located on the bank of a river or lake. If there was no natural water body, then a system of dams was organized on a stream or springs. A manor house, as a rule, was built of brick, and household buildings were laid out of local limestone on a rubble stone foundation and had brick openings.

Guseva in her thesis (2008) writes the following: "The concept of 'rural noble manor' is not localized currently. This is the name for not only large complexes that included a manor house, household buildings, and a park, but also for small ones that consisted only of a house and a front garden, a house and a service and manufacturing area, or even just a house."

During the Soviet period, almost all manors of the nobility in this region were destroyed: manor houses and household buildings were demolished, parks and farmsteads were damaged. The remaining heritage sites are on the books of district administrations. In Lomonosovsky District, numerous meetings were held on the issue of transferring the heritage of manors to private ownership, in connection with the appearance of a law on the possibility of acquiring heritage sites.

Architectural and park complexes of rural noble manors of the 18th – 19th centuries on Izhora Plateau of Saint Petersburg Governorate were considered in the master's thesis. In general, ten manors were studied: San-Susi (Bezzabotnaya), Novye Gorbunki village; Voronino manor, Voronino village; Gostilitsy manor, Gostilitsy village; Grevov's manor, Koporye village; Kotly manor, Kotly village; Kummolovo manor, Koporskoye rural settlement; Lapin's manor, Vilpovitsy village; Lopukhinka manor, Lopukhinka village; Orzhytsy manor, Orzhytsy village; Ropsha manor, Ropsha settlement.

The objective of the master's thesis was to identify the functional and planning features and specifics of the development of manors of the nobility on Izhora Plateau of Saint Petersburg Governorate, develop a conceptual proposal for their adaptation to modern conditions of economic activity. Three manors with medium indicators of investment appeal were selected for the solution (Krasnobaev 2009, Skuridina 2012).

The manors are located along Koporsky highway (tract) towards the Ingrian Finnish lands. Lapin's manor is the closest to Saint Petersburg, it is located in Orzhitsky rural settlement of Lomonosovsky municipal district in Leningrad region.

In 1727, the first buildings of Lapin's farmstead appeared on the land plot under consideration. Like many manors in this region, the residential area is located on a high clint of Izhora Plateau. All the buildings were timbered and lined with planks. Extensive orchards and vegetable gardens were laid out around it. In 1746, M.I. Felten, the owner of the manor, constructed a complex of stone buildings, which included a cattle yard, a barn, and a greenhouse. All of them were made of local limestone combined with rubble masonry and brickwork. Apparently, this very complex of buildings was constructed under the project of Yury Felten, a nephew of the owner. Now it is the only manor building that has been preserved since the 18th century (Murashova 2005, Murashova & Myslina 1999). It includes household buildings, two manor houses in Classical and Neo-Gothic styles, a system of entrances and paths in good condition (Figure 1).

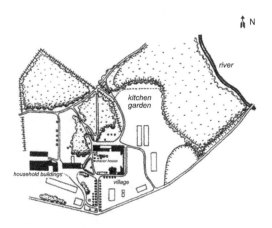

Figure 1. Layout of Lapin's manor. From the book by Murashova & Myslina (1999) "Manors of the nobility in Saint Petersburg Governorate. Lomonosovsky District".

Figure 2. Lapin's manor, 2014.

The manor house (which was constructed first) is still inhabited, household buildings are partially used as sheds and warehouses of Spirinsky farm. The buildings located around the courtyard still have roofs (Figure 2), therefore, they have preserved the internal walls, ceilings, and wooden stairs in addition to the external walls, but there are no window or door frames or casings. The external walls of all uninhabited buildings are damaged badly. Some of the original openings were blocked up, and new ones have been made. But all these buildings are in disorder. The road network is in fairly good condition. There is nothing left of the manor park, it was rather small. The owners always attached great importance to gardening, therefore, the trees were mostly horticultural. The garden was replaced with vegetable gardens of local residents.

Further along the road to Koporye and passing the village, we turn right. Kummolovo manor is located in Koporsky rural settlement of Lomonosovsky municipal district in Leningrad region. The first owner of the land plot (since 1719) was I.L. Blumentrost. The manor consisted of a house with two private rooms, a stable, and a wing for servants. In 1731, the owner fell into disgrace and was sent to Kummolovo manor, which had a utilitarian character. Later, this land property had several owners more, but the most noticeable trace in the history of its development was left by the family of Livonian noblemen von Gersdorffs, who owned it for 125 years (Murashova 2005, Murashova & Myslina 1999). By 1829, the center of the manor was a new large manor house with impressive architecture, built by architect V.I. Beretti in the Classicism style (Murashova & Myslina 1999). The main facade of the manor house is decorated with a massive six-column portico of the Doric order, flanked on the first floor by triple windows-doors and semicircular windows above them.

Fruit and landscape gardens were laid out on the sides, as well as a stable, barnyard, residential wings, and other household buildings. All of them were made of local limestone combined with rubble masonry and brickwork. On the bank of a pond fed by spring water, a distillery operated. The landscape of the manor is very picturesque, the house stands on

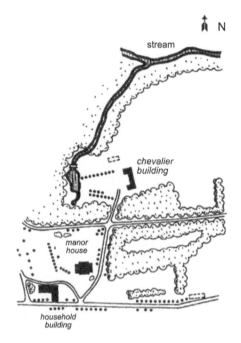

Figure 3. Layout of Kummolovo manor. From the book by Murashova & Myslina (1999) "Manors of the nobility in Saint Petersburg Governorate. Lomonosovsky District", 2014.

the edge of the Izhora Plateau clint; from it to the river flowing down the slope, a system of terraces and retaining walls was arranged (Figure 3).

The revolution marked the beginning of the collapse of the solid manor; due to the war and desolation during the last decades, today there is little left of it. The manor house preserved the porticos by the main entrance and from the park side, pediments above the porticos, magnificent three-part windows with arched ends on the main facade (Figure 4). On the back side of the house, at the foot of the slope, the pond with the stream that flowed out of it became swampy and dry. Only the ruins of the foundations remain from the mill and other household buildings that were once located here. There was a landscape park behind the pond, but now there are only partially overgrown glades.

Figure 4. Kummolovo manor, 2014.

32

Then we can go to the next manor of the Albrecht family in Kotly village. Kotly manor is located in Kotelsky rural settlement of Kingisepp municipal district in Leningrad region. The first owner of the land plot (since 1730) was I.I. Albrecht. The manor was owned by this family during the entire time of its existence. In 1742, it consisted of several log buildings. They included a residential building, household buildings, and St. Nicholas Church. In 1768, the main house was built of brick, and household buildings were laid out of local limestone on a rubble stone foundation and had brick window and door openings. In 1836, the manor house was rebuilt according to the project of architect A.I. Melnikov (Murashova & Myslina 2003). The building was designed in the classical style, had a six-column portico located on the southern side of the facade, in the center of which the Albrecht family coat of arms was depicted, as well as a swing ramp-entrance (Figure 5).

On the northern side facing the park, there is also a portico, in the second level of which there is a balcony. In the 1930s, the manor house was used as a village school. During the war, the German headquarters was located there and the Germans blew up the building during the retreat. After the war, the manor was restored and passed to the disposal of a military unit. From 1969 to 1979, Kotelskaya secondary school was in this building. In 1985, the entire complex became empty.

Today, the building is ruined. There are no internal walls, the external walls are partially preserved, and the six-column portico with remarkable columns made of chiseled local limestone still stands on its place (Figure 6).

The portico on the park side was lost during the war and was not restored. The terrace as well as the staircase leading to the lower park are ruined.

To the left of the manor house, there is a two-story building of stables, built in 1860. It is in better condition. The external and almost all internal walls are preserved, but there are no floors. To the right of the main building, there is a coach house and an ice

Figure 6. Kotly manor, April 2014.

cellar, built in 1870. From 1969 to 1979, a school canteen, an assembly hall, a gym, as well as workshops were located in the coach house.

The fourth building of the manor is a cowshed. It is a one-story building with two risalits on the edges. In the post-war years, the building was used as a warehouse.

The garden and park area of the manor is also in bad condition. There are remains of the staircase descending into the valley, and the outlines of a terraced slope and an oval pond on one of the terraces are still visible. The garden is littered with weeds, fallen trees, and wild underbrush. Special attention should be paid to the location of the manor house. It is located at the very edge of the Izhora Plateau clint. The house once had a beautiful view of the valley. Now only the unmanaged forest is visible. Old trees planted in the 19th century ennoble the surrounding landscape. Part of the estate is built up with personal farmsteads of villagers and summer residents. In the lower part of the site, partially swamped ponds have been preserved.

A full-scale survey of the sites under consideration makes it possible to draw the following conclusion: all the manors considered above are in different stages of destruction. Kotly manor, where, in fact, there is already nothing to protect, is in the most difficult situation. Almost nothing remained of the once remarkable example of the architectural heritage of the rural manor. The descendants of the family wanted to buy back and restore the manor, but they could not do that since it is considered an architectural monument of federal significance.

The condition of the manor house in Kummolovo is relatively better than that in Kotly. The manor can be restored (in the late 1980s, Signal Research and Production Association made such an attempt) and adapted to the conditions of modern economic activity.

Lapin manor can be restored and adapted to the conditions of modern economic activity as well.

During the analysis of the adaptation of similar objects abroad and in Russia, museum function turned out to be the most highly-demanded. 70% of

Figure 5. Kotly manor. Manor house. Photo of the 1910s.

such facilities in Russia and 40% abroad are converted into museums. The residential function ranks second in international practice (30% of facilities). Perhaps, after a while, this function will become more widespread in Russian practice since manors consisting of several buildings are quite common. By remodeling the premises and adapting them to be used as individual apartments, it is possible to establish homeowners' associations. According to Krasnobaev (2008), "...organizational issues of such projects can be solved since in Russia there is a number of organizations engaged in the development of manor of the nobility, in particular, the National Center of Heritage Trusteeship, the National Foundation of Russian Manor Renewal, Russian Manor non-profit partnership, and others". The PhD thesis by Malinina (2007) addresses the development of cultural heritage sites and their territories.

3 RESULTS AND DISCUSSION

When conducting research on the nature, structure, and possibilities of adapting manors of the nobility on Izhora Plateau of Saint Petersburg Governorate, we should mention the great potential of this territory. This is due to the close location of Ust-Luga port, an airfield, and A-180 federal highway. Bus transportation is provided from Gatchina, Volosovo, Kingisepp, Sosnovy Bor, Krasnoye Selo, and Saint Petersburg (Figure 7). Having assessed the potential of conservation and use (Belyankina 2008, Krasnobaev 2009), we can conclude that these manors occupy medium positions. Albrecht manor is the most promising one despite its ruined state. After reviewing the possible functions of manors' utilization, we came to the understanding that disjointed functions would not be relevant and in demand. Kotly manor takes a leading position in our combined functional complex. Having given it a central function, we have placed additional ones in other manors. Thus, we have a functional hierarchical system as a result.

Integrated adaptation can be carried out in two ways:

– by the development of the territory directly around the manor, maintaining the dominant position of the manor house, introducing new objects of modern construction aimed at its further development;
– by creating a complex based on several manors with the introduction of functions complementing each other.

The creation of a common complex on the basis of several manors meets the requirements and demands of our time.

Kotly manor is located 25 kilometers from the Ust-Luga port. The population of Kotly village is 685 people (2017). The neighboring villages — Gorodok and Kotelsky — already have a certain port infrastructure. Therefore, the appearance of a representative site related to the port on the

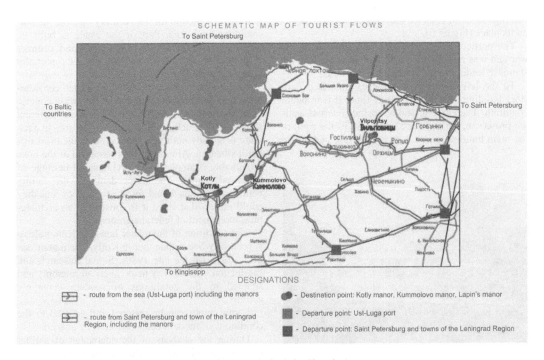

Figure 7. Schematic map of tourist flows (from the master's thesis by Shangina).

Figure 8. Rendering of Kotly manor (from the master's thesis by Shangina).

territories that are already being developed seems obvious to us. Kotly manor, after its reconstruction, may well claim the status of an object of the national importance of (Congress Hall) of Ust-Luga commercial sea port (Figure 8).

Thus, it is possible to hold negotiations, congresses, and sessions related to logistics in the context of the 19th century. It is possible to arrange a Reception House in the palace; administrative and household premises can be set in the two-story building, and a restaurant can be opened in the one-story stable. A Sunday school can be opened at St. Nicholas Church in the former barnyard.

To maintain the function of the Congress Hall, temporary accommodation, recreation, and entertainment shall be arranged. A "work – accommodation – recreation" system can be created. The accommodation can be organized in Kummolovo manor.

Kummolovo manor is located 20 kilometers from Kotly village. In Kummolovo, the main building will provide comfortable luxury rooms, and three detached cottages with 16 rooms each can be constructed. Kummolovo manor could be used for an active family holiday. It is possible to clean the park and the pond, build a sports complex with tennis courts and a bathing zone. Since pre-war times, one of the largest airfields in the area has been located next to the manor. Those who want quiet recreation can arrange a mini-vacation, a team retreat, staying in a cottage for a day or two, and visit the museum and entertainment complex in Lapin manor.

Lapin manor is located 50 kilometers from Kummolovo manor. It seems a little far, but it is worth traveling this distance to visit the place. It consists of six buildings of different styles. The main part of the manor is built in the Neo-Gothic style, which is uncommon for this region. The entire complex is nice and cozy, reminiscent of a medieval European manor. It is possible to set up an entertainment and leisure center here like Shuvalovka or Mandroga — a Russian folk village. Since the manor consists of seven buildings, it is possible to combine sightseeing and tourist functions. A museum of noble life can be opened

in one of the manor houses; in one of the household buildings, a hostel for 75 people can be set up. The other manor house can be connected with the household building with a two-story structure, and a cafe for 60 seats, a library, and a souvenir shop can be arranged there. It is possible to establish a folk and children's center with museums, workshops, and clubs in the complex of the household buildings, and construct an exhibition hall with artists' workshops, and a tour desk. For the local population, Lapin manor can become a cultural and educational center where national holidays and events can be held. The appearance of such a multi-functional complex in the western part of Leningrad region can help to revive the economic and cultural activities in the region, create new jobs, and, thus, improve the level of life of the local population.

In this regard, the article by Shulgin seems relevant since it addresses the economic aspect of the development of manors (1994): "A comprehensive approach, incorporating the historically determined territory, orienting towards not only the preserved manor building but the traditions of manufacturing and environmental management in conjunction with the challenges of tourism and the system of small businesses, creates an economic flywheel of self-development and simultaneously allows us to talk about the vitality of the manor."

4 CONCLUSIONS

This method of integrated functional association can be applied both to similar and different typological objects. With the goal of preserving historical buildings, architectural monuments (in this case, manors), it is proposed to consider these sites not separately, but to build functional links between them. Moreover, it can be not only cultural and educational tourism, but also other functional relationships between sites of cultural heritage, having tourist, social, representative, and other functions. The principles and patterns of the behavior of an architectural tourist and recreational system are addressed in the PhD thesis by Azizova-Poluektova (2015).

The main idea of functional content should be based on a specific situation, a specific place of design. When searching for a function in demand, it is necessary to consider the adaptable site in perspective since even if it is located in an accessible place and included in a tourist route, it may not be demanded in the absence of appropriate infrastructure.

We must admit that half of the manors have low investment appeal, and their adaptation is under discussion. In such cases, it is possible to include individual sites less profitable in terms of investments in functional associations, build hierarchical functional models (Azizova-Poluektova 2015) with the main significant site, and then we can hope for long-term demand for the new function of the monument, and, therefore, its long life.

ACKNOWLEDGMENTS

The author expresses her gratitude to Anastasia Shangina for the materials of her master's thesis "Integrated solution of the functional adaptation of Izhora Plateau manors".

REFERENCES

Ageev, S.A. 2005. *Preservation of local historic complexes by methods of urban planning regulation. Author's abstract of PhD Thesis in Architecture.* Moscow: Moscow Architectural Institute.

Aksenova, I.V. & Klavir, E.V. 2014. Problems of conservation and modern use of country manorial estates. *Vestnik MGSU* 9 (11): 14–25.

Azizova-Poluektova, A.N. 2015. *Systemic principles of the formation of a regional tourist and recreational environment. Author's abstract of PhD Thesis in Architecture.* Nizhny Novgorod: Nizhny Novgorod State University of Architecture and Civil Engineering.

Belyankina, N.A. 2008. *Countryside manors of Kostroma Governorate in the late 18th — early 20th centuries (functional and planning aspect). Author's abstract of PhD Thesis in Architecture.* Nizhny Novgorod: Nizhny Novgorod State University of Architecture and Civil Engineering.

Guseva, S.E. 2008. *Park and garden complex of countryside manors of the Saint Petersburg Governorate nobility (typological aspect). PhD Thesis in Architecture.* Saint Petersburg: Saint Petersburg State University of Architecture and Civil Engineering.

Krasnobaev, I.V. 2008. Experience of country estates preservation and its usage in Great Britain. *Observatory of Culture* 1: 77–83.

Krasnobaev, I.V. 2009. *Architectural heritage of rural manors of the nobility in the Kazan Volga Region: preservation and utilization potential. Author's abstract of PhD Thesis in Architecture.* Nizhny Novgorod: Nizhny Novgorod State University of Architecture and Civil Engineering.

Malinina, K.V. 2007. *Methodology of managing the development of territories with cultural heritage sites (case study of Saint Petersburg). Author's abstract of PhD Thesis in Economics.* Saint Petersburg: Saint Petersburg State University of Architecture and Civil Engineering.

Murashova, N.V. 2005. *One hundred manorial estates of Saint Petersburg Governorate: a historical reference book.* Saint Petersburg: Vybor.

Murashova, N.V. & Myslina, L.P. 1999. *Manors of the nobility in Saint Petersburg Governorate. Lomonosovsky District.* Saint Petersburg: Blitz Russian-Baltic Information Center.

Murashova, N.V. & Myslina, L.P. 2003. *Manors of the nobility in Saint Petersburg Governorate. Kingiseppsky District.* Saint Petersburg: Vybor.

Molodykh, M.S. 2019. *Historic and cultural heritage of countryside manors of the nobility in the Voronezh Region: system approach to the analysis, preservation and utilization. Author's abstract of PhD Thesis in Architecture.* Nizhny Novgorod: Nizhny Novgorod State University of Architecture and Civil Engineering.

Molodykh, M.S & Enin, A.Ye. 2014. Suburban nobility estate as a functionally integrate architectural and town-planning object. *Scientific Herald of the Voronezh State University of Architecture and Civil Engineering. Construction and Architecture* 4 (24): 85–89.

Molodykh, M.S. & Yenin, A.Ye. 2010. Problems of architectural heritage maintenance by the example of country estates of the Central Chernozem region. *Scientific Herald of the Voronezh State University of Architecture and Civil Engineering. Construction and Architecture* 3 (7): 59–73.

Shulgin, P.M. 1994. Economical aspects of manor complexes development in new conditions of economic management. *Russian Manor* 1 (17): 186–188.

Shulgin, P.M. 2004. Historical and cultural heritage as a special resource of a region and a factor of its social and economic development. *Universe of Russia. Sociology. Etnology* 13 (2): 115–133.

Skuridina, Yu.B. 2012. *Organizational and economic mechanism to manage a regional fund of immovable cultural heritage. Author's abstract of PhD Thesis in Economics.* Tomsk: Tomsk State University of Architecture and Building.

Vedenin, Yu.A. 2018. *Geography of heritage. Territorial approaches to heritage studies and preservation.* Moscow: Novy Khronograf.

Reconstruction and Restoration of Architectural Heritage – Sementsov et al (eds)
© 2020 Taylor & Francis Group, London, ISBN 978-0-367-65357-6

Specifics of forming a social and business complex in Kronstadt

O. Kokorina, D.A. Zinenkov, A.F. Eremeeva & S.A. Bolotin
Saint Petersburg State University of Architecture and Civil Engineering, Saint Petersburg, Russia

ABSTRACT: Kronstadt is a unique city in terms of architecture as well as natural and climatic characteristics. However, due to historical peculiarities, a significant part of its territory is not used, and the city is degrading. It is possible to change the situation by creating a poly-centric model in the city, i.e. complex changes contributing to the development of surrounding territories, as well as rethinking of the value of coastal and abandoned territories. The main potential for development is coastline territories as the intersection of economic, social, business, cultural, and tourist flows. The social and business sphere, based on the features of functional saturation and the uniqueness of modern architectural solutions, can attract investments and give an impetus to the economic growth of the city, providing a comprehensive approach to its exit from the stagnation state.

1 INTRODUCTION

At the moment, many cities that are part of the Saint Petersburg agglomeration are in a state of decline. In modern conditions, such settlements cannot develop, relying only on one sphere of activity. Single-industry towns that were formed in the industrial era no longer meet the needs of residents, and the structure of employment has changed, where the service sector has become predominant (Moskalenko 2011). The adaptation to the new conditions needs a comprehensive approach, including urban planning and social research, and the study of the experience of other countries. For successful further development, it is necessary to include elements of the post-industrial economy, modernize the infrastructure, and work on the image and ecology of the city.

To implement a comprehensive renovation program, it is necessary to adhere to the following principles, taking into account the existing urban environment and new urban development initiatives (Starodubrovskaya 2011):

- determination of the uniqueness of a place, its difference from nearby ones, identification of the main resources and potential areas of growth;
- creation of new unique districts included in the overall structure of the city, creation of a continuous green framework that unites all the important places in the city;
- adaptation of residential areas to modern needs, especially in disadvantaged areas;
- revival of the city center, rebranding, renovation of former industrial territories;
- construction of unique architectural, natural and landscape objects or "magnet" buildings that would attract the population, business, tourists, and, as a result, ensure the economic growth of the territory (Kokorina et al. 2020).

Kronstadt is a city in the Saint Petersburg agglomeration, located on Kotlin Island. The city is surrounded by water on three sides. However, this potential is not fully used — most of the coastline is inaccessible to visitors. Embankments are very attractive places for people, one of the main city symbols; therefore, great attention is paid to their arrangement all over the world (Marshall 2001). Besides, coastal territories can accommodate many functions that help attract people and develop the economy. Building a new "face of the city" becomes not only an important part of the marketing strategy, but also an impetus for the development of economic, social, cultural, and tourist potential (Doroshchuk 2016). Therefore, the development of coastal territories is a priority when reorganizing former industrial and military cities, such as Kronstadt.

The experience of reviving depressive territories all over the world has proved the feasibility of launching the process of revitalization based on the starting points of development — this does not require such huge economic costs. Nodes of strategic development of the city are determined after conducting a thorough analysis of the territory and identifying its resources. Historical, architectural, and natural resources have the greatest potential (Nefedov 2015). They can form growth points around themselves, attract investments, and encourage transformation of surrounding areas.

2 PURPOSE OF THE STUDY

A territory that can potentially form a center of social activity around itself should be the territory for the formation of one of the starting points of development, namely, the placement of new unique residential complexes and a multifunctional social and business cluster with additional functions. The priority is the location in the coastal zone, accessibility for residents and visitors of the city, inclusion in the existing urban structure, and the presence of identification signs.

The site identified for consideration is located in Kupecheskaya (Merchant) Harbor, west of the Italyansky (Italian) pond. Due to its location, openness of the coastline and proximity to the historical center, it is considered by the city government for the development of a business and tourist-and-recreational cluster. The site is a rectangle measuring 250 × 258 m with an area of 6.45 ha, surrounded by the Amazonka canal. The development area also includes an adjacent territory with a total area of 11.3 ha. Currently, in the north of the site, there is a complex of barracks and military buildings, as well as warehouses in poor condition. Communication with the city is carried out through the continuation of Lenina Prospekt (the main shopping street of Kronstadt) as well as Zosimova street.

This territory has a rich history, primarily related to trade. Wood was one of the main products for foreign exports of Kronstadt. In 1793, the city allocated a special place for a warehouse of timber. Fires often occurred here, so for fire safety purposes, this area was surrounded by an artificial canal, which later became known as Amazonka (Stolpyansky 2011). In 1827–1828, stone Fish Stalls were erected next to the Italyansky pond on the site of wooden shops of the Wood Exchange that burned down during a fire. They played an important role in the commercial life of the city, being a major market. One of the main attractions of Kronstadt — the building of Dutch cuisine — is located here. The building was intended for cooking food for the crews of ships stationed in the harbors of Kronstadt because cooking on the ships themselves was forbidden due to fire safety. Later, the exchange was replaced by a military town, which in the future would be transferred to the city.

The combination of factors that have developed in this area allows us to argue that the site most fully corresponds to the research topic and is suitable for the organization of a residential complex with the development of a multifunctional social and business complex, supplemented with cultural and leisure functions.

3 MATERIALS AND METHODS

Until the mid-20th century, the topic of renovation in industrial areas of cities in coastal areas was almost not discussed in the scientific and architectural community. However, at the moment, there are many examples of such territories returning to life all over the world. As a part of the study of foreign and domestic experience, the following projects were considered (Table 1):

Table 1. Analysis of experience in the development of projects for the revival of former industrial coastal areas of cities.

Object	Information	Ratio and distribution of functions
Paper Island Copenhagen, Denmark	COBE Mixed-use 45,000 m² 2016	
Aker Brygger reconstruction Oslo, Norway	Space Group Ghilardi & Hellsten Mixed-use 100,000 m² 2010–2016	
Tjuvholmen Oslo, Norway	NIEL-STORP+ Mixed-use 80,000 m² 2005–2010	
Nordhavn Copenhagen, Denmark	COBE Mixed-use 400,000 m² 2016 – present time	
Aarhus Docklands Aarhus, Denmark	BIG, Cebra JDS Architects, etc. Mixed-use 300,000 m² 2008 – present time	
Output / average value		

The territories under consideration in the cities of the Northern European countries have a number of similar features and solutions for reviving territories. Most often, these were places where powerful industrial zones were concentrated — shipyards, docks, and facilities of the mechanical engineering industry were located here. Starting from about the 1980s, enterprises gradually ceased their activities, and territories lost their practical significance. Over time, due to their location and proximity to the central parts of the city, the significance of these zones was rethought, these territories acquired value. In most cases, these areas are now developed as commercial and business ones, with the inclusion of residential development, as well as elements of other areas of activity. Historical industrial buildings are partially demolished, valuable and unique objects are preserved, and they are given a new function. Currently, the districts consist mainly of office and business centers, residential buildings and apartment complexes, and shopping centers. There are elements of service infrastructure — street cafes and eateries, beaches and pools, sports grounds, recreation areas, cultural spaces, cinemas, exhibition spaces aimed at different consumers; boat harbors and ferry terminals are preserved.

It is important to note that the projects are constantly being updated and adapted to the needs of users. The master plans are being reorganized; the main focus is on the development of embankments as public spaces — landscaping and lighting are being updated, vehicle and pedestrian flows are being separated, event venues are being added, buildings with a bright image and attractive architecture are being introduced (Baydzhanov 2018). Thus, there is a constant influx of people, including tourists, job positions are created, and investments are attracted.

When analyzing these examples, it can be concluded that most countries are creating multifunctional public clusters based on public-and-business and residential functions. Full-fledged urban polycenters are created when institutions with other functions (that can be used both by city residents and tourists — visitors of these areas) start gravitating towards the complexes with residential, public-and-business, business travel functions.

Multifunctional social and business complexes that set the overall direction of the cluster are often considered the cores of such centers. They are the center of concentration of social and business activity, a place of attraction for city residents and tourists (Eremeeva 2018). The experience in designing such structures allows us to identify the main structural elements typical for most types of multifunctional public and business complexes (Table 2).

Table 2. Analysis of the experience in creating projects of public and business complexes in the former industrial coastal territories of cities.

Object	Information	Ratio and distribution of functions
BLOX/DAC Copenhagen, Denmark	OMA / Ellen Van Loon Mixed-use 28,000 m² 2018	housing / offices / culture / retail
Timmerhuis Rotterdam, Netherlands	OMA Mixed-use 45,000 m² 2015	housing / offices / culture / retail
Città del Sole Rome, Italy	Labics Mixed-use 18,000 m² 2016	housing / offices / culture / retail
PXP Paris, France	OMA Mixed-use 50,000 m² 2016	housing / offices / culture / retail
Output / average value		housing / offices / culture / retail

4 RESULTS AND DISCUSSION

The identified structural elements of social and business complexes formed the basis for building a functional model of a social and business complex based on the principles of combining groups of premises, functional zones, and blocks into a multifunctional center (Figure 1).

Based on the studied experience, the following functional zones are identified:

– office,
– trade,

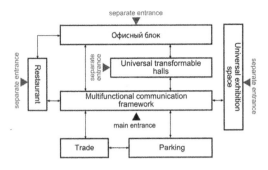

Figure 1. Functional model of a social and business complex.

- exhibition,
- transformable conference rooms,
- catering,
- recreation (aquatic center),
- accommodation (hotel),
- culture (museum and historical).

Functional office, public, exhibition, shopping, and restaurant blocks, as well as transformable conference halls, are characterized by autonomy — they have separate entrances and groups of service rooms (auxiliary and technical) but can be combined by a single communication framework of multifunctional atriums, halls, recreation areas, corridors that combine them into a single complex (Gelfond 2006).

In this case, components that perform functions for the main events (office sections, sections of universal conference halls and exhibitions able to perform a public function) refer to the main sections; that was identified based on the analysis of the Kronstadt infrastructure, the needs and demands of residents. Optimal ratios of the main functional elements are identified on the basis of the research of foreign and domestic design experience (Table 2) and also based on the analysis of the Kronstadt infrastructure.

The office and business section will be the dominant main functional section. Additional retail and catering sections are located in the multi-functional space of the complex. Additional autonomous sections are located in the adjacent territory and have a non-business purpose: hotels, cultural institutions, an aquatic center, etc. They can be combined with the main complex by an underground communication framework.

The office section consists of administrative offices, meeting rooms, individual offices, common areas for work and recreation, and a group of technical rooms.

The conference center section includes multifunctional transformable conference rooms, meeting and negotiation rooms, presentation and multimedia rooms that can be used for various scenarios with different numbers of participants, including for city events (Eremeeva 2018).

The exhibition center block includes a transformable exhibition space, terraces, storage rooms, service and technical rooms.

The trading section includes a transformable sales area, storage rooms, service and technical rooms.

Thus, the cooperation of a unique residential complex, a multifunctional public-and-business complex with historical and cultural objects, a recreational aquatic center, several autonomous (or included in business complexes) hotel sections, retail and catering facilities creates a multifunctional cluster on the basis of the public-and-business and residential functions in the territory under consideration. It not only solves the existing problems of the city by developing empty territories, creating jobs, improving the appearance and image of the city but also contributes to the development of social infrastructure, attracting the urban population and tourists.

5 CONCLUSIONS

The transformation of public spaces is possible with a proper arrangement of relationships in the created environment, including buildings, premises, and equipment connected by a single functional scheme. Multifunctional social and business complexes are often the central objects of such an environment. The typology and functional content of a complex should be formed in accordance with the following principles (Eremeeva 2018):

- The design area must meet the requirements of uniqueness, inclusiveness, availability of resources that determine the potential of the territory. The choice of a site depends on a thorough analysis of historical, cultural, socio-economic, climatic, environmental, and urban factors.
- The organized functional program of the complex should meet the needs and requirements of the city or its designated part, and focus on residents and other potential consumers. The composition and size of the main and additional functional blocks should be determined by a comprehensive analysis of the existing situation and infrastructure, taking into account future development, as well as match the portrait of consumers.
- The balance and proportion of functional sections and premises in the overall structure of the complex, options for their cooperation with each other should be formed based on the analysis of the experience in implementing such structures and the technology of processes occurring in each section. This determines the overall orientation of the complex, the dominance of a certain functional section, or their equivalence.
- The size and composition of premises and functional sections are adopted taking into account a number of requirements:

1. the set of premises must correspond to the quantitative characteristics and scale of events;
2. grouping and joint operation of premises should be carried out depending on their technical capabilities – structural, engineering, etc.;
3. it is necessary to optimize the parameters of premises based on internal processes, taking into account the possibility of their transformation and universal use.

The practical and scientific significance of the research lies in the development of an architectural solution for the coastal territory of the city of Kronstadt, as well as in the formation of principles and recommendations for working with such spaces. The research can become a basis for developing projects for new public spaces and transforming existing public spaces on embankments.

The result of this study is to develop proposals for the comprehensive development of the coastal zone of Kronstadt District through the design of a public and business cluster, forming the center of social and business activity, and attraction of tourists.

REFERENCES

Baydzhanov, I.S. 2018. *Principles of the transformation of urban public spaces*. Moscow: Izdatelskiye Resheniya.

Doroshchuk, N.R. 2016. Town planning features of coastal areas. *International Scientific Review* 21 (31): 90–92.

Eremeeva, A.F. 2018. *Design of multi-functional public complexes using business tourism centers as an example: study guide*. Saint Petersburg: Saint Petersburg State University of Architecture and Civil Engineering.

Gelfond, A.L. 2006. *Architectural design of public buildings and structures: study guide*. Moscow: Arkhitektura-S.

Kokorina, O. et al. 2020. New public spaces as the basic nodes for development of new city areas (case study of Kronstadt, Russia). *E3S Web of Conferences* 164: 04026.

Marshall, R. (ed.) 2001. *Waterfronts in post-industrial cities*. London, New York: Taylor & Francis.

Moskalenko, I.O. 2011. General features, problems, and prospects of the development of single-industry towns in Russia. In O.A. Shulga (ed.), *Topical issues of economics and management: proceedings of the international virtual scientific conference, Moscow, April 2011, Vol. 2*: 157–160. Moscow: RIOR.

Nefedov, V.A. 2015. *How to return the city to people*. Moscow: Iskusstvo XXI Vek.

Starodubrovskaya, I. (ed.) 2011. *Development strategy of old industrial cities: international experience and prospects in Russia*. Moscow: Gaidar Institute Publishing House.

Stolpyansky, P.N. 2011. *Guide to Kronstadt: historical essays*. Moscow, Saint Petersburg: Tsentrpoligraf, Russkaya Troyka.

Reconstruction and Restoration of Architectural Heritage – Sementsov et al (eds)
© 2020 Taylor & Francis Group, London, ISBN 978-0-367-65357-6

Formation features of high nobility estates in neighboring districts of Saint Petersburg governorate on example of Peterhof district

E.A. Kozyreva
Saint Petersburg State University of Architecture and Civil Engineering, Saint Petersburg, Russia

ABSTRACT: The article is dedicated to research of the situating of representative and prestigious suburban estates – "estates of high nobility" on the territory of Peterhof district before 1917. These unique estates got special functional and spatial-landscape characteristics that significantly distinguish them from the "ordinary" noble estates. The relevance of the work is due to the fact that an active search is currently underway for solutions to adapt and preserve estate heritage of Russia. The aim of the research is to identify a separate type of objects, called "estates of high nobility." For the purposes of the research the following methods were used: archival-bibliographic, cartographic, analytical, field studies. Results: the preserved and pre-existing estates of the high nobility are analyzed.

1 INTRODUCTION

During the development of Saint Petersburg and Saint Petersburg governorate, a belt of suburban estates belonging to the nobles of the highest ranks was formed. Such estates differed from hundreds of ordinary urban, suburban and rural estates by significant functional, spatial, compositional features, especially created under the orders of their owners by well-known Saint Petersburg architects. Hundreds of noble estates were studied in sufficient detail in the works of I.V. Barsova, S.E. Guseva, T.E. Isachenko, O.V. Litvintseva, indicated in the general reference works on the theory and history of architecture and landscape art by T.B. Dubyago, A.V. Ikonnikov, M. A. Ilyin, T.P. Kazhdan, V.Ya. Kurbatov and others; in works devoted to Russian estates I.A. Bondarenko, T.P. Kazhdan, A.Yu. Nizovsky and others; in books and articles on the history of Saint Petersburg that contain information about the estates by Georgi, M.I. Pylyaev, N.N. Wrangel, N.V. Murashova, L.P. Myslina, S.V. Sementsov and others.

This article is aimed at identifying patterns of placement and search for features of estates of high nobility around Saint Petersburg. It is based on the study of numerous complex historical materials, which include extensive collections of cartographic and descriptive data stored in various archives and libraries of Saint Petersburg.

2 MATERIALS AND METHODS

2.1 Territorial features of the placement of high nobility estates

At the beginning of the XX century there were about two thousand estates in the Saint Petersburg governorate. Among them were unique imperial palace and park residences as well as the modest dwellings of the village landowners (Murashova 2005). Nowadays hundreds of noble estates are still preserved in the contemporary center of Saint Petersburg, in the areas of its near and distant suburbs, in the distant regions of modern Leningrad Region (until 1917 – districts (uyezdy) of the Saint Petersburg governorate), although in a different state. They were studied in sufficient detail by I.V. Barsova (Barsova 1971), S.E. Guseva (Guseva 2008), T.E. Isachenko (Isachenko 2003), O.V. Litvintseva (Litvintseva 2006). All the works identified the estates that were owned by the most influential people and families of Russia, the estates of a particularly large size with diverse "set" of buildings, natural and man-made landscapes.

In my research, such unique estates are referred to as "high nobility estates" ("nearby high nobility estates") (Kozireva 2016).

Previous research of N.Ya. Tikhomirov dedicated to the estates near Moscow made it possible to identify estates in Moscow that were similar in size and structural diversity and show how, the number of buildings, and the internal planning characteristics of the main estates with greenhouses, pavilions, sculptures, bosquets, etc. in park compositions depend on the social status and socio-political responsibilities of the estate's owner as a nobility (Tihomirov 1995).

Taking into account the peculiarities of the formation of nearby estates of high nobility around Saint Petersburg, the countries of Europe were analyzed in order to identify similar unique estates. As a result of the analysis, we can conclude that the appearance of pompous, large and presentable objects that were built for themselves close to the ruling elite is also characteristic of European countries.

As a result of the research the following characteristics of European states (with the exception of England) were revealed: most of the possessions are state-owned, there are museums, national galleries, higher educational institutions and other state institutions situated in them. The exceptions are estates in England, where, unlike the rest of Europe, most are privately owned. All objects are recognized as the cultural heritage of those countries where they are located, and where the state is trying to preserve this heritage for posterity. (Lavrov 2015, Victoria Jenkins 2018, Tony Matthews 2017, Mihailo Grbić 2016, Rudi Hartmann Heritage 2020)

"High nobility estates", which appeared and developed around Saint Petersburg, are a unique type of possessions of the upper strata of the Russian nobility, created and operating until 1917 - 1920s. Distinctive features of such estates include them being much larger in size in comparison with estates of the ordinary noblemen - more than 5.0 hectares (in most cases - tens, even hundreds of hectares), as well as the presence of various auxiliary and office buildings. An integral attribute in the "high nobility estates" was the presence of non-trivial elements that might not be functional, but it is necessary to show the elitism of its owner ("Friendship Temples", "Islands of Love", "Parnassus Hills", greenhouses for growing exotic plants, stables for especially thoroughbred horses and cattle, gardens and parks in a particularly artistic execution, ponds and lakes with special landscape outlines, complex systems of paths and viewing platforms, etc.). In parallel, the systems of estates of the high nobility were formed in suburban and distant districts (uyezdy).

As a result of the analysis, it is possible to reveal that the estates of the high nobility can be divided into categories depending on the area of the estate: ordinary - up to 30 hectares, medium - from 30 hectares to 60 hectares, large - up to 200 hectares. It is worth noting that with the increase in the area of the estate's land plot, the number of outbuildings does not increase, however, park "frolics" appear: man-made ponds, grottoes, parnassuses, curtain walls and other entertainment facilities. For a more detailed examination of each category, depending on the area, we consider specific examples of estates.

2.2 Estates of Peterhof district (uyezd)

In total, by the year 1917, there were about 100 different estates in Peterhof district (uyezd). Of these, 27 estates of the high nobility were identified as a result of the research. Representatives of the highest 4 ranks socially belonged to the high nobility (according to the "Table of Ranks", begun as early as 1719, approved by the Highest Act of January 24, 1722) (No. 3890. 1722) and according to the Manifesto of Succession (introduced by Paul I as the "Establishment of the Imperial Family" dated April 5, 1797) (No. 17906. 1797). For example, these are: the manor complex in Ropsha, in

Gostilitsy, in Lopukhinka, the Grevov estate in Kopor`ye and others (Figure 1).

They can also be typologized according to several characteristics, taking into account the fact that the majority of owners had not lower than the fourth rank according to the table of ranks:

According to the stages of prosperity: I - 1703 - 1725. Dylitsy, palace and park ensemble of Ropsha, Gostilitsy palace, park and garden ensemble, Medushi, Grevov estate in Kopor`ye, Kummolovo, Voronino, Alutin`s estate, Oranienbaum, Sergievka palace and park ensemble, Znamenka palace and park ensemble, ensemble of Mikhaylovskaya dacha. II - 1725 - 1761 Ust-Ruditsa, Orzhitsy, Lopukhinka, Vlasovo, Novoivanovskaya. III - 1762 - 1800 Gomonovo, Seltso, Wolfhounds, Wells, Terpilitsy, Sumino. IV - 1801 - 1836 Kaskovo, Volgovo.

V – 1837 – 1900 гг. Bezzabotnaya, Torosovo.

By building area: up to 30 hectares - Gomontova, Sumino, Alutino, Orzhitsy, Volkovitsy, Medushi, Volgovo, Novoivanovskaya, Seltso, Lapino, Ust-Ruditsa, Kummolova; Kolodezi, Terpilitsy; from 30 to 60 hectares - Vlasovo, Torosovo, Grevov's estate in Kopor`ye, Kaskovo, Bezzabotnaya; to 200 ha - the palace and park ensemble of Sergievka, the palace and park ensemble of Znamenka, the palace and park ensemble of the Mikhailovsky dacha, the palace and park ensemble of Ropsha, the palace and park ensemble of Gostility, Oranienbaum, Lopukhinka, Voronino.

By preservation: only the park (part of the park) was preserved: Ust-Ruditsa, Orzhitsy, Kaskovo, Bezzabotnaya; the park and the manor house (or buildings on the territory) have been preserved: the Ropsha palace and park ensemble, the Gostilitsy palace and park ensemble, the Grevov's estate in Koporye, Lopukhinka, Medushi, Kummolovo, Voronino, Alutin`s, Vlasovo, Novoivanovskaya, Gomonovo, Seltso, Terpilitsy, Sumino, Volgovo, Torosovo; the park, the manor house and buildings on the territory are preserved): the courtyard-park ensemble of Sergievka, the palace and park ensemble of

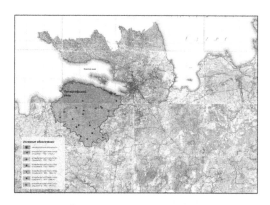

Figure 1. The layout of estates of the high nobility on the territory of Peterhof district (uyezd).

Znamenka, the palace and park ensemble of the Mikhailovsky dacha, Oranienbaum, Dylitsy.

2.3 Unified spatial structure of high nobility estates in the territory of Peterhof district (uyezd)

Estates on the territory of the governorate were located along the main highways departing from Saint Petersburg, as well as along the largest land and water highways. In the Saint Petersburg governorate a dense road network was developed, in which several types of roads can be distinguished: railways are single-track and double-track, highway, stone, dirt country roads.

Estates of the high nobility in Peterhof district were located along highways.

The following differences can be distinguished: linear situation on one side of the road in the direction of travel from Saint Petersburg; the two-sided location of the estates along the Peterhof road to Oranienbaum (some stretched from the highway to the Gulf of Finland, others went from the highway deep into the territory of district) (Figure 2, 3, Table 1).

The geographical features of the area influenced the appearance of the characteristic features of the placement of estates of the high nobility.

According to A.G. Isachenko the following landscape areas are located on the territory of St. Petergof: Izhora Plateau (plateau-shaped on Ordovician limestones) and Pre-Glint areas (lake-glacial and water-glacial swampy sandy) (Isachenko 2003).

Given these characteristics, most of the estates under consideration are located on the territory of the Izhora Plateau, as in a more favorable area. The estates of the high nobility in the Pre-Glint areas gravitate towards the Peterhof highway and the Imperial residences, essentially decorating the way along the highway.

The study of the laws of spatial localization of estates of the higher nobility around the capital city

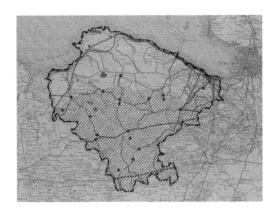

Figure 3. The layout of estates of the high nobility in the territory of Peterhof district (uyezd) with the designation of the Izhora plateau.

of Saint Petersburg shows the unevenness of the distribution. Moreover, up to the outer borders of the neighboring districts, such estates were placed in almost a single tight ring, concentrating as much as possible around the city imperial palace ensembles and suburban imperial palace and park residences. Such a dense and interconnected spatial localization of estates of the higher nobility in the vast spaces around Saint Petersburg suggests that during the XVIII - early XX centuries. a well-ordered network of estates was formed, which became in fact a special representative layer (layer of representative estates) of the historical Saint Petersburg (Sementsov 2019).

As an example, let's take a look at three different estates of the Peterhof district (uyezd) - from the Pre-Glint area, situated in a linear manner (along the Gulf of Finland) - Znamenka, from the Izhora plateau a singular estate along the district council roads (dead-end) - Orzhitsa; located along the highway, which is in the possession of the governorate council - Gomontovo.

Znamenka

The palace and park ensemble of the Znamenskaya dacha within the modern borders is a small part of the vast estate that existed here from the 1770s to the 1860s (located between Alexandria and the Mikhailovsky dacha) (Figure 4).

The main facade of the estate is facing the lower Peterhof road. In front of the palace on the lower terrace is a regular garden-parterre with three alleys leading to the gulf. Between the alleys there are three artificial rectangular reservoirs, at the slope on the upper terrace – two oval reservoirs. In the first third of the XIX century, the foundation was laid for the formation of the Upper Landscape Park with regular elements in the form of three rays - alleys diverging from the central manor buildings. Along the construction of the Upper Peterhof Road (1780-1784), the

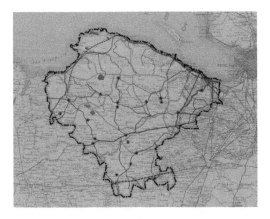

Figure 2. The layout of the estates of the high nobility in the territory of Peterhof district (uyezd) in regard to transport arteries.

Table 1. Classification of the high nobility estates of Peterhof district (uyezd) according to the peculiarities of accommodation.

Pre-Glint area

Linear (along the coast of the Gulf of Finland)	Linear (going deeper into the territory of the district (uyezd)	Railway	Single estates along the Primork-saya road
Mikhaylovka	Sergievka	Oranienbaum	Alutin`s
Znamenka			
Oranienbaum			

Izhora plateau

Highway owned by the governorate council (zemstvo)	A short distance from the road	The roads of the district council (zemstvo)	Singular estates along the roads of the district council (dead-end ones)
Gostilitsy	Sumino	Lapino	Ust`-Ruditsa
Ropsha	Seltso	Lopukhinka	Dylitsy
Bezzabotnaya	Kummolovo	Medushi	Orzhitsy
Kaskovo		Voronino	
Gomontovo		Kopor`ye	
		Vlasovo	
		Novoivanovskaya	
		Valgolovo	
		Torosovo	

Figure 4. Inventory plan of park of the Znamenka ensemble. 1957. Archive of the Committee for the state preservation of historical and cultural monuments, Saint Petersburg.

Ropshinskaya clearing became the main entrance to the estate.

Orzhitsy

Small estate of 6.3 hectares. Manor buildings were located on both sides of the stream (Мурашова 2005). The manor complex was located on the road from Ropsha, an entrance alley led to the buildings.

The landscape gardening complex was divided into two parts - regular and landscape. It is located on a flat terrain, with small differences, surrounded by forests, meadows and arable lands. (Guseva 2007)

Gomontova

The estate is located along the Narva tract (Tallinn highway) (Figure 5).

A small estate with an area of 5 acres (8.2 hectares). An entrance linden-birch alley led from the tract to the manor house (Murashova 2005). Around is a garden with trees of various species. The garden is surrounded by a moat. The garden is divided into various sections of household complexes. Located on a flat terrain, surrounded by forests. (Guseva 2007)

3 CONCLUSIONS

These materials analyze a special "layer" of the most prestigious estates located around the capital city of Russia - Saint Petersburg, including Peterhof district (uyezd), and by 1917 formed a unified spatial system focused on the territorial and social interaction with the imperial court (imperial residences), as well as - on the periphery - for especially close interaction with each other. The processes, features, patterns of identifying such a territorial dislocation of the largest estates in the spaces of hundreds of square kilometers in size require additional research.

The analysis of the processes of further development of the Saint Petersburg agglomeration throughout the XIX - early XX centuries showed a rather rapid

45

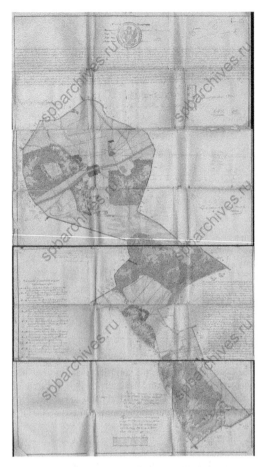

Figure 5. Petersburg governorate. Peterhof district (uyezd). Begunitsy county.

landscape elements - summer settlements, summer zones and summer belts, created on the principles of regular "garden cities" and "garden suburbs". All this led to the continuation of the processes of transforming the emerging agglomeration into "ideal agglomeration" on the principles of "regularity".

REFERENCES

Guseva, S.E. 2008. Garden and Park complex of rural noble estates of the Saint Petersburg province (typological aspect). Saint Petersburg.

Barsova, I. V. 1971. Estate parks of the Leningrad region and the principles of their use.

Litvintseva, O. V. 2006. Formation of rural noble estates of the Novgorod province of the end of the XVIII-XIX centuries. Saint Petersburg.

Isachenko, T. E. 2003. Interrelation of natural and cultural complexes of noble estates and landscapes. Saint Petersburg.

Kozyreva, E. A. 2016. Nearby estates of the highest nobility as a phenomenon of historical and urban planning life of Saint Petersburg (on the example of Ekateringofsky Park). Bulletin of civil engineers, (54): 5–11.

Guseva, S.E. 2007. Alleys in the estates of the Saint Petersburg province. Industrial and civil construction 7: 49–50.

Tikhomirov, N. Ya. 1995. The architecture of suburban estates, Gosstroiizdat.

Murashova, N. V. 2005. One hundred noble estates of the Saint Petersburg province: historical reference.

Sementsov, S., Akulova, N. 2019. Foundation and development of the regular Saint Petersburg agglomeration in the 1703 – 1910-s. *Advances in Social Science, Education and Humanities Research* Volume 324: 425–433.

Lavrov, L., Perov, F. 2015. The landscape development of Saint Petersburg (1853 – 2003). *Proceedings of the institution of civil engineers: urban design and planning* 168 (5): 259–266.

Victoria Jenkins, 2018. *Protecting the natural and cultural heritage of local landscapes: Finding substance in law and legal decision making*, 73: 73–83.

Tony Matthews, Deanna Grant-Smith, 2017, Managing ensemble scale heritage conservation in the Shandon Architectural Conservation Area in Cork, Ireland. 62: 152–158.

Mihailo Grbić, Aleksandar Čučaković, 2016. Biljana Jović, Miloš Tripković Garden cultural heritage spatial functionalities: The case of anamorphosis abscondita at Vaux-le-Vicomte Journal of Cultural Heritage Volume 18 March–April 2016: 366–369.

Rudi Hartmann Heritage and Economy, 2020. International Encyclopedia of Human Geography (Second Edition) 2020 pages 369–372.

expansion of the territory of the agglomeration under the influence of a number of factors and after the reformation of the transport system outside Petersburg, which included, along with the land- and waterways, traditional, from the XVIII century, railway lines from the middle of the XIX century. One of the factors that influenced the development of the Saint Petersburg agglomeration is the emergence and development of a network of nearby estates of the high nobility, which were the center of attraction and the center of development of the surrounding area. These transformations in the territories of the significantly expanding near agglomeration belt led to the birth of new functional

Reconstruction and Restoration of Architectural Heritage – Sementsov et al (eds)
© 2020 Taylor & Francis Group, London, ISBN 978-0-367-65357-6

Architect J.-B. A. Le Blond and his role in construction of Grand Peterhof Palace

A.G. Leontyev

Saint Petersburg State University of Architecture and Civil Engineering, Saint Petersburg, Russia

ABSTRACT: In the course of studying archives, conducting exploratory research and restoration works, the author of the article has studied the history of construction and restoration of the largest facility of the Peterhof ensemble and describes unique facts based on historical data and his own experience, making a conclusion that architect J.-B. A. Le Blond had a decisive role in the early period of the construction of the Grand Peterhof Palace.

1 GENERAL INSTRUCTIONS

Many authors have dedicated their research to the history of the construction of Nagorny (Upper) Chambers and subsequent reconstruction that shaped the final image of the Grand Peterhof Palace. Papers by A.N. Petrov (1949), N.V. Kalyazina and E.A. Kalyazin (Kalyazina 1984, Kalyazina & Kalyazin 1997), N.I. Arkhipov (Arkhipov 2016, Arkhipov & Raskin 1961, 2012), A.G. Raskin (Arkhipov & Raskin 1961, 2012), B. Lossky (1932, 1934, 1936) cover a great deal of archive recordings. The majority of the papers characterize the works by architect Jean-Baptiste Alexandre Le Blond in Peterhof assessing positively his impact on the construction of the palace and park residence as a whole as well as of the main facility — the regal Upper (Nagorny) Chambers that subsequently became the Grand Peterhof Palace. Publications by O. Medvedkova can be distinguished among the newer studies (1999, 2003, 2007). A biographical dictionary published by V.S. Rzheutsky and D.Yu. Guzevich in consultation with A. Mezen (2019) is considered the fullest publication that includes all known and new information on foreign specialists in Russia at the beginning of the 18th century. It clarifies and corrects earlier, widely used data. The article on Le Blond focuses on the personality of the architect and his work in France and Russia.

A talented architect, theoretician and practician of the garden art of the late 17th — early 18th centuries, Jean-Baptiste Alexandre Le Blond became famous in France thanks to his buildings, theoretical works, and numerous drawings (Rzheutsky & Guzevich 2019). He came to Russia upon invitation from Peter the Great. By that time, he was already a full-fledged expert with practical experience in architecture.

2 INITIAL WORKS BY LE BLOND IN PETERHOF

When he arrived at Peterhof in September 1716, Le Blond found a number of serious violations in the organization of the works: simultaneously with making the foundations and erecting the walls of the future palace, a great deal of soil was being excavated at its foot and loaded under the cascade walls. Groundwater filled the basement and eroded the foundation of the palace, which led to the deformation of the walls and the arches and the formation of dangerous cracks (Arkhipov & Raskin 1961, 2012).

The architect submitted a report to A.D. Menshikov, the Governor-General, and included a number of criticisms and important proposals. First of all, Le Blond, without waiting for the Tsar's approval, immediately proceeded with the construction of an underground "aqueduct" (a collector drain) for the drainage of groundwater and rainwater from the building. Installation of the drainage radically changed the situation at the construction site.

2.1 Le Blond's "aqueduct"

The early history of the construction of the Grand Peterhof Palace is closely related to a unique engineering structure — a collector drain called "aqueduct" by Le Blond. It collects groundwater that accumulates under the palace and then it is drained through underground tunnels of the Grand Cascade to a bucket and then, to Morskoy Canal.

Over time, the underground hydraulic engineering structure deteriorated gradually. In the late 19th century, architect A.I. Semenov partially changed the operation of the collector drain. This is evidenced by the "drawing of the works for the drainage of dirty water and backfilling of underground tunnels of the Grand Palace in Novy Peterhof" with an inscription "checked on-site by architect Semenov, 8/XII 1897" (Peterhof State Museum-Reserve Archive). The drawing shows a layout of the area of the Upper Gardens to the south of the Grand Palace with the route and profiles of the new drainage made of ceramic pipes (diameter — 6–15 inches (15–38 cm), total length — 95.76 fathoms (204.4 m)). The section shows the "aqueduct" part under the Grand Palace. Installation across the whole cross-section of the collector drain is shown under its southern wall and at the outlet to a tunnel of the Grand Cascade. Thus, Semenov excluded the southern area located in the Upper Gardens from the works.

In his manuscript about the Grand Palace, N.I. Arkhipov says that "according to some palace's long-term residents, in the 1890s, there appeared a long sinkhole that was being filled with sand for a long time in the Upper Gardens, along the palace facade", and he makes a conclusion "…that the sinkhole was due to the collapse of the archways of Le Blond's aqueduct" and, thereafter, after the war, "a narrow and long depression in the ground was visible" at the place where the aqueduct passed (Arkhipov).

On one of the drawings made by architects V.M. Savkov and E.V. Kazanskaya in the course of the restoration of the Grand Palace after World War II, there is a sketchy dashed designation of the Western section of Le Blond's "aqueduct" (immediately below the palace) with an inscription "an old tunnel". So, probably, there was access to it. The Eastern section on the same drawing symmetrical to it was not shown; its outlet was closed at the end of the 19th century and there was no more access to it (Archive of A.G. Leontyev).

In the late 1980s, during the restoration of the Grand Cascade according to the design of architect M.A. Dementyeva (LENNIIProyekt Institute), by efforts of PKZ workshops (Poland) specializing in the restoration of historic landmarks, works on the restoration of the cascade and the northern terrace in front of the Grand Palace were performed. Access to both sections of the underground structure was opened. The brickwork of the Western section of the collector drain was partially restored. At the time, the author was engaged in the development of the design documentation for the Grand Palace and, in the course of the designer supervision, he participated in the work of the commission on the restoration of the Grand Cascade as the basement section of the palace was closely connected to it. The research materials of those years and subsequent personal research are provided in this paper.

In the course of the examination of accessible Eastern and Western sections, the structure of the collector drain was identified. The collector drain corridors are about 90 cm in width and about 1.9 m in height; the masonry of the side walls has chink-shaped vertical holes in two rows at an equal distance from each other. Groundwater accumulated in the soil mass outside the collector drain is drained through these holes. An inverted brick arch serves as a chute

for water; the corridors are crossed with cylindrical brick arches. A comparison of known descriptions with the on-site investigation has shown that the existing "aqueduct" matches the data given in Le Blond's notes. The "aqueduct" surrounded Peter the Great's palace at a distance of 4 fathoms (8.5 m) from its sidewalls (Arkhipov 2016). Currently, its Western and Eastern sections passing under the buildings of the palace built in the mid-18th century and granite stairs of the 19th century exit in the tunnels under the steps of the Grand Cascade (Figure 1).

Accessible sections of the collector drain were measured and described in the cadastral plan and the technical passport of the "Grand Cascade with waterfall stairs, water jets, Upper and Lower grottos" in 2011 at the initiative of the Service of the Chief Architect of the Peterhof State Museum Reserve (A.G. Leontyev was the head of the Service from September 2009 till May 2012) by the design and inventory bureau of Petrodroverts District, a branch of GUION state federal enterprise (Municipal Department for Real Estate Appraisal and Inventory). The unique engineering structure of the early 18th century is registered in the register of federal property and has been included in the list of sites in accounting records of the Peterhof State Museum-Reserve (Figure 2).

In 2015, the author conducted research on finding the Southern section of Le Blond's collector drain when engineering and geotechnical survey was conducted in the Upper Gardens by Profil OOO for the purposes of developing design and estimate documentation for the restoration of the site and replacement of utility facilities. A brick arch was found at a depth of 1.5 m in accordance with the indicative layout of the Southern section of the collector drain. The condition of the brickwork opened in the area of less than one square meter was assessed as satisfactory. The section found confirmed the author's calculations concerning the location and the layout of Le Blond's "aqueduct". The Eastern and Western sections of Le Blond's "aqueduct" are connected by the Southern section.

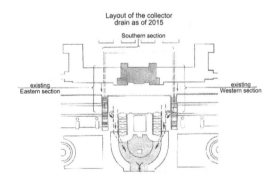

Figure 1. Layout of the collector drain. Restoration. Architect A.G. Leontyev, 2019.

Figure 2. Collector drain of the early 18th century. General view of the Western section after restoration in the 1990s Photo, 2015. A.G. Leontyev.

It seems very expedient to conduct restoration works with a comprehensive examination of the unique structure, its strengthening and recovery of its full operation in the nearest future.

3 RADICAL ADJUSTMENTS IN THE PALACE MADE BY LE BLOND

After Peter the Great approved some of the proposals for the palace, Le Blond proceeded with their implementation: he reinforced the foundations and arches of the basement with brick, introduced buckstave into the walls by partially laying it again, and therefore prevented the destruction of the building (Arkhipov 2016).

Decisive and professional actions of the architect eliminated the consequences of incompetent construction works within the shortest time. After that, the architect proceeded with significant changes to the palace facade and interior solutions.

3.1 Facades

In 1717, the architect increased the central risalit of the building by one story and completed its Northern and Southern facades with triangular gables. After Le Blond's death (he died on February 27, 1719), the carved bas-reliefs he envisaged were installed. The architect made three doorways in each of these facades. These doorways were joined by wide porches, thus, creating front doors in the center of the palace. A large balcony with a chiseled wrought lattice installed in 1720 was built on the facade facing the sea.

3.2 Interior

Due to the fact that another story was added in the central part of the building, a great hall with two tiers of windows, an Italian saloon, appeared. Arched door windows and windows of the second tier situated in the opposite end walls filled the hall with light. High intrados with polychrome paintings and plaster friso with rhythmically paired consoles, a profiled coving enframing a decorated ceiling, panels at the baseline of the high walls bore features of an architecture that was new to Russia. The exquisite decoration of the hall was completed with marble fireplaces and the floor made of marble slabs (later it was replaced with parquet flooring). The hall in the center of the palace highlighted the main axis of the whole ensemble. The hall commands a view of the Lower Park with the Grand Cascade and Morskoy Canal facing the sea and the Upper Gardens. Subsequently, the hall also became embellished with new decorations, but its harmonious proportions and plasterwork remained unchanged.

Le Blond's adjustments also involved a wooden staircase leading to the second floor made before him. According to the architect, it was too small and nasty" to lead to such a "decent and grand room as an Italian saloon". A new wide staircase was installed where the disassembled staircase and the kitchen rooms were situated. It was lit with large windows facing the southern side. The steps and the finishing of the flights were made of oak. Oak rails with carved balusters were installed in 1721 according to the sketches and with the participation of skilled master Nicolas Pineau who delivered on Le Blond's idea.

The architect joined several rooms of the first floor to be a single large entrance hall with rhythmically located load-bearing poles and a wide arched doorway leading to the Grand Staircase; the floor was made of marble slabs (Medvedkova 1999).

By implementing into life the ideas of a comfortable, rational and distinguished architecture of the palace chambers, Le Blond introduced a study for Peter the Great in the list of the grand halls. It was situated in the eastern risalit facing the southern facade into the Upper Gardens. The blueprint of the interior was implemented after the death of the architect. During 1719, Philippe Pillement, a decoreographer, painted the ceilings in four rooms of the palace including Peter the Great's study according to his sketches. A fire destroyed the fresco, and it was never restored. However, in 1721, French cutters Follet, Ruste, Farsure, Taconait, under the guidance of Michel, made fourteen oak murals with elaborate carvings of allegorical images of the forces of nature, seasons, the four winds, sciences and arts glorifying the acts of the reigning monarch for the good of Russia. A marble fireplace on the northern wall and glued-laminated parquet with geometrical pattern gave magnificence to a small room.

4 CONCLUSIONS

Le Blond corrected obvious and glaring faults of the first stage of construction and therefore ensured promising development of the palace. In the mid-18th century, carrying out instructions of Elizabeth of Russia, a daughter of Peter the Great who zealously venerated the memory of her father and a great monarch, Francesco Rastrelli preserved Peter the Great's part of the building and included it into a new enlarged palace. One-story galleries with pavilions that became adjacent to Peter the Great's palace in the 1720–1730s were disassembled, side risalits of the northern facade were enlarged by one story (within one ledge with the central hall), and two window axes were added. Rastrelli joined "the old stone house" (the old middle palace) with new buildings by a developed crowning ledge and made changes to the roofing. He made a wavelike pyramidal roof in the central part and decorated it with carved gilded sculptures and embellishments. The architect delicately cited the patterns of the balconies with delicate lattices and decorative parts of Le Blond's palace facades in the new parts of the building. The palace, stretching over 300 meters along the northern terrace, united the Grand Cascade, Morskoy Canal with a bucket, Samson Fountain and pattern beds into a single grandiose ensemble. The size of Peter the Great's Upper Chambers is currently hidden in the structure of the Grand Palace enlarged by Rastrelli but it can be visualized by means of a projection on its northern facade of the side walls of the Grand Cascade that preserves the initial width — 41 meters (Figure 3).

N.V. Kalyazina justly noted that F. Rastrelli not only used what Le Blond had done, but also consistently developed his principles. Thus, in his Grand Staircase (Kupecheskaya) in the new part of the palace, Rastrelli used the principles applied by Le Blond during the design of grand staircases (Kalyazina & Kalyazin 1997).

Rastrelli enlarged the palace using properly what had been done before him, and what was preserved naturally merged with the new building.

In assessing Le Blond's work, N.V. Kalyazina says: "The amount of his contribution to the Russian architecture should not be limited to what was actually done by him: it is measured by the influence he had on his colleagues — architects and the followers of his ideas, projects and knowledge, with which he arrived and which he sought to distribute in a number of ways" (Kalyazina & Kalyazin 1997).

Figure 3. Grand Cascade and the central part of the Grand Peterhof Palace with a projection of the dimensions of the Upper (Nagorny) Chambers. A photo by A.G. Leontyev, 2008.

However, the hopes of Peter the Great he held out for the wealth of knowledge and experience in various areas of architecture, civil engineering and park arts of Le Blond, which made him a very valuable and significant person in the implementation of the great plans, did not come true. The early death of the architect interrupted the good influence his gift had on civil engineering in the Russian capital. However, those solutions of the architect that were implemented are characterized by high professionalism and creative vision, thanks to which radical enlargement of the palace and development of its architectural and artistic features became possible, as well as long-term operation of the unique palace ensemble due to the competent technical drainage solution.

REFERENCES

Arkhipov, N.I. 2016. *Studies on the history of Peterhof.* Saint Petersburg: Peterhof State Museum Reserve.

Arkhipov, N.I. & Raskin, A.G. 1961. *Petrodrovets.* Leningrad, Moscow: Iskusstvo.

Arkhipov, N.I. & Raskin, A.G. 2012. *Peterhof.* Saint Petersburg: Abris.

Arkhipov, N.I. Grand Peterhof Palace. Peter the Great's section (1714–1750). Historical note. Chapter I (Entrance Hall, Grotto, Le Blond's Staircase and adjacent rooms of the first floor). Manuscript: 8. Archive of A.G. Leontyev.

Drawing "An integrated layout of original and existing walls of the Grand Palace with places of soil excavation for the analysis of the condition of the foundations, tunnels of fountain pipelines and the traces of old brickwork" with a stamp of LENPROYEKT Institute 1959. Plan of earthwork for the analysis of the foundations and archaeological excavations. Blueprint. Archive of A.G. Leontyev.

Kalyazina, N.V. 1984. Architect Le Blond in Russia (1716–1719). In T.V. Alekseeva (ed.), *From Medieval times to the New Time. Materials and studies on Russian arts of the 18th – first half of the 19th centuries:* 94–123. Moscow.

Kalyazina, N.V. & Kalyazin, E.A. 1997. Jean Le Blond. *Architects of Saint Petersburg. 18th century*: 67–111. Leningrad: Lenizdat.

Lossky, B. 1932. Le sejour de Pierre le Grand en France. *Le Monde Slave* 8: 278–303.

Lossky, B. 1934. L'architecte Jean-Baptiste-Alexandre Le Blonde et son oevre en France. *Bulletin des Musees de France* 8: 167–168.

Lossky, B. 1936. *J.-B.A. Le Blond, architecte de Pierre le Grand: Son oevre en France*. Praha: Université libre russe à Prague.

Medvedkova, O. 1999. Le Blond Jean-Baptiste-Alexandre (1679–1719). Encyclopaedia Universalis: Dictionnare des architectes: 364–366. Paris: Albin Michel.

Medvedkova, O. 2003. Les architectes et les artistes français à Saint-Pétersbourg au XVIIIe et au début du XIXe siècle: présences directes et indirectes. Jean-Baptiste Alexandre Le Blond. In *Les Français à Saint-Pétersbourg. Catalogue de l'exposition*: 27–40, 111–114.

Medvedkova, O. 2007. *Jean-Baptiste Alexandre Le Blond: architecte 1679–1719. De Paris à Saint-Pétersbourg*. Paris: Alain Baudry & Cie.

Peterhof State Museum-Reserve Archive, PDMP 291 ar.

Petrov, A.N. 1949. *The Grand Palace in Petrodvorets. Historic reference. Revisiting its restoration*. Leningrad: KGIOP.

Rzheutsky, V.S. & Guzevich, D.Yu. (eds) 2019. *Foreign specialists in Russia during the reign of Peter the Great. Biographical dictionary of specialists from France, Wallonia, French-speaking Switzerland and Savoy*. Moscow: Lomonosov.

Fundamental differences of ensemble system of historical center of Saint Petersburg from European city centers

S.O. Markushev
Research and Design Center of the General Plan of Saint Petersburg, Saint Petersburg, Russia

ABSTRACT: The study, taking into account the historical urban planning principles of Saint Petersburg and the existing problems of maintaining the integrity of the ensemble of its historical center, determines the need for a special study of the urban planning compositional system of the historical center of Saint Petersburg and reveals its features in comparison with the systems of historical centers of Venice, Amsterdam, London, Paris and Rome. A graphical comparative analysis was carried out on the basis of historical fixation plans and developed graphic models of urban planning and compositional systems of cities. It was determined that the vast majority of the compositions of the historical center of Saint Petersburg are regular, the ensemble is characteristic of the entire territory of the historical center of Saint Petersburg, the scale of the integral ensemble of the historical center of Saint Petersburg significantly exceeds the parameters of the historical centers of European cities.

Keywords: ensemble, urban planning and compositional system, historical center of Saint Petersburg, Venice, Amsterdam, London, Paris, Rome, regularity, scale

1 INTRODUCTION

The urban planning and architectural character of the city center of Saint Petersburg was laid in the first decades of its existence. Since 1712, the principle of regularity has been introduced by the joint efforts of Peter I and the Italian engineer D. Trezzini (Sementsov, S.V. 2012, 2013, 2014). Since 1717, the principle of ensembles had been introduced by Peter I and J.-B.-A Leblond and adapted in relation to the territory of grandiose scale. The features of Saint Petersburg were laid specially taking into account the existing European achievements in comparison with the leading European capitals. Based on these principles, a well-developed regular and extensive ensemble system was formed in the city center; the real urban planning situation was fixed by Siegheim's plan in 1738.

In modern conditions, the urban compositional system of Saint Petersburg, the principles and central ensembles of which had been formed by 1738, is threatened by destructive processes. The Admiralty Square, Razvordnaya Square, Kollezhskaya Square have been lost, and they used to be important open spaces in the city center (Lavrov, L.P. 2016).

The radiation patterns of the Krestovsky, Kamenny, Petrovsky islands are degrading, modern irregular land surveying "makes" new buildings chaotic within the boundaries of the plots (Nikonov, P. N. 2013), new transport interchanges destroy the regularity of the planning structure. The problem of the negative impact on urban planning compositions

that had developed by 1917 has a 150-year history and continues to remain relevant.

Comparisons of Saint Petersburg with European cities and capitals are popular as Saint Petersburg is assigned their characteristics. Saint Petersburg is Venice, Amsterdam, Rome, Paris, Berlin, London, Athens, Palmyra. Saint Petersburg has incorporated the features of many European cities, while its very special qualities and striking differences are obvious, some of which have already been recognized by the world community by including the ensemble of its center in the list of UNESCO sites.

To prevent the destruction of the integrity of the ensemble of the historical center, the substitution of its metropolitan qualities for ordinary provincial ones, the author performed a special study of its urban planning and compositional system and the identification of its features against the background of other outstanding European capitals, with which Saint Petersburg is often compared.

An impressive amount of domestic works dealing with the problems of history, development theory, issues of preserving individual objects, urban ensembles and an integral ensemble of the historical center (Sementsov, S.V. 2004, Slavina, T. A. 2016, Shvidkovsky, D. O. 2003) are devoted to the historical center of Saint Petersburg. Despite a large number of works on the topic, there is no consolidated study considering the entire territory of the historical center as an interconnection of urban planning compositions of various scales and types (Markushev, S. O. 2019, Markushev, S.O. 2020).

The results can form the basis for supplementing the existing system of protection of the historical center of Saint Petersburg and for revising approaches to the use and development of the urban development heritage.

2 METHODS

A graphical comparative analysis of urban planning and compositional systems of cities was carried out on the basis of historical fixation plans that have a fairly high cartographic accuracy. For comparison, the cities of Venice, Amsterdam, London, Paris, Rome (Hall, T. 2005, Arkin, D. Paris. 1937, Tussenbroek, Gabri van 2019, Boerefijn W. 2016) and their plans, contemporary to Siegheim's plan of 1738, were chosen (Figure 1).

On Siegheim's plan, the central space of the Neva, uniting the Admiralty, the eastern tip of Vasilievsky Island and the zone of the Peter and Paul Fortress, finally became the epicenter of the entire composition of Saint Petersburg (Sementsov, S.V. 2004). On the plan, Saint Petersburg is presented as the interconnection of individual compositions of various scales and types, united by the central open space of the Neva.

Giuseppe Giuliano on the "Iconographic Image of the Famous City of Venice, Dedicated to the Government of the Magnificent Venetian State" in 1728 shows the existing irregular network of rivers and the central space of the Grand Canal, which has similar parameters to the Fontanka River.

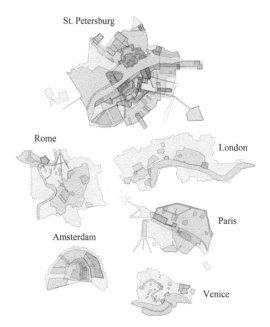

Figure 1. Comparison of graphical hierarchical models of ensemble systems of the centers of the cities of Venice, Amsterdam, London, Paris, Rome for the period of 1728-1746.

"A Very Accurate Map of the Famous Trading City of Amsterdam" by Hendrick de Let in 1735 shows the central composition of the Amstel River, the irregular medieval core of the city by the 14th century, the regular structure of the 16th century quarters surrounded by ramparts; the whole city is comparable to the scale of the regular composition on Vasilievsky Island.

In the center of the "Plan of the Cities of London and Westminster and the Southwark Area" by John Rock of 1746 there is a composition of the City of London on the territory of Roman Londinium, in the area similar to the composition of the defensive system of the Peter and Paul Fortress.

On the "New Map of Paris and Environs" by Jean Delagriv in 1728, the parameters of irregular quarters with dozens of plots and houses correspond to one plot and a house in Saint Petersburg.

"The New Topography of Rome in Honor of Our Holiness Pope Benedict XIV" by Jambattista Nolly in 1748 shows the composition of the three rays of Piazza del Popolo with an area of 17 hectares, which is one of the prototypes of the Saint Petersburg radiation system of the Admiralty with an area of 370 hectares.

3 RESULTS

The regularity features of the urban planning and compositional system of the historical center of Saint Petersburg in comparison with the centers of European capitals are:

– the regularity of most of the compositions of the historical center of Saint Petersburg. The exceptions are the quarters of the first city center in the Troitskaya Square area, parks and gardens of the English type, and special irregular intra-quarter compositions;
– the basis of the urban planning and compositional system is regular sections and quarters, regulated riverbeds and canals, regulated roads and paths of pre-Peter time;
– the regularity is ubiquitous within the boundaries of the historical center as regular compositions are located both in the city center and on the periphery of the center;
– the regularity was introduced purposefully by the joint efforts of Peter I and the Italian engineer and architect Trezzini since 1712 with the aim of creating a completely new type of capital city for Russia (S.V. Sementsov);

The features of the ensemble style of the urban planning and compositional system of the historical center of Saint Petersburg in comparison with the centers of European capitals are:

– the ensemble character of the entire territory of the historical center of Saint Petersburg. Many ensemble city-planning compositions of different scale and level, intersecting and overlapping each other, form a single ensemble of the historical center;

- on the territory of the historical center there are special areas that have no analogues in Europe, formed by the intersection and overlap of a large number (up to 10) of urban ensembles;
- within the boundaries of the territory of the historical center there are examples of ensembles that have no analogues in terms of their compositional structure (composition of the central space of the Neva, the system of the Palace, Admiralteyskaya, Petrovskaya, Isaakievskaya squares);
- the ensemble style was introduced purposefully by strong-willed joint efforts of Peter I and French architect J.-B.-A. Leblond since 1717, taking into account the unique Russian political, geographical and social conditions and the territorial scale (Sementsov S.V.);

The features of the scale of the urban planning and compositional system of the historical center of Saint Petersburg in comparison with the centers of European capitals are:

- the scale of the integral ensemble of the historical center of Saint Petersburg exceeds the parameters of the historical centers of European cities;
- in the urban planning and compositional system of Saint Petersburg there are compositions that do not have European counterparts in terms of their scale. For example, the layout of Vasilievsky Island, the Five Rays of the Admiralty, the central space of the Neva River;
- in the urban planning and compositional system of Saint Petersburg there is a phenomenon of scale displacement. Large (not the most important) urban planning compositions of Saint Petersburg correspond to single major urban compositions of European centers and so on. For example, the Fontanka River corresponds in scale to the Venetian Grand Canal;
- the especially large scale of the urban planning and compositional system of Saint Petersburg was specially laid down under the leadership of Peter I by engineer and architect Trezzini in 1712 and architect Leblond in 1717. Based on this principle of space exploration, it was possible to unprecedentedly organize a vast territory of 5300 x 8700 m in a relatively short time. (Sementsov S.V.).

4 DISCUSSION

The study considers the historical centers of cities as a hierarchy of urban planning compositions. The comparison is carried out within the framework of the selected characteristics: regularity, ensemble and scale, corresponding to the principles that were specially introduced for the design of Saint Petersburg. The article for the first time reveals the features of the urban planning and compositional system of the historical center of Saint Petersburg against the background of the selected cities. It is possible to further refine the results of the study by expanding the geography of comparison and including new cities in the comparative analysis, expanding the source base, and using new archival materials.

5 CONCLUSIONS

The research determined the fundamental differences of the urban planning and compositional system of the historical center of Saint Petersburg in comparison with the centers of European capitals.

The urban planning and compositional system of European capitals is based on the system of irregular curved medieval roads and streets. The ensembles of European cities are separated from each other by irregular medieval quarters; the ensembles of their centers are local and contrast with the chaotic urban planning and compositional system. The conditions for the development of the European urban center, its intra-fortress origin and expansion became a limitation on the parameters of the scale of the central urban planning and composition system.

Against the background of European cities, the vast majority of the compositions of the historical center of Saint Petersburg are regular, the ensemble is characteristic of the entire territory of the historical center of Saint Petersburg; the scale of the integral ensemble of the historical center of Saint Petersburg significantly exceeds the parameters of the historical centers of European cities.

6 RECOMMENDATIONS

The realization that Saint Petersburg has always existed as an urban planning and compositional system should entail a serious revision of value guidelines in the issues of preservation and development of its historical center. Changes should concern urban planning regulating documents focused on resolving the functional and transport problems of the historical center. Regional and federal laws that consider the conservation of the historical city center not as a system of ensembles, but as a "set" of individual cultural heritage sites with protection zones may be adjusted. Incorporating urban planning compositions into the system of urban development and architectural heritage protection may be in the form of:

- especially valuable objects of cultural heritage in accordance with the Decree of the President of the Russian Federation of November 30, 1992 No. 1487;

- elements of a urban planning and compositional system in the Law of Saint Petersburg of January 19, 2009 No. 820-7 "On the boundaries of the zones of protection of cultural heritage objects";
- special components of the historical center of Saint Petersburg as a UNESCO site The Historic Center of Saint Petersburg and Related Groups of Monuments.

ACKNOWLEDGEMENTS

The author expresses deep gratitude to his leader S. V. Sementsov, Doctor of architecture, professor of Saint Petersburg State University of Architecture and Civil Engineering, for help in understanding the center of Saint Petersburg as a system of ensembles on the basis of which the study was conducted.

REFERENCES

Arkin, D. Paris. 1937. *Architectural ensembles of the city.* Moscow: All-Union Academy of Architecture Publishing House.

Boerefijn W. 2016. About the ideal layout of the city street in the twelfth to sixteenth centuries: the myth of the Renaissance in town building. *Journal of Urban History* 5: 938–952.

Hall, T. 2005. *Planning Europe's capital cities.* London: E & FN Spon.

Lavrov, L.P. 2016. Lost transparency of historical open spaces of Saint Petersburg. *Bulletin of Civil Engineers* 6: 35–48.

Markushev, S.O. 2019. Typology of urban planning compositions of the historical center of Saint Petersburg. *Architecton: University News* 4: 1–9.

Markushev, S.O. 2020. The hierarchy of urban planning compositions of various scales in the structure of the historical center of Saint Petersburg. *Bulletin of BSTU named after V.G. Shukhov* 3: 68–75.

Nikonov, P. N. 2013. *The problem of maintaining the boundaries of historical land surveying* [Electronic resource]. Access mode: http://www.nipigrad.ru/publ/6/probl_meg.php. (Date of access: May 26, 2017).

Sementsov, S, V. 2013. Formation of the principles of preservation of the architectural and urban planning heritage of Saint Petersburg on the basis of the laws of its three-century urban development. *Bulletin of Saint Petersburg State University* 2: 190–211.

Sementsov, S. V. 2013. Mythology of urban development of Saint Petersburg. *ARDIS* 2: 11–15.

Sementsov, S.V. 2004. *Saint Petersburg on the plans and maps of the first half of the 18th century.* Saint Petersburg: Eklektika Tourist and cultural center.

Sementsov, S.V. 2012. Saint Petersburg historical agglomeration - a unique urban development site of a global scale. *Internet-vestnik VolgGASU* 1: 1–16.

Sementsov, S.V. 2014. The beginning of the creation of a regular Saint Petersburg agglomeration under Peter the Great. *Bulletin of Civil Engineers* 3: 46–55.

Shvidkovsky, D. O. 2003. *The great city of Peter the Great.*

Slavina, T. A. 2016. *On the issue of the strategy of conservation and development of the Russian architectural heritage:* 79–83.

Tussenbroek, Gabri van 2019. The great rebuilding of Amsterdam (1521–1578). *Urban History* 3: 419–442.

Reconstruction and Restoration of Architectural Heritage – Sementsov et al (eds)
© 2020 Taylor & Francis Group, London, ISBN 978-0-367-65357-6

Historic urban landscape in Saint Petersburg: To question of determining attributes of outstanding universal value of world heritage objects

N.V. Marushina

Saint Petersburg State University of Architecture and Civil Engineering, Saint Petersburg, Russia
NIiPISpecrestavratsia LLC, Saint Petersburg, Russia
Likhachev Russian Research Institute of Cultural and Natural Heritage, Moscow, Russia

ABSTRACT: The discussion on the permissibility and extent of changes, which has become a turning point in the history of international heritage protection, is still relevant for historical cities that belong to one of the most dynamic categories of protected sites. With the introduction of such a concept as "historic urban landscape", the concept of the urban territory and its values has expanded significantly, and management systems have been supplemented with a new approach that takes into account the most diverse components of the human environment: from natural to man-made, from material to socially determined. Historic Center of Saint Petersburg is the core of the historically formed urban agglomeration and is characterized by unique cultural and natural features. One of the urgent tasks is to determine the limits of permissible changes and potential impacts on the heritage of a particular development project. Under these conditions, the identification of attributes of outstanding universal value, which are the subject of protection of the World Heritage property in the terminology of Russian legislation, is the key to preserving one of the most complex urban systems in the world. The article, highlighting copyright developments in the field of analysis of the value characteristics of World Heritage sites, offers methodological approaches to identifying attributes of outstanding universal value.

1 INTRODUCTION

As a result of the ever-accelerating growth of cities and an increase in their population, the name of the urban era was firmly entrenched in the current century, as reflected in the documents of the UN Habitat Program (Guzmán et al. 2014).

Urbanization in its various manifestations (from the development of new territories and migration to the revitalization of the existing urban planning environment and its adaptation to new conditions) is of particular importance for historic cities that are still populated and continue to develop under the influence of socio-economic and cultural processes (UNESCO 2019). It is for them that uncontrolled development is the factor that significantly affects both the integrity of urban fabric and the sense of place expressing the specifics of a particular city (Grishina & Grishin 2012). In this regard, the tasks of preserving the identity of historic settlements and the sustainable development of their territories can be attributed to one of the most urgent ones in the modern cultural agenda.

The international community has come a long way before recognizing the special status of cities as heritage sites. This led to the emergence of such a concept as "historic urban landscape", which significantly expanded the idea of the urban territory and its values, and also contributed to the emergence of

a new management approach (HUL approach). Based on the interdisciplinary nature of urban planning, this approach takes into account the most diverse components of the human environment (Bandarin & Van Oers 2012, Pasechnik & Marushina 2019, Ripp & Rodwell 2015, Rodwell 2018).

Today, changes in the broadest sense are recognized as an integral part of a person's life, and conservation has received a new interpretation and is understood as the management of ongoing changes (Feilden 2003, Tomaszewski 2012). Under these conditions, the question of the permissible limits of transformation remains relevant, especially when it comes to the preservation of urban heritage (Lipp 2012, Petzet 2012). In this regard, to determine potential impact that historic urban areas may be exposed to, the international community uses such a tool as heritage impact assessment. It is recommended by the World Heritage Committee as a measure to prevent threats or minimize their consequences. The analysis of the degree of potential influence is based on the value characteristics of the heritage object.

2 METHODS

Given the multilayer structure of the cultural and natural landscape values of Saint Petersburg agglomeration, the study of its components was based on the methods of system and structural analysis. They made it possible to consider the elements that form the environment and image of Saint Petersburg as a single system, which is characterized by complex internal connections. A comparison of the Statement of Outstanding Universal Value and the characteristics of the unique urban development object protected at the national level made it possible to identify both various aspects of its value and specific material characteristics (attributes) that are their expression.

3 RESULTS AND DISCUSSION

The expansion of the concept of heritage, the complexity of its typology and the recognition of the importance of interconnection with the surrounding context marked the transition from separate protection of heritage sites and the idea of conservation as an independent process, primarily related to providing physical protection, to understanding the role of cultural, natural, social, economic and other factors. However, despite the commonality of the problems and challenges that the heritage faces today, including historic cities, the set of characteristics expressing the value of each object is unique.

Historic Center of Saint Petersburg and Related Groups of Monuments, included in the UNESCO List in 1990 in accordance with the decision of the 14th session of the World Heritage Committee (Banff, Canada), became the first Russian World Heritage Site. The concept of the nomination was bold and innovative: not only the city center or the most famous of its suburbs was proposed for inclusion, but the entire historically developed agglomeration as a single unique urban planning and historic and cultural monument.

The value of this object, recognized at the time of entering into the World Heritage List, lies in the continuity and sequence of urban development of the territory, a distinctive feature of which is the integrity of the historic urban environment: monuments, ensembles and ordinary buildings belonging to different eras and styles that were combined by the efforts of generations of architects and urban planners into a unique architectural and urban ensemble. Saint Petersburg, a unique socio-cultural phenomenon, is associated with the names of artists, poets, writers, architects, statesmen and scientists, as well as with events that influenced the course of world history. These features of the object are embodied in the Retrospective Statement of Outstanding Universal Value adopted in 2015 by the World Heritage Committee (Bonn, Germany).

Historic Center of Saint Petersburg and Related Groups of Monuments is a serial object, combining 36 components and 86 elements belonging to different typological groups. In addition to the historical center of Saint Petersburg, the object includes palace and park and estate ensembles, fortification and engineering structures, scientific and memorial components, natural and cultural landscapes, elements of the historical planning structure, as well as historical centers of small towns and summer cottages.

The features of the outstanding universal value of a serial object are that each of its components and elements contributes to the formation of the outstanding universal value of the object as a whole. That is why it is fundamentally important to determine this contribution to the overall-resulting-outstanding universal value.

Despite the fact that the wording of the outstanding universal value should contain the most complete description of the World Heritage Property, as a rule, it is not sufficiently detailed from the point of view of indicating the quality of the object, which must be preserved and received the name of value attribute in international practice (ICOMOS 2011). First of all, this is relevant for objects with a complex structure, which undoubtedly include Historic Center of Saint Petersburg and Related Groups of Monuments.

Attributes can be physical characteristics or materials, but can also be processes associated with the object and capable of affecting its physical qualities (ICOMOS 2011). Their selection is an integral part of the process of conservation and managing the object. At present, the international and domestic practice of working with value attributes is becoming more and more stable. However, a general algorithm of work has not yet been proposed. In any case, the list of attributes is determined based on the individual characteristics of the object.

Historic cities and urban centers, which include the historic center of Saint Petersburg as the main component of the World Heritage Site, are interpreted by the Guide to the Implementation of the World Heritage Convention as a "group of urban buildings". The key values of such objects, according to the Convention, are: the value from the point of view of art (artistic value, aesthetic value), the value from the point of view of science (scientific value) (Jokilehto et al. 2008).

Each of these values, to one degree or another, is reflected in the criteria approved by the World Heritage Committee to determine the outstanding universal value of the object The Historic Center of Saint Petersburg and Related Groups of Monuments (Table 1).

One of the most interesting, in our opinion, methods that can be used to determine value attributes was proposed by T. Nypan (2013). Despite the fact that the researcher focuses on visual integrity, the principle of highlighting valuable characteristics of an object allows you to structure outstanding universal value and achieve maximum detail when determining the features of a World Heritage property to be protected. In this regard, three groups of components are distinguished: attributes (the most general characteristics of an object), components (describe significant aspects or parts of each attribute) and features/elements (lower level characteristics). Such an approach to highlighting

Table 1. The analysis of the key characteristics of the object Historic Center of Saint Petersburg and Related Groups of Monuments based on the criteria approved by decision 14 of the COMVIIA of the World Heritage Committee (1990).

Criterion	Key characteristics	Values
(i) masterpiece	a unique artistic achievement in the field of urban development: – ambitiousness of the program, – plan consistency – execution speed (1703-1725)	urban planning landscape historic aesthetic associative
(ii) values/ influences	ensembles created in Saint Petersburg and its environs: – influence on the development of architecture and monumental art of Russia and Finland – normative significance of Saint Petersburg as the capital – the urban planning model of Saint Petersburg	architectural urban planning
(iv) typology	architectural ensemble of Saint Petersburg with outstanding examples of imperial residences: – mainly baroque and classicism – reconstruction after World War II	architectural artistic/ aesthetic historic
(vi) associations	connection with world events: – the foundation of Saint Petersburg (opening of Russia to the Western world) – victory of the Bolshevik revolution – (formation of the USSR) associative	associative historic memorial

the value characteristics of a serial World Heritage Site seems to be most appropriate for its complex hierarchical structure (Table 2). For ease of perception, the terminology of T. Nypan's methodology has been adapted taking into account the structure of the object Historic Center of Saint Petersburg and Related Groups of Monuments and, thus, value characteristics can be divided into three levels: attributes of the 1st, 2nd and 3rd levels.

The uniqueness of the object Historic Center of Saint Petersburg and Related Groups of Monuments lies in the fact that, despite the enormous area, all its components developed in a single rhythm and on the basis of general urban planning principles. With this in mind, attributes of the 1st level describe the values of the object as a whole, attributes of the 2nd level characterize typological groups of components and elements of the object (i.e., they refine and specify the parameters of each attribute of the 1st level), attributes of the 3rd level describe the valuable characteristics of a specific component or element of an object. Thus, the identification of attributes of the 3rd level is the result of the maximum detailing of each value attribute of the World Heritage property as a whole.

Integration of the principles of analysis of outstanding universal value into the ICCROM system of "values - value attributes" allows for a comprehensive study of all the characteristics of an object, going beyond exclusively outstanding universal values and considering a wider urban, landscape, cultural and social context.

Not only the unique features of the historical, urban and natural environment, but also the authenticity of its main components determine the status of Saint Petersburg as the World Heritage Site. At the same time, authenticity has a complex multilayer structure, since the value characteristics inherent or attributed to an object can be considered in the context of various categories of values and expressed through a variety of characteristics (form and design, location and environment, materials and substances, etc.) (The Nara Document on Authenticity 2014, UNESCO 2019, Van Balen 2008).

The approaches to highlighting the value attributes proposed in this work are also based on many years of experience of domestic experts in determining the subject of protection of cultural heritage objects. The subject of protection, in accordance with Federal Law of June 25, 2002 No. 73-FZ "On Objects of Cultural Heritage (Monuments of History and Culture) of the Peoples of the Russian Federation", represents the

Table 2. The structure of the outstanding universal value of the World Heritage Site Historic Center of Saint Petersburg and Related Groups of Monuments.

World heritage site	Typological groups of object components	Specific component or elements of the world heritage site
Attribute 1	Attribute 1.1	Attribute 1.1.1 Attribute 1.1.2
	Attribute 1.2 and etc.	Attribute 1.2.1 Attribute 1.2.2 and etc.
Attribute 2	Attribute 2.1	Attribute 2.1.1 Attribute 2.1.2
	Attribute 2.2 and etc.	Attribute 2.2.1 Attribute 2.2.2 and etc.

features of the object of cultural heritage that are grounds for inclusion in the register and subject to mandatory preservation. Thus, it is possible to establish a correspondence between the subject of protection and the attributes of the outstanding universal value of the World Heritage Site, which are the basis for inclusion in the World Heritage List and need protection, management and monitoring (UNESCO 2019).

Table 3 presents the most common attributes that can be applied to various typological groups of components and elements of the object Historic Center of Saint Petersburg and Related Groups of Monuments.

Table 3. Attributes of outstanding universal value of the World Heritage Site Historical Center of Saint Petersburg and Related Groups of Monuments (1st level attributes).

Value category	Attribute	Attribute type (in the context of authenticity)
Town-planning/ Historical/Socio-cultural/ Functional/Ecological	Historical planning framework of the Saint Petersburg agglomeration	Form and design; location and surroundings
	The role of components in the overall urban structure of the object	Location and environment; usage and functions
	The system of historical settlement of the Saint Petersburg agglomeration	Location and environment
Landscape/Natural/ Historical/Aesthetic	Historical relief of Saint Petersburg and its environs	Form and design, mass and scale
	Historical green spaces of Saint Petersburg and its environs	Location and surroundings; mass and scale
	Historical hydrosystems of Saint Petersburg and its environs	Location and environment
	The historical relationship of green, free and built-up spaces within the Saint Petersburg metropolitan area	Location and surroundings; mass and scale, spiritual and physical perception
	The main visual directions for the architectural and urban development dominants of Saint Petersburg and its environs	Form and design; location and surroundings; mass and scale; spiritual and physical perception
	Unobstructed view of natural dominants and panoramas from them	Location and environment
	Features of the perception of species and panoramas	Spiritual and physical perception
Architectural/Functional/ Aesthetic	Spatial structure of the historical buildings of Saint Petersburg and its environs (ensembles, monuments, objects of ordinary historical buildings)	Form and design; location and surroundings; mass and scale; usage and functions
	Spatial and planning structure of the historical buildings of Saint Petersburg and its environs (ensembles, monuments, objects of ordinary historical buildings)	Form and design; location and surroundings; usage and functions
	Architectural and artistic solution of the historical buildings of Saint Petersburg and its environs (ensembles, monuments, ordinary historical buildings)	Form and design; materials and substances; mass and scale; color and texture
Archaeological/Historical	Archaeological heritage as a source of information about the history of an object	Location and environment
Memorial/Scientific/ Artistic/Aesthetic	Connection with events of world and domestic history, science and culture	Language and other forms of intangible heritage; spiritual and physical perception
	Associations with works of art and literature, scientific achievements and the names of scientists and cultural figures	Language and other forms of intangible heritage; spiritual and physical perception
Sociocultural	Traditions, legends, beliefs, myths	Language and other forms of intangible heritage; spiritual and physical perception; traditions, methods and management systems

4 CONCLUSION

Listing on the World Heritage List means recognition by the international community of the outstanding universal value of the object: the value that such an object has for all of humanity, its present and future generations. It also means the willingness of the state where the monument is located to assume obligations to protect, preserve and popularize it in the present and future.

The implementation of the recommendations of international experts aimed at maintaining the value, integrity and authenticity of the World Heritage Site, requires a complete and accurate idea of the outstanding universal value and attributes expressing it - key characteristics of the historical cultural and natural landscape of Saint Petersburg. In this regard, the determination of attributes of outstanding universal value and the creation of conditions for their preservation also turns out to be the key to preserving the object Historic Center of Saint Petersburg and Related Groups of Monuments, which is one of the most complex urban systems in the world.

The analysis methods proposed in this article to determine outstanding universal value were tested in the framework of heritage impact assessments for typologically different components of the World Heritage Site Historic Center of Saint Petersburg and Related Groups of Monuments, it was also used to determine the value attributes of several Russian World Heritage Sites as part of the work carried out by the Russian Scientific Research Institute of Cultural and Natural Heritage named after D.S. Likhachev.

REFERENCES

Bandarin, F. & Van Oers, R. 2012. *The historic urban landscape: managing heritage in an urban century.* Oxford: Wiley-Blackwell.

Feilden, B.M. 2003. *Conservation of historic buildings.* Oxford: Architectural Press.

Grishina, O.A. & Grishin, A.I. 2012. Historical and cultural heritage in the context of sustainable development. *Vestnik of the Plekhanov Russian University of Economics* 5 (47): 16–24.

Guzmán, P.C. et al. 2014. Bridging the gap between urban development and cultural heritage protection. In *IAIA14 Conference Proceedings. Impact assessment for social and economic development. 34th Annual Conference of the International Association for Impact Assessment, April 8–11, 2014.* Accessed May 02, 2020. https://conferences.iaia.org/2014/IAIA14-final-papers/Guzman,%20P.C.%20Bridging%20the%20gap%20between%20urban%20development%20and%20cultural%20heritage%20protection.pdf.

ICOMOS 2011. *Guidance on heritage impact assessments for cultural world heritage properties.* Accessed June 06, 2020. www.icomos.org/world_heritage/HIA_20110201.pdf.

Jokilehto, J. et al. 2008. *The World Heritage List. What is OUV? Defining the outstanding universal value of cultural world heritage properties.* Berlin: Hendrik Bäßler Verlag.

Lipp, W. 2012. Dimensions of limits – philosophical and cultural aspects. In W. Lipp et al. (ed.), *Conservation turn - return to conservation: Tolerance for change, limits of change: Proceedings of the International Conferences of the ICOMOS, International Scientific Committee for the Theory and the Philosophy of Conservation and Restoration, Prague, 5–9, May 2010, Florence, 3–6 March, 2011:* 135–142. Florence: Polistampa.

Nypan, T. 2013. On a method for indicator based monitoring of visual integrity. In *International Conference on contemporary architecture in historic settings. UNESCO recommendation on historic urban landscapes.*

Pasechnik, I.L. & Marushina, N.V. 2019. Value category in theory and practice of conservation of historical urban environment. *Vestnik Tomskogo gosudarstvennogo arkhitekturno-stroitel'nogo universiteta. Journal of Construction and Architecture* 3: 9–19.

Petzet, M. 2012. Conservation/preservation: limits of change. In W. Lipp et al. (ed.), *Conservation turn - return to conservation: Tolerance for change, limits of change: Proceedings of the International Conferences of the ICOMOS, International Scientific Committee for the Theory and the Philosophy of Conservation and Restoration, Prague, 5–9, May 2010, Florence, 3–6 March, 2011:* 261–263. Florence: Polistampa.

Ripp, M. & Rodwell, D. 2015. The geography of urban heritage. *The Historic Environment: Policy & Practice* 6 (3): 240–276.

Rodwell, D. 2018. The historic urban landscape and the geography of urban heritage. *The Historic Environment: Policy & Practice* 9 (3-4): 180–206.

The Nara Document on Authenticity 1994. Nara Conference on Authenticity in Relation to the World Heritage Convention, Nara, Japan, 1–6, November 1994, http://unesco.urbanismosevilla.org/unesco/sites/default/files/02.TerjeNypan-Ponencia.pdf

Tomaszewski, A. 2012. Conservation between "tolerance for change" and "management of change". In W. Lipp et al. (ed.), *Conservation turn - return to conservation: Tolerance for change, limits of change: Proceedings of the International Conferences of the ICOMOS, International Scientific Committee for the Theory and the Philosophy of Conservation and Restoration, Prague, 5–9, May 2010, Florence, 3–6 March, 2011:*43–46. Florence: Polistampa.

UNESCO 2019. *The operational guidelines for the implementation of the World Heritage Convention.* Accessed May 1, 2020. https://whc.unesco.org/en/guidelines/.

Van Balen, K. 2008. The Nara Grid: An evaluation scheme based on the Nara document on authenticity. *APT Bulletin: The Journal of Preservation Technology* 39 (2): 39–45.

Reconstruction and Restoration of Architectural Heritage – Sementsov et al (eds)
© 2020 Taylor & Francis Group, London, ISBN 978-0-367-65357-6

Saavedra's contribution to modern methodology of study of Roman Roads in Spain

I.F. Menendez Pidal de Navascues
Escuela de Ingenieros de Caminos, Canales y Puertos, UPM, Madrid, Spain

M. Almagro-Gorbea
Real Academia de la Historia, Madrid, Spain

I. Moreno Gallo
Ministerio de Fomento, Madrid, Spain

I. Aguilar Civera
Universidad de Valencia, Madrid, Spain

J. Mañas Martínez
Fundación Juanelo Turriano, Madrid, Spain

J. Alonso Trigueros & A. Arcos Álvarez
Escuela de Ingenieros de Caminos, Canales y Puertos, UPM, Madrid, Spain

ABSTRACT: Paper means to set and clarify the roll of Don Eduardo de Saavedra as the precedent of the methodological and scientific study of the Roman Roads in Spain. His widely humanistic and engineering condition contributed significantly to promoting the study of Roman Roads from a multidisciplinary point of view. The work shows a complete vision of the study of these Roman engineering works in Spain, reflecting on the importance of their examination to base the young civil engineering as opposed to the military and architecture, and their situation nowadays, where their study remains a reference for the idiosyncrasy of the *ingeniero de caminos* or civil engineer.The paper reveals, compiles and analyses the historical environment lived by Eduardo Saavedra that push him to understand the roots of the technical humanism of his time, which leads on the one hand to the search for the Roman roots of engineering and, on the other, to the consolidation of civil engineering as opposed to the military one.

1 INTRODUCTION

Don Eduardo Saavedra y Moragas (27.2.1829-Madrid, 12.2.1912), born in the city of the former imperial Tarraco in 1829, is one of the great figures of Spanish society and culture of the second half of the 19th century (Figure 1). In addition to being a very remarkable engineer and architect, he was "an archaeologist-engineer and an Arab-historian", "… as a wise person of the Renaissance", as Julio Caro Baroja said of him, and "one of the most complete and encyclopedic people of our country and the most important Spanish humanist engineer of the twentieth century", in the words of José Mañas, since he added to his excellent professional quality an extensive humanistic training, which is reflected in his brilliant skills as an archaeologist and an excellent arabist. In a word, he was an intelligent multidisciplinary fellow in current terminology, and is therefore considered one of the most distinguished Spanish archaeologists of the second half of the nineteenth century.

It is not easy to explain what the figure of Saavedra represents as an archaeologist in 19th-century Spain. This activity was surely inspired by the memories of his native Tarragona and because circumstances led him to carry out his activity as a civil engineer (ingenieros de caminos) in the province of Soria, where he finally located Numantia and where he carried out his study "Description of the Roman way between Uxama and Augustobriga", published in 1879 in the Memoirs of the Royal Academy of History, which is his best known work, because he owes his unquestionable reputation as an archaeologist.

However, Eduardo Saavedra, in addition to a very brilliant engineer, was a true humanist, able with his intelligence to deal deeply with very diverse fields of letters or science. In this regard, it has been pointed out that it represents a humanist trend that positively influenced an entire generation of engineers and a model for successive generations, because, in addition to be a great engineer, he felt interest in

Figure 1 and 2. Portrait at Royal Academy of History. Bernesga River Bridge built in November 1863.

numerous fields, as evidenced by his publications, in addition having completed the architectural career in three years, studies that finalized in 1870. Besides, he joined his ability for technical drawing as an engineer, but he also become skilled at artistic drawing, as evidenced, for example, by the sheets drawn by him for his publication about the church of San Juan de Duero and that of San Nicolás, in Soria, Castille, Spain.

2 METHODOLOGY

2.1 Saavedra and his background

The After eight years of teaching at the School of Civil Engineers (Escuela de Ingenieros de Caminos), where he consolidates his professional prestige, he joins the Palencia Railroad Company in Ponferrada. At this time implantation of the first railroads in Spain were carrying out, and the best engineers are hired by the railway companies. Saavedra could not be left out of this historical moment.

As Chief Engineer of the Company, he directs the construction of the Palencia-León track, inaugurated in November 1863, and designs the León–Astorga track with all its rail stations. His project is the beautiful bridge over the Bernesga River (Fig. 2), to link the León station with the city (the bridge was replaced in the 50s by the current one). Saavedra will also carry out the draft project of the Torralba to Soria (Soria province, Spain) railway line, in 1863, which was not built until many years later, in 1892. In this line it is worth highlighting his design of viaduct at the Golmayo River, Soria, Spain.

It was usual practice of the time that the numbers one of the promotions of the School were claimed as teachers of the same. For this reason, Saavedra was part of the faculty staff from 1854 to 1862 and would still be teaching years later. This was a stage in which he alternated teaching with an intense publicist activity. From the arrival to the School until his departure he is in charge of Applied Mechanics. He also teaches Rational Mechanics and Construction.

Saavedra worried about writing texts for his lessons, thus relieving the shortage of texts in Spanish in the Engineering Schools. His texts were widely disseminated and, in particular, his Theory of Hanging Bridges. In addition to his original works, he translated another authors making available to the students, for the first time in Spanish, lectures that were traditionally studied in foreign languages. A multitude of technical news published in magazines abroad were translated by his personal initiative to the *Revista de Obras Públicas,* Public Works Journal, main engineering and professional journal in Spain for civil engineers. In this way he did an important work of scientific dissemination. During his teaching period at the Escuela de Caminos, Saavedra is responsible for carrying out projects with different students. In the summer of 1857, he moved to Andalusia and designed the lighthouses of Chipiona, Salmedina and Trafalgar. Carry out the project of the road from Cudillero to Cornellana through Pravia, in Asturias, North Spain. Another project that he carries out is some tracks of Garray-Villar road in Soria, Spain.

2.2 Statement of the civil engineering identity question in the XIX century

We know that the nineteenth century seeks its roots, its regional and national identity, an aspect of some impact on the consolidation of professions dedicated to the broad field of construction art.

The Civil Engineer, *ingeniero de caminos*, will be a key figure in this century, not only because as a professional was born and configured itself in this century, from the foundation of the Corps of Engineers in 1799, *Cuerpo de Ingenieros de Caminos*, but because it responds to social, economic and political needs. He is a professional who identifies with the progress of society, who knows and manages technological advances, new construction arts, communication systems, acting on the territory, on the city and on society itself. Well, in civil engineering, the objective is public utility, social welfare and responds to certain promotion policies to organize and put the territory into production. This kind of professional will be consolidated throughout the century, as the numerous engineering professionals must assume a new role in the administrative structure of the State.

The reign of Isabel II (1844-1868) will be an exemplary period in forming administrative structures of government, restructuring the powers of the engineer in its various branches (military, roads, canals, ports, mines, mountains and industrial). In this sense, educational training becomes a fundamental factor, from which specialized schools and the different professional bodies are created.

This reorganization creates a series of conflicts over the powers of each specialized body, and will be a vast field to develop a desire for brand, identity, as if it were a product in a competitive market.

Therefore, as in any marketing study, it is essential to define, know and make yourself known, differentiate yourself from other products, other professions, reflect their high quality, communicate both internally and externally, and disseminate their works and works, etc.

In this area and in the second half of the 19th century, in relation to the figure of the road engineer, Eduardo Saavedra (1829-1912) will be an active part of that exciting project of being an *ingeniero de caminos,* Civil Engineer.

In 1896, an article by Manuel Pardo in the Public Works Journal called attention to the convenience of celebrating the centenary of the creation of the Corps of Civil Engineers on June 12, 1799 to pay tribute to the profession that, not in small scale, had contributed "to the development of industry and commerce and the welfare of all social classes" and to serve to highlight their valuable services. In the brief article, the engineer recounted the difficulties of the profession in his first decades, but he also praised the results of the works carried out throughout the century. At the end of the century we can observe a consolidated profession and proud to belong to a Corp, to a Body. Pardo encouraged solemnly to celebrate the first centenary "serving as a stimulus for contemporaries to continue vigorously the work begun by their ascendants".

2.3 *Analysis of Saavedra's contribution*

On the other hand, to contextualize Saavedra's work, one must understand that Rome was the reference for many reflections or investigations, the look of the past for its comparative analysis. In the engineering world, Rome was a top period in public constructions or public utility buildings. A moment of great infrastructures, such as the road network, the hydraulic network, territorial planning, a period of large constructions in new port facilities. Someone could check, among many other texts, for example one of Toribio de Areitio the "Essay on water legislation" published in the "Memories and Documents" collection of the Public Works Journal, *Revista de Obras Publicas*, in 1858. In its long first chapter, entitled "About water considered in the physical order of nature", introduces us to the history of its uses and uses (irrigation, inland navigation, water-driven artefacts, population supply) since ancient times.

Other interesting text is the speech at Manuel Pardo's reception ceremony at the Royal Academy of Exact, Physical and Natural Sciences in 1894 that dealt with the "Importance of chemistry in construction". A speech full of praise for ancient architecture and engineering, but which, in turn, serves to affirm the importance of modern science, applied chemical industry and the great advantages it brings to the art of construction.

Regarding to roads, "Memory on the pavement or road construction or road sign", by Francisco Xavier de Barra, is one of the main texts that we can analyze and study from its historical side and from its present side. It was published in 1826 in full reinstatement of the absolutism of Fernando VII and in an immobilistic environment with a weak public works policy and when it was closed the School of Civil Engineers, *Escuela de Caminos*. This text was conceived as a treaty, Barra dedicates the first part of the book to developing the method followed by the Romans to build the pavement of the roads and explain the method that had been followed in Spain during the 18th century.

The rest of the chapters refer to the mistakes that had been made in these constructions and the way in which the pavements were to be made: convexity of the surface, construction process, materials and different uses of stones. It is the part of the text that can be considered as treated or manual. Finally, it concludes with a comparative analysis between the Roman firms and those that were made in their time.

3 DISCUSSION

His humanistic training led him to undertake many different activities in this field. He founded in 1876 Geographic Society of Madrid, of which he was President, which became the Royal Geographic Society. He made the Report on the Study Plan of Alejandro Pidal y Mon (1846-1913), a plan that remained in force for more than half a century, in 1868 he was part, together with Aureliano Fernández Guerra, of the commission to examine the designs for the peseta wedges as a new currency and managed the Spanish delegation at the inauguration of the Suez Canal, in 1869, in addition to being a member of the Royal Academy of History (1861), of the Royal Spanish Academy (1878) and of the Royal Academy of San Fernando (1894).

In this academic field his activity is framed to a large extent as an erudite and as an archaeologist. Eduardo Saavedra was one of the most distinguished numeraries of the Royal Academy of History at the time (1861-1912), elected academic in 1861, at thirty-two years old, when he won the National Antiquities Award in 1861 for his Description of the Roman Road from Uxama to Augustobriga, Descripción de la vía romana de Uxama a Augustobriga, which was published in 1879 and is its best known archaeological study and a model to imitate. His entry speech dealt with public works in ancient times, which was answered by the Perpetual Aureliano Fernández-Guerra. He was Numerary over 50 years (1861-1912), Treasurer (1878-1895), Dean (1898-1912) and became Director (1908-1909), a very remarkable fact in the history of this institution, but it is also a fact to point out in the cultural life of the Restoration period in Spain. In addition, as an architect, he was in charge of the works for the rehabilitation of the building of the *Nuevo Rezado*, built by the famous Spanish architect Juan de Villanueva in the 18th century (Figure 3), which is the current official location of the Royal Academy of History. It was a laborious work that ended in 1874 and was elected senator by the Royal Academy of History in 1895 and re-elected for seventeen years.

His studies as a civil engineer, *ingeniero de caminos*, provided him with special professional and technical training to contextualize archaeological finds in the landscape, which was especially useful for studying Roman Roads, such as the one mentioned from Uxama to Augustobriga (Figure 4 (a) (b) and (c)), that is traced from Burgo de Osma to Muro de Ágreda, Province of Soria, Spain. The interest in this Roman way allowed him to locate the famous, ancient and immortal Celtiberian city of Numantia, whose situation was discussed since the Middle Age.

Saavedra confirmed that it corresponded to the hill of Garray, near Soria city, on the Duero River, a place already proposed previously, but without definitive arguments, by Florián de Ocampo, Fr. Flórez and Juan Loperráez in his "Historical Description of the Bishopric of Osma" of 1788. To prove it, he started excavations in 1853 and carried out various campaigns from 1861 to 1867 under the patronage of the Royal Academy of History, which unfortunately were unpublished, so we only have brief reports and a plan, since the original memory is lost.

Some of the objects appeared, including two fragments of ceramic tubas and bowls and painted Celtiberian jugs, were deposited in the Royal Academy of History, where they are currently kept next to a small bottle with a label, which could be of Saavedra letter, which indicates that it contains "Ashes of the fire of the immortal Numancia" (Figure 4), remains that reflect the post-romantic nationalism of the time, which has been closely associated with this site.

We cannot forget that, as we had previously pointed out, Soria had been Saavedra's first destination as a Civil Engineer, being appointed Delegate of Public Works, a position he held with great activity for two years until 1853. When he was building the road from Soria to Burgo de Osma, the cities and Roman vestiges of the *Meseta* (Spanish central geographical natural region) had to refresh his memories about their native Tarragona city and decided to study the Roman road that from Uxama, the current Burgo de Osma, goes to Muro de Ágreda, the old Augustobriga, which passed also by Numantia, whose disputed situation was definitely confirmed, dismantling the arguments about the alleged location in Zamora.

The Description of the Roman Way between Uxama and Augustóbriga, *Descripción de la Vía Romana entre Uxama y Augustóbriga*, is a model and innovative work in many aspects, especially for its precise topographic data due to an engineer, so it reminds an memorable precedent in Spanish Archeology as the works of the also engineer Joaquín de Alcubierre in the excavations of Carlos III that discovered Pompeii.

Figure 3. Royal Academy of History, Madrid, Spain.

Figure 4. Ashes of Numantia city, supposed after its destruction in 133 b.C. by Rome.

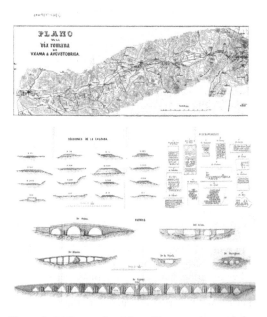

Figure 5. (a) Roman Road from Uxama to Augustobriga. General ground plan. (b) Cross Section (c) Bridges.

This work of Saavedra enthused at the Academy for its technical rigor, since then rarely exceeded, and for its knowledge of humanistic erudition by including a broad appendix with the news of geographers and classical historians, for which it deserved the 1861 Antiquities Award of the Royal Academy of History, recently established, which was a silver medal (Figure 6), the sum of 3,000 reals and to be elected as a member of the Royal Academy of History.

With great practicality, he knew how to accompany the memory with a remarkable chase-suitcase (Figure 7), to expose the multiple objects and coins he had collected when studying the Roman road, which he generously donated to the Royal Academy of History, where it is currently preserved. These findings include bell-shaped assegai tips and axes from the Bronze Age and various Celtiberian fibulae, several Visigoth brooches and other objects gathered when studying the Roman Road, although not all belonging to it.

In those years Saavedra also recovered an important Roman bronze helmet of Montefortino type appeared in 1863 in Quintana Redonda, Soria (Figure 8), found with several amphorae full of silver coins that have been lost, although Saavedra managed to save recovered some who donated, next to the helmet in Figure 8, to the Royal Academy of History in 1888.

Figure 6. Medal. Royal Academy of History.

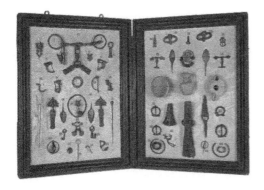

Figure 7. Special closet chase belonged to Saavedra's antiquities collection.

Figure 8. Celtiberian helmet and silver coins. Quintana Redonda, Soria.

4 CONCLUSIONS

In this context, the brilliant and original personality of Eduardo Saavedra y Moragas, which stands out among the archaeologists of his generation, must be outlined. Saavedra was, without a doubt, who covered more fields of knowledge in antiquarian studies. This diversity of orientations and themes, which he dominated with his prodigious intelligence, provided him with an exceptional wideness of vision, which today we would call interdisciplinary, and a much higher capacity than usual to associate the data he knew in many different fields.

His heritage, push the scientific, organizational, methodological, systematic study of roman roads in Spain till nowadays. An uncountable group of scientists, academics, professors, engineers, somehow, follow his multidisciplinary approach: Antonio Blázquez, Gonzalo Menéndez Pidal, Claudio Sánchez Albornoz, Federico Wattenberg, Gonzalo Arias, Raymond Chevallier, Pierre Sillières, José Antonio Abásolo, Juan Manuel Abascal, Xosé Manuel Caamaño, Manuel Durán, Segundo Alvarado, Carlos Nárdiz, Enrique Cerrillo, Santiago Palomero, etc.

He cultivated Archeology, together with Epigraphy and studies of the Hispanic Islamic world with his deep humanistic vocation, which reflects the tradition of antiquarian erudition originated in the Renaissance and enriched in the Enlightenment. But to these humanistic knowledge he added the technician knowledge with a practical engineering efficiency that shows his precise descriptions, far superior to those of his time. Although his work, so varied, is consequently dispersed and within his numerous archaeological publications, monographs such as lost memory on his excavations in Numancia are missing, as well as the corresponding part of the Roman Epigraphy of the Province of León that did not write.

ACKNOWLEDGMENTS

The authors would like to thank International Association of Hispano-American Roads, *Escuela de*

Ingenieros de Caminos, Canales y Puertos, Polytechnic University of Madrid, and Royal Academy of History of Spain for providing material and figures.

REFERENCES

Almagro-Gorbea, M. 2013. Zóbel de Zangróniz, Jacobo, Diccionario Biográfico Español, 50, Madrid: 852–854.

Almagro Gorbea, M. (2013) Schulten Klarenbach, Adolfo, Diccionario Biográfico Español, 46, Madrid, 2013: 387–390.

Almagro-Gorbea, M., Maier, J., 1999. El futuro desde el pasado: la Real Academia de la Historia y el origen y funciones del Museo Arqueológico Nacional, *Boletín de la Real Academia de la Historia* CXCVI, 1999, 2: 183–207.

Almagro-Gorbea, M., 2007. La Real Academia de la Historia y la Escuela Superior de Diplomática, F. de los Reyes y J. Mª de Francisco, eds., 150 Aniversario de la Escuela Superior de Diplomática (1856-2006). Reglamentos y Programas. Madrid: 13-32.

Almagro-Gorbea, M., Casado, D., Fontes, F., Mederos A., Torres J. 2004. Prehistoria. Antigüedades Españolas I. Real Academia de la Historia, *Catálogo del Gabinete de Antigüedades* I.2.1. Madrid: 321–330.

Betancourt, A. 1869. Noticia del estado actual de los caminos y canales de España, causas de sus atrasos y defectos, y medio de remediarlos en adelante, *Revista de Obras Públicas*, 1869: 54-58, 68-71, 115-116, 156-158.

De la Torre Echávarri, J. I. 1998. Numancia. Usos y abusos de la tradición historiográfica, *Complutum*, 9: 193–211.

Gadea, E. 1882. Descripción de un trozo de la vía romana de Braga a Astorga por Chaves, *Revista de Obras Públicas*1882:. 169-172, 181-185.

González Reyero, S., Pérez Ruiz, M., Bango García, C. I. eds. 2007, Una nueva mirada sobre el Patrimonio Histórico, Madrid: 79-142.

Isaac, A. 1987. Eclecticismo y pensamiento arquitectónico en España. Discursos, revistas, congresos, 1846-1919, Granada, Diputación Provincial de Granada: 107.

Loperráez Corvalán, J. 1788. Descripción Histórica del Obispado de Osma, con tres disertaciones sobre los sitios de Numancia, Uxama, y Clunia, Madrid.

Marcos Pous, A. 1993. ed., De Gabinete a Museo, tres siglos de historia (catálogo de exposición), Madrid.

Mañas Martínez, J. 1983. Eduardo Saavedra. Ingeniero y Humanista, Madrid, Turner-Colegio de Ingenieros de Caminos, 1983.

Maier, J. 2003. La Comisión de Antigüedades de la Real Academia de la Historia, en M. Almagro-Gorbea y J. Maier, eds., 250 Años de Arqueología y Patrimonio. Documentación sobre Arqueología y Patrimonio Histórico de la Real Academia de la Historia. Estudio general e índices, Madrid: 27–52.

Maier, J. 2004. La Real Academia de la Historia y la arqueología española en el siglo XIX, *Eres. Arqueología/Bioantropología*, 12: 91–121.

Pardo, M. 1896. El centenario de la creación del Cuerpo de Ingenieros de Caminos, *Revista de Obras Públicas* 1896: 103–104.

Pericot, L. 1940. Adolfo Schulten. Su vida y sus obras. Homenaje de la universidad de Barcelona a su doctor honoris causa. con motivo de su 70 aniversario. *Anales de la Universidad de Barcelona. Memorias y Comunicaciones*: 1–32.

Saavedra, E. 1856. Arquitectura. San Juan de Duero, en Soria, *Revista de Obras Públicas*, IV, 24, 1856: 277–282.

Saavedra, E. 1879. Descripción de la vía romana entre Uxama y Augustóbriga, Memorias de la Real Academia de la Historia, IX.

Saavedra, E. 1863. Discursos leídos ante la Real Academia de la Historia en la recepción pública de Don Eduardo Saavedra, el día 28 de Diciembre de 1862, Revista de Obras Públicas, Colección Memorias y Documentos, Madrid: 103 pp.

Saavedra, E. 1879. Descripción de la Vía romana entre Uxama y Augustóbriga. *Memorias de la Real Academia de la Historia*, Tomo IX, Madrid.

Sáenz Ridruejo, F. 1996. Los ingenieros de Caminos, Madrid, Colegio de Ingenieros de Caminos, Canales y Puertos: 47.

Urban, R. 2007. Schulten, Ernst Adolf (1870-1960), *Neue Deutsche Biographie*, 23, Berlin: 691–692.

Reconstruction and Restoration of Architectural Heritage – Sementsov et al (eds)
© 2020 Taylor & Francis Group, London, ISBN 978-0-367-65357-6

Historical cemeteries as element of urban planning subject of protection of Saint Petersburg

A.V. Mikhailov
Saint Petersburg State University of Architecture and Civil Engineering, Saint Petersburg, Russia

ABSTRACT: The article is devoted to the study of historical cemeteries (existing and destroyed) as an important component of the urban planning subjects of protection of a historical settlement. For clarity, the analysis of the urban planning structure of one of the largest UNESCO-protected urban planning entities - the World Heritage Site Historical Center of Saint Petersburg and Related Groups of Monuments has been carried out. The relevance of the work is due to the fact that at present in the historical part of Saint Petersburg there are practically no territories left for development, and the economic and investment attractiveness of the territories is much higher as it approaches the historical center of the city. Thus, in the zone of active development there are territories of historical industrial enterprises and territories for some reason not previously involved in the economic activity. One of these areas is sites of former or destroyed cemeteries. The aim of the study is to consider these territories precisely as an element of the urban development environment with a possible analysis of methods for its conservation and acceptable transformation. For research purposes, the following methods were used: archival and bibliographic, cartographic, analytical, on-field studies. The research has achieved the following results: historical cemeteries have been analyzed as components of the urban planning subject of protection, and existing methods for their conservation and transformation have been evaluated.

Keywords: historical cemeteries of Saint Petersburg, subject of protection of the cultural heritage object, Mitrofanievskoe cemetery, Sampsonievskoe cemetery

1 INTRODUCTION

Studying history is vital, both for the individual and for the people as a whole. The popular expression "historical roots", speaks of a deep understanding that knowledge of one's history allows one to live. Unconditionally, part of the story of man and people accumulates in burial places, to which mythology, religion and psychology pay a lot of attention. However, cemeteries are not only a place of concentration of intangible history, it is also a quite tangible space that makes up the urban planning structure of the city and, as a result, affects its image and development.

The study of historical cemeteries in Saint Petersburg was and is being performed by many researchers. However, there are still not many professional scientific papers. We would first of all note the works of A.V. Kobak and Yu.M. Pirutko (Historical cemeteries of Saint Petersburg 1993, 2009, 2011). In their monumental works, these researchers describe both the existing and the lost historical cemeteries of the city, the history of their appearance and development. Without a doubt, we should also mention the works of I.G. Georgi (Georgi 1996) and M.I. Pylyaev (Pylyaev 1996, 2000) that describe in detail Saint Petersburg in different historical periods. Modern city researchers A.L. Punin (Punin 2014), S.V. Sementsov (Sementsov 2007) pay attention to the architectural

and urban development of the city. However, the topic of the significance of the cemetery as an important and not always static element of the urban planning structure of the city, due to a number of reasons, is not fully disclosed by any of these works.

We believe that now, despite the complexity of the topic of identifying, preserving and modern use of the territories of historical cemeteries and, in the first place, destroyed historical cemeteries, it is already necessary to deal with it practically. It is important to work on this topic putting aside all conventions and prejudices, because if we delay, we will either lose part of our history (as it once was with the deliberate destruction of historical necropolises), or we will destroy the historical environment (as it once happened during a spontaneous industrial or construction development of former cemeteries), or we will get a poor-quality urban environment of a modern city (if we abandon these territories by providing them for uncontrolled use).

One of the approaches to the modern study of such territories should be the historical and urban planning approach, representing historical cemeteries as a territorial and planning element of the natural historical development of the urban environment to be preserved. The study should answer the question about the form of conservation of this component of the subject of protection of the historical environment and the degree of its transformation. Given the volume of the question, this article is not in a position to give comprehensive answers. Its task is to determine the main approaches to the study of the identified problems.

The article was preceded not only by the study of historical scientific textual and cartographic materials stored in archives and libraries, but also by practical experience in creating a system of protection zones for cultural heritage sites of Saint Petersburg (Mikhaylov 2017), preparing tasks for historical, cultural and archaeological research of destroyed historical cemeteries and consideration of the results of these works.

2 MAIN PART

2.1 Urban planning subject of protection of Saint Petersburg and the role of historical cemeteries in it

In the article on the territorial and planning components of the urban planning subject of protection (using the example of the World Heritage Site Historical Center of Saint Petersburg and Related Groups of Monuments), the author tried to consider various factors affecting the territorial and planning subjects of protection of such a historic settlement as Saint Petersburg. In fact, the same factors influenced the appearance of historical cemeteries (location and spatial development) in the structure of the urban environment. The first and probably the main one is a geographical factor. As you know, cemeteries were located on the outskirts of the city, preferably in the not flooded parts of it, but at the same time not so far off (taking into account the presence of highways). Estate or class and economic factors (the proximity of settlements and the application of labor of various segments of society), as well as to some extent mythological and religious factors (habitual places of worship and the presence of religious shrines) also influenced the location of the cemetery. We suppose it would be wrong to assume that historical cemeteries appeared due to urban planning of creators of our and most of the other historical cities, although the presence of such objects is a prerequisite for the modern design of the city structure.

However, the city is growing and what once was located on its outskirts gradually turns out to be part of its urban structure and even at some point part of its historical urban structure. This fate befell most of the historical cemeteries of our city: the Alexander Nevsky Lavra, Lazarevskoye, Nikolskoe, Smolenskoe Orthodox, Smolenskoe Lutheran, Smolenskoe Armenian, Volkovskoe Orthodox, Novo-Dyevitchiye, Malookhtinskoe, Bolsheokhtinskoe, Gromovskoe, Krasnenskoe, In Memory of the Victims of January 9, Hebrew, Novo-Volkovskoe, Kinovievskoe, Porokhovskoe, Serafimovskoe, Severnoe, Bogoslovskoe, Sampsonievskoe, Mitrofanievskoe. As a rule, moral

and ethical and sanitary-epidemiological factors affect the nature of the development of such territories by contemporaries, and cemeteries are surrounded at the beginning with non-public buildings, and over time, with public and even residential facilities. Thus, based on the territorial planning characteristics of the facility, and not vice versa, the surrounding urban planning situation is formed. However, the emerging urban development environment can still not be called spontaneous. The territorial development of a cemetery is usually influenced by natural (e.g. water) or artificial (e.g. roads) objects, in some cases restricting, and in some cases correcting it. So, for example, along the peripheral areas of the territory of historical Sampsonievskoe cemetery (the first burial sites discovered date back to 1703), already in the 19th century a number of roads were laid that served as the starting line for the development of adjacent residential and non-residential quarters (Figure 1).

Thus, it can be stated that quite often the cemetery is a consciously conceived element of the historical urban planning composition (excluding memorial burial sites), but at the same time it has rational reasons for its origin and development. Also, the Necropolis of the Alexander Nevsky Lavra, Volkovskoe and destroyed Mitrofanievskoe cemeteries, despite being close to the city, were historically located on the periphery of the urban environment. It should be clarified that this article does not discuss individual burials within the boundaries of religious sites, since their classification as cemeteries is not entirely objective.

We can draw a preliminary conclusion that historical cemeteries become part of the urban environment as the developed territory of the urban settlement expands. Urban development, despite moral, ethical, religious and sanitary-epidemiological restrictions, due to the limited land resources, gradually structures

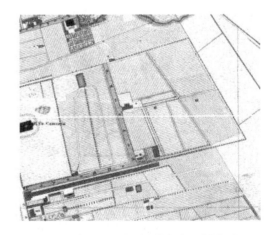

Figure 1. The fragment of a detailed plan of Saint Petersburg in 1828, Major General Schubert. National Library of Russia, K2-Pb 4/58 with the image of Sampsonievsky Cathedral and the cemetery.

Figure 2. Historical and cultural reference plan of Sampsonievskoe cemetery.

and, as a rule, reduces the territories occupied by burial sites, but does not fully involve them in circulation. The exception was the Soviet period, when historical cemeteries, along with objects of religious purpose, were destroyed purposefully to break the connection between generations (similar historical examples are repeated on a global scale throughout the entire history of mankind).

Summing up the definition of historical cemeteries as an element of the territorial and urban planning subject of protection, it is necessary to distinguish their characteristics: these are open or semi-open geometrically designed spaces in the style of historical traditions combined with the architectural dominants of religious objects and occupied by various types of green spaces; these spaces are included in the urban structure as a rule after their creation and even after their main use and are surrounded by non-residential buildings; in the process of their development, these objects receive either a memorial function (Necropolis of the Alexander Nevsky Lavra, museum-necropolis Literatorskie Mostki of Volkovskoe cemetery, Piskarevskoe memorial cemetery), or a recreational one (former Sampsonievskoe cemetery, the park near Lomonosovskoye metro station on the site of Transfiguration cemetery) (Figures 1 and 2), or become industrial or warehouse facilities (former Mitrofanievskoe cemetery, part of former Bolsheokhtinskoe cemetery) (Figures 3 and 4).

2.2 Features of the identification, conservation and modern use of territories of historical cemeteries

It has already been mentioned that the identification of this type of objects is important, as for the historical self-identification of an individual person and the people as a whole, as well as for the correct formation of the urban environment. The indicated characteristics and their functional use, which usually develops, can be recognized as logical, with the exception of those territories where industrial and warehouse facilities were formed due to historical and political features. In addition to historical, bibliographic and cartographic studies of the location of historical cemeteries, it is important to carry out

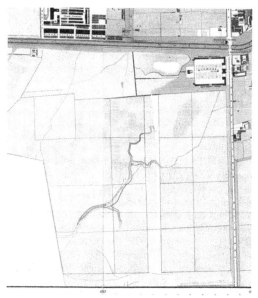

Figure 3. The fragment of a detailed plan of Saint Petersburg in 1828, Major General Schubert. National Library of Russia, K2-Pb 4/58 with an image of Mitrofanievskoe cemetery.

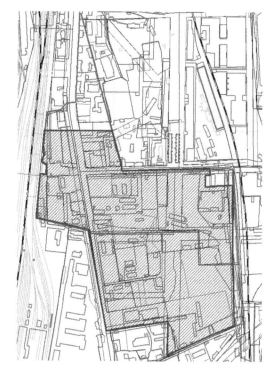

Figure 4. Historical and cultural reference plan of Mitrofanievskoe cemetery.

archaeological research, as practice has shown that historical boundaries do not always correspond to the urban planning situation, even if the cemetery was formed in the 19th century. Archaeological research on Neyshlotsky Lane 3 revealed multi-layered burials in the urban development territory beyond the accepted historical boundaries of former Sampsonievskoe cemetery. Furthermore, investigations of the territory adjacent to the garden near the Church of the Annunciation of the Blessed Virgin Mary on Maly avenue of Vasilievsky Island showed that under the foundations of the kindergarten (it sounds strange but there was also a kindergarten on Neyshlotsky Lane) there were multi-layered burial sites, which made it possible to establish that the boundaries of such objects were not always historically reliably recorded (Figure 2). It can be assumed that in fact there was a violation of historical boundaries and burials were carried out behind them, but due to religious traditions this was permissible only for certain categories of the deceased and was unlikely to be wide-spread as we are now finding.

If there are no methodological contradictions with the identification and preservation of historical cemeteries, and it's rather a matter of economic, social and political will, which now exists, although it may seem to someone that it's not very strong, as they would like it to be. However, the modern development of such territories is not so smooth. A particular example of this is the process of transforming the territory of former Mitrofanievskoe cemetery. Purposefully destroyed during the Soviet period along with the destruction of the entire complex of religious objects, this territory was essentially given over to spontaneous industrial and warehouse development and seemed to be practically lost. Acquisition of the status of an object of cultural heritage to most of the territory in 2014 and subsequent inclusion in the list of green spaces of common use gave a chance for a proper transformation. However, due to civil law regulation, it did not completely relieve it of its existing use. Further use of this territory, given the approaching development is obvious. The memorial and recreational function should and will be in demand both by residents of the neighborhoods under construction and by residents of the city, and with proper information support it is possible to include this object in tourist routes, especially since the history of this place is connected not only with Russian prominent figures, but also with foreigners (for example, the crypt of the Polish citizen Maria Szymanowska is there).

3 CONCLUSIONS

Thus, in spite of the apparent absence of a city-planning architectural and urban plan with regard to the location and configuration of historical cemeteries, it is necessary to state that their characteristics are fixed in urban planning and are usually associated with architectural dominants and have an evolutionarily rational system of transformation, including functional purpose. The town-planning qualities of these territories, along with historical and cultural ones, make it possible to attribute them to one of the important urban planning subjects of protection of the historical urban environment to be preserved.

At the same time, a detailed historical, archival and field study of these objects is necessary in order to establish historical and actual borders, as well as borders on which irreversible urban transformations have not occurred. These studies will determine the prospective functional use of the territories of historical cemeteries and establish an objective prohibition or restriction on further urban planning transformation of those parts on which historical burials can be found. The memorial and recreational functions with the preservation and restoration of architectural dominants are the priority for territories of historical cemeteries that have not undergone irreversible urban development.

These measures will allow us to preserve the memory of the history of our city, people and country as a whole, not to lose our roots and rationally preserve and develop the urban planning environment of our magnificent city that successfully combines historical heritage and modern megalopolis.

REFERENCES

Kobak, A.V., Pirutko, Yu. M. 1993. *Historical cemeteries of Saint Petersburg*. Saint Petersburg: Chernyshev Publishing House.

Kobak, A.V., Pirutko, Yu. M. 2009. *Historical cemeteries of Saint Petersburg*. Saint Petersburg: Centerpolygraph, MiM-Delta.

Kobak, A.V., Pirutko, Yu. M. 2011. *Historical cemeteries of Saint Petersburg*. Saint Petersburg: Centerpolygraph.

Georgi, I.G. 1996. Description of the Russian imperial capital city of Saint Petersburg and memorials in the vicinity of it, with a plan/I.G. Georgi; entry art. By Pirutko. Saint Petersburg: Liga.

Punin, A.L. 2014. *The architecture of Saint Petersburg in the middle and the second half of the 19th century. Volume 2. Saint Petersburg of the 1860-1890s in the context of urban development of post-reform Russia.* – Saint Petersburg: Kriga.

Pylyaeva, M.I. 2000. *Old Petersburg: Stories from the past life of the capital*. Moscow: Svarog and K.

Pylyaev, M.I. 1996. *The Forgotten Past of the Surroundings of Peterburg*. Leningrad: Lenizdat.

Sementsov, S.V. 2007. *Urban development of Saint Petersburg in the 1703-2000s*. Saint Petersburg: SPbGASU.

Mikhailov, A.V. 2017. The definition of subjects of protection for objects of archeology, on the example of the Lower Cottage in the Alexandria Park. *International Research Journal* 10 (64): 125–128.

Mikhailov, A.V. 2019. Definitions of subjects of protection for objects of cultural heritage on the example of hospital complexes of Saint Petersburg. *Bulletin of Tomsk State University of Architecture and Civil Engineering* 3: 20–37.

Mikhailov, A.V. 2020. Urban planning objects of protection as a component of the system of protection of the world heritage object. *E3S Web of Conferences*, vol. 164.

Sementsov, S., Akulova, N. 2019. Foundation and development of the regular Saint Petersburg agglomeration in the 1703 – 1910-s. *Proceedings of the 2019 International Conference on Architecture: Heritage, Traditions and Innovations (AHTI 2019). Advances in Social Science, Education and Humanities Research* 324: 425–433.

Ptichnikova, G. 2016. New Century High Risers in the Core Areas of Historic Cities in Russia. *Procedia Engineering Volume*: 1903-1910.

On the boundaries of the zones of protection of cultural heritage on the territory of Saint Petersburg and the land use regimes within the boundaries of these zones and on amendments to the Law of Saint Petersburg "On the General Plan of Saint Petersburg and the boundaries of the zones of protection of cultural heritage on the territory of Saint Petersburg". The law of Saint Petersburg No. 820-7 of December 24, 2008. Saint Petersburg: Petrocentr, 2009.

Federal Law of June 25, 2002 No. 73-FZ (as amended on May 7, 2013) "On Objects of Cultural Heritage (Monuments of History and Culture) of the Peoples of the Russian Federation". *Parliamentskaya Gazeta* no. 120–121.

Territorial-planning constituents of urban planning components to be protected

A.V. Mikhaylov

Saint Petersburg State University of Architecture and Civil Engineering, Saint Petersburg, Russia

ABSTRACT: The article is dedicated to the research of the territorial-planning constituents of the urban planning components to be protected of both individual objects of cultural heritage and complexes of such objects. For clarity, the territorial-planning structure of one of UNESCO's largest urban development entities, the World Heritage site "Historic Centre of Saint Petersburg and Related Groups of Monuments," is analyzed. The research is relevant is due to the fact that in recent years the Russian Federation has been actively analyzing the accumulated so-called city-planning mistakes that have invaded the historical fabric of the development of both small and large historical cities and attempts are being made to prevent such mistakes. The purpose of the research is to identify the causes of the destruction of the historical environment of cultural heritage objects, consisting either in the imperfection of the protection system itself, or in an insufficient methodological study of the methods for its application. For the purposes of the research, the following methods were used: archival-bibliographic, cartographic, analytical, field studies. Results are: the territorial-planning constituents of urban planning components to be protected are analyzed and existing methods for their conservation are evaluated.

1 INTRODUCTION

What is the secret of harmony in the historical city environment of Saint Petersburg? A professional scholarly answer was sought by many urban planners, art historians, architects. In their works, I.G. Georgi (Georgi 1996), M.I. Pylyaev (Pylyaev 2000), A.L. Punin (Punin 2014), V.Ya. Kurbatov, S.V. Sementsov (Sementsov 2007), M.A. Ilyin and many others mentioned this topic. I.G. Georgi in his work gives a description of the city at the end of the XVIII century. M.I. Pylyaev in his work describes the city from the moment of its foundation until the first half of the XIX century, while pointing out sights, estates and describing the life and special features of living. S.V. Sementsov in his dissertation reveals to us the principles and features of the formation of Saint Petersburg. And similar searches will go on. Indeed, the depth and detail of the security system of the elements, constituent of the historical structure of the urban environment – and therefore its safety – depend on how in depth such elements are studied.

The purpose of this article is to analyze the territorial planning structure of the historical part of Saint Petersburg. The article was preceded not only by the study of historical scientific textual and cartographic materials stored in archives and libraries, but also by practical experience in the formation of a historical and cultural reference plan that became the basis of the system of protection zones for cultural heritage of Saint Petersburg (Mikhajlov 2017).

2 MAIN PART

2.1 *Territorial-planning features of Saint Petersburg*

I would not break the usual course of my thinking from general to particular.

Probably the first factor that determines the territorial-planning features of the development of most cities is the geographical conditions. Of course, people did not always incorporate their buildings into landscape. During some historical and political periods, the maximum counteraction to natural environmental conditions was considered to be true, but as a rule such experiments did not last long and did not determine the architectural and urban planning outlook of world cities (Mikhajlov 2019). Saint Petersburg is partly an experiment in counteracting harsh environmental conditions, but it is harmoniously integrated into the natural landscape that determined its primary characteristics. The relief of our city is quite flat. The ledges of the Baltic-Ladoga Glint surround the city, forming separate scenic sections (Pulkovo Observatory, Peterhof's water supply system and the associated park system, etc.), but do not determine its nature (Figures 1, 2). The next geographical factor is the significant water areas that determined the nature and scale of the overall composition of the layout of the embankments and the associated road network. The rule of forming, or even filling in, large-scale spaces was transferred from the embankments to the inner territories of the

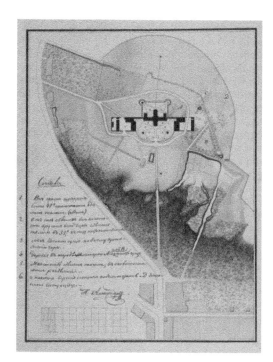

Figure 1. Bryullov A.P. General plan of the Pulkovo Observatory. МИМ РАХА-1763.

Figure 2. Plan of Peterhof. 1830s.

Figure 3 Fragment of the 1828 detailed plan of Saint Petersburg by Major General Schubert . РНБ К2-П6 4/58.

city, influencing the formation of magnificent ensembles as well.

I would consider the factor of politics as the next important one. The need in new way of thinking and representation of itself as a leading European-thinking state defined the strict following to the canons of beauty and harmony of that time. Hens the utmost geometry of the planning structure and its mathematical regularity thanks to the system of regulation of the urban-planning activity. If mathematicians came to study the urban structure, they would have found out everywhere the rational ratio of the numerical

characteristics. Initially, the regulatory width of streets of 6-10-12 sazhens (12.96-21.6-25.92 meters) often corresponds to the 1:1 or 1:2 height of buildings (fortunately, we do not live in New York and the proportions of 1:3 and 1:6 have not been applied) (Shuhana Shamsuddin 2012, Hoda Zeayter 2018, Lodovica Valetti 2019). The apogee of this rule is the famous Teatralnaya Street (now Zodchego Rossi Street) where the width of 22 meters is equal to the height of the buildings, and its length is 10 times the width and is 220 meters. Th scale of blocks and sections followed this logic: large were 20 sazhens (43.2 m), medium - 15 sazhens (32.4 m), small - 10 sazhens (21.6 m) in size. M.G. Zemtsov supplemented the rules of the exemplary building by increasing the height of buildings and introducing regulations depending on the importance of the street: 2 floors on the main streets in squares and block corners, 1.5 floors along building lines and in 1 floor on the intra-block territories. Segregation of sites has formed a modal building both along buildings lines of streets and in depth of the blocks.

Such geometrical rigor, which was, of course, not always observed, but still prevailed, has also influenced the transformation of architectural preferences. It is not surprising that the style of classicism has taken root in the city and largely determined its appearance.

The standardization, which I talked about before and which can be traced back to the historical general plans of Saint Petersburg (Figure 3), is also reflected in the historical and cultural reference plan, which is the basis of the contemporary system for standardizing of the preservation of the historical environment (Figure 4). In this way, we can talk about the continuity of systems.

These factors, among others, determined the inclusion of the historical center of the city and its suburbs as a unified urban planning, territorial and planning system in the list of World Heritage sites. (Sementsov 2019).

2.2 Accounting for territorial-planning constituents in regulatory documents

However, the purpose of the present article is not to list the the important and rather well-known facts, but to analyze the current system of regulation

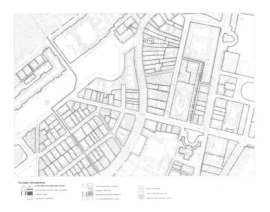

Figure 4. Fragment of historical-cultural reference plan with markings of historical surveying, historical building lines and historical planting.

territorial-planning transformations of the urban environment. The most significant among such documents are the Law of Saint Petersburg on Zones of Protection of Cultural Heritage Objects No. 820-7, General Plan of Saint Petersburg and the Rules of Land Use and Building of Saint Petersburg. In the near future the preparation of the order of the Ministry of Culture of the Russian Federation on the historical settlement of Saint Petersburg will be completed, which will finally legalize many principles of protection of historical territorial-planning elements and principles, previously mentioned in the system of regulation of protection zones.

The progress in the materials of the historical settlement is in terms of increasing the depth of the protected panoramas up to 9 km and in some areas up to 11 km is in fact only a partial and late compensation for the regress that was made in the during high-rise development of peripheral areas. The addition, made to the list of panoramas and views, including those that have been transformed, such as the view of the Epiphany Church on Gutuevsky Island, local views of temple dominants, will preserve the historical territorial planning system of street orientation on the squares and dominants.

In general, the adherence to territorial and planning genetics of the historical environment should be decisive both in relation to new inclusions and in relation to the assessment of the architectural heritage, which XX century left us. (Ptichnikova 2016) This is the goal of urban planning components to be protected. Whether they are applied to the determination of the designation of this or that object as a cultural heritage site, to the determination of the possible transformation of undeveloped territories to the architectural correspondence of the historical environment of the new objects. Thus, as time went on and the initial fear of novelty passed, such objects as the building of the North-West Department of the Central Bank on the corner of Fontanka Embankment and Lomonosova street; the building of the Gallery on Ligovsky 30, perceived in the context of the buildings of the Moskovsky Railway Station and the Oktyabrskaya Hotel, which itself in its history of transformation is an illustrative example of the both disruption of genetic foundations and following them (the unaccustomed modern facade was replaced by the project of 1887) ceased to irritate, and sometimes even began to please the city residents and specialists. However, it is impossible to get used to such buildings as the Vladimir Passage, and new glass-made buildings which richly "decorated" the embankments. And Petrovsky Island is a cataclysmic example, when neither the system of legal regulation nor the subjective factor of professional consideration of development projects did not work. Initially, the wrong maximum development parameters of 25-33 meters were interpreted as mandatory, the scale and module of development were forgotten at all, because they were not prescribed in the regulatory documents as a result of which the island is turning into a ghetto for the middle class.

3 CONCLUSIONS

This article, although very general in nature, is a part of a major work on the analysis of the system and establishment of methods for the definition of components of the objects of cultural heritage to be protected on the example of Saint Petersburg. In the course of the work it was determined that such an important factor as urban planning components to be protected and their constituents are not practically taken into account in the current system. It is possible to demonstrate the importance of urban planning characteristics on the example of the historical environment of the big city. For Saint Petersburg, these characteristics are fundamental in its recognition as a global value. And every element of this environment contributes to the perception of the whole.

Thus, the initial understanding of the importance of urban planning, territorial-planning characteristics as a whole and the degree of participation in them of separate elements representing a street, square, block, boundary area or just a separate building. The importance of understanding of their transformation in time or static character, can give an answer or part of the answer to the question what is harmony for a complex system, such as the historical environment of our city.

Taking these characteristics into account when determining the protection items for individual objects of cultural heritage and forming a system of regulation of transformation and preservation of urban environment is of primary importance for such complex urban planning formations as Saint Petersburg.

REFERENCES

Georgi, I. G. 1996. Description of the Russian-Imperial city of Saint Petersburg and its memorable surroundings, with a plan of Saint Petersburg: Liga.

Punin, A. L. 2014. Architecture of Saint Petersburg in the middle and second half of the XIX century. Volume 2. Saint Petersburg of the 1860-1890s in the context of urban planning in post-reform Russia, Saint Petersburg.

M. I. Pylyaeva 2000. Old Petersburg: Stories from the former life of the capital Moscow.

Sementsov S. V., 2007. Urban Development of Saint Petersburg in 1703-2000s, Saint Petersburg.

Mikhailov A.V. Definition of objects of protection for objects of archeology, on the example of the "Lower dacha" in the Park Alexandria international research journal no. 10 (64)2017.

Mikhailov A.V.2019. Definitions of objects of protection for objects of cultural heritage on the example of hospital complexes in Saint Petersburg Bulletin of the Tomsk state University of architecture and construction.

Sementsov, S., Akulova, N. 2019. Foundation and development of the regular Saint Petersburg agglomeration in the 1703 – 1910-s. Advances in Social Science, Education and Humanities Research, 324: 425–433.

Galina Ptichnikova 2016 New Century High Risers in the Core Areas of Historic Cities in Russia Procedia Engineering, 165.

Shuhana Shamsuddin, Ahmad Bashri Sulaiman, Rohayah Che Amat 2012 Urban Landscape Factors That Influenced the Character of George Town, Penang Unesco World Heritage Site Procedia - Social and Behavioral Sciences, 50.

Hoda Zeayter, Ashraf Mansour Habib Mansour 2018, Heritage con-servation ideologies analysis – Historic urban Landscape approach for a Mediterranean historic city case study HBRC Journal, 14.

Lodovica Valetti, Anna Pellegrino, Chiara Aghemo 2019, Cultural landscape: Towards the design of a nocturnal lightscape Journal of Cultural HeritageIn press.

The law of Saint Petersburg № 820-7 2008, "On the General plan of Saint Petersburg and borders of zones of protection of objects of cultures-tion of heritage on the territory of Saint Petersburg". Saint Petersburg: 2009.

Federal law No. 73-FZ of 2002 On objects of cultural heritage (monuments of history and culture) of the peoples of the Russian Federation"

Analysis of research on Bakhchisaray Palace

Z. Nagaeva & L. Budzhurova
Vernadsky Crimean Federal University, Simferopol, Russia

ABSTRACT: The article presents an analysis of drawings of the Bakhchisaray Palace Complex (BPC) made by Giacomo Trombaro (in 1784), William Hastie (in 1798), and I.F. Kolodin (1820–1824). The purpose of the study is to perform a comparative analysis of the designs made from 1784 till 1824 by architects G. Trombaro, W. Hastie and I.F. Kolodin for the identification of the architectural and spatial structure of the BPC in the late 18th – early 19th centuries. The study tasks are to identify spatial planning as well as compositional-and-artistic features of the BPC of the end of the 18th – the beginning of the 19th centuries, to analyze the planning pattern of the BPC reflected in the drawings of 1784–1798, 1820–1824, and to give theoretical and practical recommendations on further research with regard to the BPC. In the course of the study, the following methods were used: analysis of literature, historical and academic sources, reports on the studies and works related to BPC restoration, drawings made by G. Trombaro, W. Hastie and I.F. Kolodin. It is established that the restoration and renovation of the BPC, the sole monument of the Crimean Tatar palatial architecture, are necessary for the preservation of the architectural heritage of Crimea.

1 INTRODUCTION

The Bakhchisaray Palace Complex (BPC) is a typical allegorical Garden of Eden, which is manifested in the absence of fortification elements, the monumental style of a state building, open plan with a breakdown into ceremonial and residential sections.

The BPC was founded by Khan Sahib Giray at the same time when Bakhchisaray was founded in 1532. The palace grew and broadened over time. Each Khan made a contribution to its development. According to researchers, the first period of BPC construction ended during the reign of Khan Sahib I Giray (Ibragimova 2015, Bodaninsky 1916).

The Crimean capital was seized by the troops of the Russian Empire under the command of field marshal Burkhard Christoph von Münnich in 1736 (Bogdanova et al. 1961, 1963). After that, the BPC was repeatedly rebuilt, repaired and restored. However, the attempts to achieve historical credibility of the architectural image of the complex were not successful (Krikun 1998, Manstein 1895).

Drawings made by Italian architect Giacomo Trombaro (1784) and Russian engineer and architect of Scottish descend William Hastie (1798) kept at the funds of the Bakhchisaray State Historical and Cultural Reserve are the oldest preserved drawings of the palace complex in Bakhchisaray. Russian architect I.F. Kolodin started studying the BPC in 1820 for restoration works. He made drawings of the palace complex and estimates (Dombrovsky 1863).

2 SUBJECT, TASKS, AND METHODS

The subject of the study is archival, bibliographical and graphical sources of the Bakhchisaray Palace Complex. The study tasks are to identify spatial planning as well as compositional-and-artistic features of the BPC of the end of the 18th – the beginning of the 19th centuries, to analyze the planning pattern of the BPC reflected in the drawings of 1784–1798, 1820–1824, and to give theoretical and practical recommendations on further research with regard to the BPC. In the course of the study, the following methods were used: analysis of literature, historical and academic sources, reports on the studies and works related to BPC restoration, drawings made by G. Trombaro, W. Hastie and I.F. Kolodin (Hastie 1964, Gerngross 1912, Imperial 1910).

3 RESULTS AND DISCUSSION

II his paper "Khan's Palace in Bakhchisaray: the basic building periods", E.E. Osmanov (2015) breaks down the construction of the palace complex into five periods:

1532–1551 — construction of the square, Khan Jami Mosque, madrasa, Small Khan Mosque, residential facilities of the Old Palace, Sari-Guzel Baths;

Second half of the 16th century — construction of the mausoleum of Devlet I Giray at the Khan cemetery near the palace; from the early 17th century until 1736 —construction of the Divan Hall (first half of the 17th century), Southern tomb at the Khan cemetery, Summer Gazebo, state apartments on the second floor; a "new palace" was built, the Golden Fountain was installed near the Small Mosque; 1740–1743 — the Juma-Jami Mosque was rebuilt, the main building was restored, the Fountain Yard and the Falcon Tower were built. Probably during that time, the Portal of Italian architect Aloisio the New was transferred to the Fountain Yard of the palace from Salachik, the former residence of the Crimean Khans;

1758–1769 — construction of the Golden Study, the tomb of Dilyara Bikech; the Summer Gazebo was reconstructed, and the Western facade of the Juma-Jami Mosque was decorated with paintings.

The drawings of Italian architect G. Trombaro (1742–1838) were made supposedly in 1784 before the repair and restoration works conducted in 1784–1787. Figure 1 presents a layout depicting the first floors of the BPC, as well as the adjacent territory.

The layout by G. Trombaro is detailed: six entrances to the palace are shown — the Main Gate of the palace square (9), the Back Gate (Б) and entrances to the Old Palace (В), to the Kitchen Yard (Г), to the Ambassador Yard (two entrances) (Д, Е); the Khan Palace and a detailed layout of all structures are presented. The amenities and beautifications of the palace complex with gardens, flower gardens and fountains are depicted carefully (Kondakov 1899, Kotov 1896).

It is possible to go to the "gates to the palace square" (9) (hereinafter G. Trombaro's explanations to the drawings are given in quotes) by crossing a bridge over the Çürük Suv River; the gates are situated in the northern part of the BPC. "The guards' building" (47) is situated to the right of the gates. The entourage's building is situated to the left and to

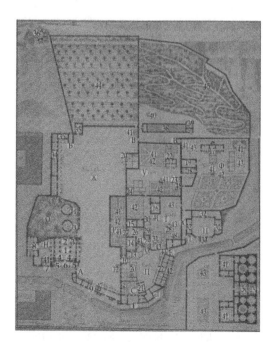

Figure 1. Layout by G. Trombaro. 1784.

the right of the gates. It is not specified in the explanations to the G. Trambaro's layout.

We arrive at the Gran Corte (Ж) through the main gates. It was an open square and the center of the palace complex during the khans' times. Meetings and various solemnities were conducted there, and the troops gathered there before military campaigns.

"The Big Mosque" (1) is located to the left of the gates, in the eastern part of the palace. The mosque was one of the largest Crimean mosques and the first structure in the BPC. The mosque premises are depicted in detail: "the mihrab" (2), the sacristy (3), two spiral staircases (4), two minarets (5). The main entrance to the mosque is to the side of the embankment of the Çürük Suv River. G. Trombaro called it a "portico" (6). To the left is the entrance to the ablutions room (7) with a fountain (41) and the entrance to the inner mosque yard (H) where a madrasah, "a school where the Quran is interpreted, where the Arabic language, religion and the Muhhamad civil law are taught", was situated (8).

To the south of the mosque, there is a Khan cemetery (O) with two tombs — "mortuary chapels marked as XX in the layout and a delicate monument with eight columns without a dome marked as X". These tombs are mausoleums built of limestone, they are eight-square and have a dome. The northern tomb (3) was built for Devlet I Giray and the southern one (И) — for İslâm II Giray. The rotunda is a tomb with a monument at the grave of Meñli I Giray (К).

Behind the cemetery, there are "stables with residential premises for stablemen" (34), "rooms for outfitting horses" (35) and the Back Gate (11).

In the southern part of the BPC, there are gardens — an orchard (44) and a forest orchard (45), vineyards with terraces and stairs. There is also a well there (40), a large swimming pool (38), and a fountain (39).

Above the gardens, there is an eight-square building with a dome — a tomb of Dilyara Bikech, the wife of Khan Qırım Giray. It is noted in the explanations to the G. Trombaro's layout as "a mortuary chapel and a cemetery for sultanesses and other Christian women of the harem" (36).

To the right of the palace square, in its western section, there is "the Front Gate to the palace" (10), which leads to the open Ambassador Yard (П).

To the left of the Front Gate, there is an entrance to the Swimming Pool Yard (P), which was drawn by the architect in detail: it depicts gardens, flower gardens, terraces, stairs, and a cascade fountain.

In the northern section of the Ambassador Yard, there are "residential premises of various sultans" (20) and a small kitchen yard (S) with a "khan and hafizes' kitchen" (21) and a "house of the keepers of the Bakhchisaray Palace". In the western section of the Ambassador Yard, there is a harem (22) with an inner garden (42).

In the southern part of the Ambassador Yard, there is a "pretorium where guards and other members of the Divan were situated" (13). There are two entrances to the Divan Hall (14). The Divan, the Council and Court Hall, is the central part of the complex. It is one of the earliest BPC structures with stone bearing walls. There is an entrance to the Fountain Yard (26) with two fountains from the Divan Hall. One of them, the Golden Fountain, is located in front of the Small Khan Mosque (15) that consists of two rooms.

In the eastern section of the Fountain Yard, there is an entrance to the Summer Gazebo (27) "surrounded by excellent comfortable sofas from three sides, and with a fountain in the middle".

In the western section of the Fountain Yard, there are entrances to the Harem Yard (T) and the Ambassador Yard. The entrance from the Ambassador Yard to the Fountain Yard is decorated with the Demir-Kapu portal (M) (translated from the Crimean Tatar language as a "steel door") created in 1503–1504 by great architect Aloisio the New. It was the grand entrance to the palace. The Harem Yard includes: a harem building (16), rooms for eunuchs (29, 30, 31), Khan baths (28, 48), fountains, and gardens.

To the south of the Harem Yard, there is the Persian Garden (У) with a "harem building with an eight-square four-story tower, which served as the sultan's pinnacle" (24), another harem building (22), and a kitchen (23). Plants, grapes and flowers grow across the entire area. It is decorated with four fountains. In the south-eastern section of the Harem Yard, there are khans' residential premises (20).

Another yard (Ф) is situated to the west of the Persian Yard. It is the largest fenced area of the palace complex with three structures: an old house with an outbuilding (17), a kitchen (18), and another building that is not described. The yard is decorated with fountains, flower gardens, orchards, and vineyards. Most probably, the Khan's personal apartments were located there. There is a way to the embankment of the Çürük Suv River through a long corridor (X) from this yard.

Another two systems of buildings are depicted in the western section of the embankment: one of them is adjacent to the BPC, and the other is situated on the other bank of the river. The first complex has a number of buildings — "large storerooms" (32) and premises for watchmen and guards (33) along the perimeter of the inner yard (Ц). The other complex is the "regal baths" (50) with a large yard decorated with gardens and flower gardens, with numerous fountains and an inner yard (Ч).

In 1798, before the planned repairs of the palace, English architect William Hastie characterized and made various drawings of the layouts, facades and the developed views of the BPC.

In W. Hastie's layout of the first floor of the palace complex (Figure 2), the BPC buildings and structures are shown in detail, but beautifications are

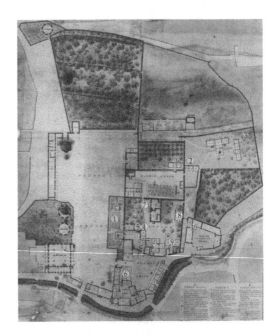

Figure 2. Layout by W. Hastie. 1798.

shown without details. Only the beautification of the Swimming Pool Yard (1) was drawn in detail. However, in comparison to G. Trombaro's layout, there are no significant changes.

The complex is divided into the main building, old palace, right and left wings, harem, mosque, tombs, stables, back yard, kitchen yard, and bakery yard.

Some premises and beautification elements shown in G. Trombaro's layout are absent: the storeroom adjacent to the kitchen (2) and the flower garden (3) with the fountain in the south-western section of the complex in the yard of the Old Palace, one of the eunuchs' houses (4), and the right room of the Small Khan Mosque (5). New premises are depicted: a room (6) adjacent to the Khan's apartments in the Persian Yard, a room (7) adjacent to the harem building, and a separate building (8) in the western section of the Harem Yard. The configuration of the baths (9) and large storerooms (10) differs a bit. In W. Hastie's drawings, there is no depiction of the "regal baths" shown in G. Trombaro's layout (No. 50).

In 1820, architect Ivan Fyodorovich Kolodin started preparing the BPC for a visit of emperor Alexander I of Russia. Upon assignment from the Construction Committee of the Ministry of Interior, the architect made layouts (Figures 3–4) and developed views with explanations, and made two estimates. The results of the works were submitted to the Ministry.

The Construction Committee decided not to preserve all the buildings of the palace. As a result, the following buildings were demolished: the small harem building (1), bathing zone (2) in the Harem Yard, Persian Palace (3), old palace (4) (only the

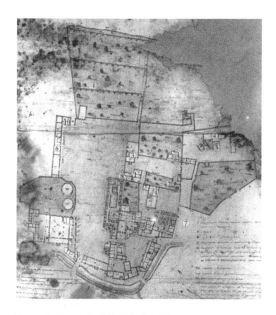

Figure 3. Layout by I.F. Kolodin. 1820.

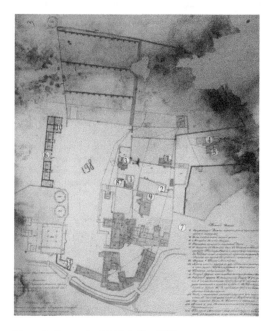

Figure 4. Layout by I.F. Kolodin. 1824.

gazebo near the palace building later called the "gazebo of Selâmet I Giray" was preserved (5)), stables (6), "large storerooms" (7) marked in G. Trombaro's layout. The delicate paintings by Iranian master Omer who worked for Khan Qırım Giray were painted over. The Falcon Tower (8) and the large Harem building (9) were repaired. "A building for the palace keeper and 12 invalids" (10)

(instead of the Khan's stables) and a wing to the east of the Front Gate (11) were built according to I.F. Kolodin's design.

The works carried out in the course of the repairs that permanently changed the original image of the palace caused objections of architects and artists. As a result, Novorossiysk Governor-General, count M.S. Vorontsov, suspended the repair works. I.F. Kolodin was dismissed, and architect Philipp Elson was appointed. During 1825–1831, he redid almost all the works carried out by I.F. Kolodin.

Numerous changes made by the architect in the Bakhchisaray Palace Complex are not depicted in I.F. Kolodin's drawings. The essence of the changes is not explained. The places of the works are shown only with color (Figure 4) with the "description of colors: a building to be demolished — with ink, a demolished building — with blue color, a new building — with pink, a changed building — with yellow, an unchanged building — without color"(the words I.F. Kolodin used in the descriptions to the 1824 drawings are in quotes). No color is used only for the Big Khan Mosque, the building for ablutions near the mosque, and the madrasah.

The works carried out by I.F. Kolodin led to a permanent loss of the unique architectural image of the BPC, damage to its interior, and destruction of the beautifications and amenities of the complex.

The study conducted represents a framework for further studies and analysis of the materials for the purpose of the restoration of the BPC image with all the characteristics to the greatest degree.

1. The layouts made by G. Trombaro and W. Hastie provide an indication of the architectural and artistic image of the BPC at the end of the 18th century.
2. When comparing the drawings made by G. Trombaro and W. Hastie, the author has found differences in the images, which confirms the need for further research.
3. The architectural image of the BPC underwent significant changes during a relatively short period of time (1784–1798). The reorganization was chaotic.
4. To restore the original historical architectural image of the BPC, it is necessary to conduct meticulous research, which is of great interest and importance for the preservation and restoration of the cultural heritage of Crimea.
5. The drawings made by I.F. Kolodin give an indication of the architectural and artistic image of the BPC at the beginning of the 19th century and of the changes in the architectural ensemble after the repairs in 1820–1824.
6. A comparative analysis of two sets of drawings made by I.F. Kolodin shows the destructive results of the works carried out by the architect in the 1820s.

4 CONCLUSIONS

The restoration of the BPC during different periods led to the distortion of its original image. Studying of phase-by-phase changes in the layout and the facades of the palace complex restores the original image of the BPC. To restore the BPC, it is necessary to study carefully its architecture, the interior and the decorations on a phase-by-phase basis.

Further research on the literature of various periods of the palace existence is necessary: the Khanate period, the period of the annexation of Crimea by the Russian Empire, the Soviet period, and the modern period.

REFERENCES

Bodaninsky, U. 1916. *Bakhchisaray monuments.* Simferopol: Printing Office of the Taurida Governorate Council.

Bogdanova, N.A. 1963. *History of reconstruction, repairs and restoration of the former Khan Palace in Bakhchisaray.* Bakhchisaray.

Bogdanova, N.A. et al. 1961. *Bakhchisaray historical and archaeological museum.* Simferopol: Krymizdat.

Broniewski, M. 1867. A description of the Crimea (Tartariae descriptio). *ZOOID — Proceedings of the Imperial Odessa Society for History and Antiquities* 6: 337–367.

Çelebi, E. 2008. *A book of travels. Crimea and surrounding areas: fragments from the book of a Turkish traveler of the 17th century.* Simferopol: Dolya Publishing House.

Craven, E. 1795. *A journey through the Crimea to Constantinople (1786).* Translated from French. Saint Petersburg: University Printing House of Ridiger and Claudia.

De Ségur, L.-P. 1865. *Notes by earl de Ségur about his stay in Russia during the reign of Catherine the Great.* Saint Petersburg: V.N. Maykov Printing House.

Dombrovsky, F.M. 1863. Crimean Khans' Palace in Bakhchisaray. *Tavricheskiye Gubernskie Vedomosti.*

Hastie, W. 1964. Layout of the Khan Palace in Bakhchisaray. A copy of the layout of 1798. In A.L. Yakobson (ed.) *Medieval Crimea.* Moscow, Leningrad: Nauka.

Gerngross, V. 1912. *Khan Palace in Bakhchisaray.* Saint Petersburg: Sirius Printing House.

Ibragimova, A.M. 2015. *Bakhchisaray Khan Palace in the 16th – 18th centuries.* Kiev: Vidavets Oleg Filyuk.

Imperial Archaeological Committee 1910. Reports of restoration meetings. Bakhchisaray Palace. *Bulletin of the Imperial Archaeological Committee* 34: 43–50.

Kondakov, N.P. 1899. Bakhchisaray Palace and its restoration. *Art and Art Industry* 6: 435–452.

Kotov, G.I. 1896. Bakhchisaray Palace. *Zodchiy* 1: 1–5.

Krikun, E.V. 1998. *Crimean Tatar landmarks (13th – 20th centuries).* Simferopol: Krymuchpedgiz.

Manstein, C.H. 1895. Description of Bakhchisaray capital city and Khan chambers made by Captain Manstein after field marshal von Münnich seized this peninsula, 1736. *Otechestvennye Zapiski* 31: 32.

Markevich, A. 1895. Revisiting the history of Bakhchisaray Khan Palace. *Izvestiya Tavricheskoy Uchenoy Arkhivnoy Komissii* 23: 130–176.

Münnich, E.B. 1891. *Russia and Russian court in the first half of the 18th century.* Saint Petersburg: V.S. Balashev Printing House.

Murzakevich, N.N. 1837. A travel to Crimea in 1836. *Journal of the Ministry of Public Enlightenment* 13: 625–691.

Nagaevskaya, E.V. 1976. *Bakhchisaray: travel guide.* Simferopol: Tavria.

Osmanov, E.E. 2015. Khan's Palace in Bakhchisaray: the basic building periods. *Crimean Historical Review* 3: 149–164.

Pallas, P.S. 1999. *Travels through the southern provinces of the Russian Empire, in the years 1793 and 1794.* Moscow: Nauka.

Prokopenkov, V.N. 2011. *Bakhchisaray on lithographs, engravings and picture cards: an album.* Simferopol: Firma Salta Ltd. OOO.

Vlasyuk, A. 1958. Concerning the work of architect Aloisio the New in Bakhchisaray and Moscow Kremlin. *Architectural Heritage* 10: 101–110.

Zaytsev, I.V. 2014. Establishment of the Crimean Khanate. In R. Khakimov (ed.), *History of the Tatars from the ancient times in seven volumes. Vol. IV. Tatarian states of the 15th – 18th centuries:* 130–146. Kazan: Mardzhani Institute of History of the Academy of Sciences of the Republic of Tatarstan.

Reconstruction and Restoration of Architectural Heritage – Sementsov et al (eds)
© 2020 Taylor & Francis Group, London, ISBN 978-0-367-65357-6

City image in protection system of historical environment

A.Yu. Nazarova
Saint Petersburg State University of Architecture and Civil Engineering, NIiPISpetsrestavratsia LLC, Saint Petersburg, Russia

ABSTRACT: The image of the city combining material elements and a wide range of associations is a reflection of ideas about the values of planning, development and urban planning composition that dominate at different stages of the formation and development of the historical urban landscape. At present, at the national and international level, the value of the entire set of elements of the historical environment and the importance of the integrity of urban landscapes in the process of maintaining cultural identity has been recognized. Based on the analysis of the tools for historical environment conservation, the article identifies the most effective methods for preserving the visual image of a historical city, draws conclusions regarding the development of urban planning legislation in the field of protection of visual and compositional relationships of environmental elements, continuous development of historical principles for creating a visual image of a city.

Keywords: city image, silhouette, views and panoramas, historical and urban environment, historical settlement

1 INTRODUCTION

The category of the image of the city refers to complex concepts denoting a set of interconnected material and intangible elements. The natural landscape, planning and development are the most significant, visually perceived elements of the urban landscape. Associations of the city with significant historical events and cultural phenomena complement the visual image.

Scientists and philosophers of the 18th century, comprehending the relationship between the visible and the realized in the urban landscape, determined the characteristics of urban space that most strongly affect the observer. They are horizontals, verticals and contrasts (Belyaeva, E.L. 1977, Savarenskaya, T.F., Shvidkovsky, D.O., Kiryushina, L.N. 2004). Throughout the 19th and 20th centuries, theorists of architecture and urban planning developed issues of urban development composition. Particular attention was paid to the study of the interconnections of the main components, such as planning, development and landscape (Bunin, A. V. 1940, Baranov, N.N. 1980). At the present stage of the theory of urban development, the concept of the vertical composition of the city is being studied (Chudinova, T.S. 1985).

The perception of the urban landscape of historical cities is a complex process of revealing the value of visible elements and awareness of their relationships (Vedenin, Yu.A. 2019). The focal points in the structure of perception of the urban image are the frame of architectural dominants, nodes and landmarks, directions (paths and communications), borders (Lynch,

K. 1982). Evaluation of the image as pleasant or unpleasant for residents and tourists depends on the integrity (harmony) of the architectural and urban planning composition (Nasar, J.L. 1998).

The theme of the image of the city is extremely relevant at the present stage of development of the theory and practice of the protection of urban development heritage. In the context of the implementation of the concept of sustainable development of large cities, maintaining the integrity of the historical environment contributes to the maintenance of cultural identity. City panoramas, perspective views and silhouettes are unique characteristics of the urban image, which determine the recognition of a particular city.

The expression of the historical value of the urban landscape in the structure of views and panoramas is ensured by the preservation of historical buildings and the ratio of background buildings and dominants, as well as the tactfulness of new inclusions. The artistic (aesthetic) value of the image of urban space is determined by the compositional completeness of views and panoramas, silhouette activity, the expressiveness of the contrasts of horizontals and verticals, the configuration of the completion of the dominants (Bunin, A. V. 1940).

The perception of intangible value is associated with the comprehension of eidos and genius loci of urban space, the understanding of the general cultural context broadcasted in the image of the city (Stepanov, A.V. 2016).

Awareness of the integrity of the historical environment testifies to the need of society to preserve the whole variety of types of historical information of urban landscapes, the carriers of which are not only individual monuments, but their spatial and compositional relationships.

2 METHODS

The study is based on the integrated approach, including general cultural theoretical methods (system-structural analysis, comparative analysis and theoretical generalization), as well as special methods such as landscape-visual analysis of objects of urban development heritage. In the system of the city's image, the forming elements were identified, which included landscape, planning, development and their interconnections. Based on the study of the tools for the protection of historical urban environment, conclusions are drawn about the most effective tool for protecting the key characteristics of the urban image. They are the silhouette, panoramas and views of the urban landscape, experiencing a strong impact in the conditions of dynamic urban development.

3 RESEARCH RESULTS AND DISCUSSION

The legal basis for the protection of the image of the city was formed on the basis of ideas about the inseparability of historical and cultural monuments from the environment in which they are located (*Venice Charter,* 1994; *Washington Charter,* 1994). According to the International Charter on the Conservation of Historic Towns and Urban Areas (The Washington Charter, 1987), the expression by a historic city of the values inherent in traditional urban civilizations is considered a necessary condition for classifying historic cities as protected urban landscapes. The image of historic cities is defined as an expression of the historic nature of the city, a combination of material and spiritual elements. The basic characteristics of the image include the configuration of the city plan, the relationship between different urban spaces, the shape and type of structures, the relationship between the city and its surroundings, various functions acquired by it during historical development. The ideas of the Washington Charter found successive development in the provisions of the Vienna Memorandum (*Vienna* 2005) and the Recommendation on Historic Urban Landscapes (*UNESCO,* 2011).

The approach currently adopted in international practice, focused on the conservation and management of historic urban landscapes (HUL), develops ideas of the value of the wide context of the urban landscape, which is a layering of natural and cultural values, including, inter alia, perception and visual relationships of material and intangible elements (*UNESCO,* 2011).

Based on many years of research, experts in the field of preservation of historical urban landscapes came to the conclusion that special studies of the silhouette and view disclosures are necessary, as well as the introduction of high restrictions within visibility zones as the most effective method of view protection (*World Heritage* 2008).

In order to protect the visual image and silhouette, a number of cities adopted documents of a strategic nature that ensure the protection of the most significant views and panoramas, the characteristics of the visual image of the city. Strategies for the conservation of views and silhouettes are accepted for official use in London, Paris, Vancouver, Ottawa (Karaga, K. 2020). Such documents contain an assessment, description and mapping of the most significant views and viewpoints, silhouette compositions, a description of the factors affecting the value characteristics of these views, and measures to preserve and maintain their value.

In Russian legislation, security zoning is the instrument for protecting the visual integrity of the urban environment. Protection zones of cultural heritage objects are established on the sites of the historical and natural environment of cultural heritage objects; they can cover territories that are significant in area if the protection zones of several monuments are combined (united protection zones). The system of restrictions on urban development in the protection zones includes special land use regimes and requirements that limit urban development and preserve the compositional and visual connections of cultural heritage objects in their historical and natural environment.

At present, at the national level, a new "tool" is being developed for the comprehensive protection of the historical environment - the "historical settlement". The status of a historical settlement provides protection of valuable elements and parameters of the environment: city-forming objects, planning and spatial structure, composition and silhouette of buildings, relationships between different urban spaces, relationships between composition and views (panoramas), relationships between natural and man-made environments" (Law No. 73-FZ, Article 59).

The historical center of Saint Petersburg serves as a unique example of a holistic historical and urban planning environment. Many years of experience in protecting the historical environment of the city center through the establishment of joint zones for the protection of cultural heritage objects have shown insufficient protection zoning in terms of maintaining the visual integrity of panoramas and views of the urban landscape and silhouette. The inclusion of a large architectural module with a disordered configuration of crowning parts in the view zones of buildings has a negative impact on the integrity of panoramas and silhouette compositions of the Neva that is the main planning and compositional axis of the historical center of the city. River panoramas and silhouettes are sensitive to high-rise construction beyond the borders of the historical center. Complexes of multi-storey buildings erected at considerable distances from the historical center, due to the bends of the river bed, invade the panorama structure of the Neva, violating the existing structure of perception of the visual image and silhouette. The significant impact of new construction on the perception of panoramas and the silhouette of buildings leads to the loss of historical and cultural value of the visual image of the city, the loss of the cultural identity of the historical urban landscape.

A new tool for urban development protection - the status of a historical settlement - opens up the possibility of establishing detailed requirements for urban development, ensuring the protection of valuable panoramas and views of the urban landscape and the continuous development of sustainable principles for organizing the urban environment outside the historical center.

The determination of the content of requirements for urban development activities within the boundaries of a historic settlement aimed at protecting city panoramas and views, the silhouette and composition of buildings should be based on a comprehensive analysis of the urban planning composition and perception conditions, as well as the determination of the vulnerability of individual elements. The result of the research is the identification of the most valuable panoramas and views, silhouette compositions to be protected, areas of perception of views, within which urban development is subject to special regulation (Figure 1).

Figure 1 The analysis of urban planning composition of the territory of the historical center of Saint Petersburg.

Figure 1. The analysis of urban planning composition of the territory of the historical center of Saint Petersburg.

4 CONCLUSIONS

Currently, the conservation of the historical environment is considered among the priority tasks for the development of historic cities. In the field of cultural heritage protection, the most pressing issue is the problem of maintaining cultural identity, the expression of which is the historical and urban heritage as a whole and its image, expressed in the aggregate of urban views and panoramas, silhouette compositions.

The protection of the visual image of historic cities is provided for in Russian legislation within the framework of the implementation of a new conservation status - the "historical settlement" - the subject of protection of which is the silhouette and composition of buildings, relationships between the composition and views (panoramas), the ratio of the natural and the man-made environment.

General directions of maintaining the silhouette and composition of buildings as a subject of protection of a historic settlement include the study of value features, determination of visibility pools, zones of compositional influence and characteristic points of perception of valuable silhouette compositions, regulation of urban development within the perception zones with the aim of maintaining and continuing development of the main compositional and formative principles of organization of urban perspectives and panoramas, elimination of negative impact on the value features of the silhouette.

The methods of preserving the visual image, based on a comprehensive study of valuable elements and characteristics, managing changes in the historical environment, meet the urgent tasks of sustainable development of historical urban landscapes.

REFERENCES

Baranov, N.N. 1980. *City silhouette*. Leningrad: Stroyizdat.
Belyaeva, E.L. 1977. *The architectural and spatial environment of the city as an object of visual perception*. Moscow: Stroyizdat.
Bunin, A. V. 1940. *Architectural composition of cities*. Moscow: Academy of Architecture.
Chudinova, T.S. 1985. *The formation of a vertical composition of the historical centers of riverine cities (on the example of the Volga cities)*. PhD thesis of PhD in Architecture Chudinova, T.S. Moscow: Moscow Order of the Red Banner of Labor Architectural Institute.
Karaga, K. Urban skyline planning strategy. Casestudy. London 2014-2015 [Electronic resource] Access mode: http://www.academia.edu/22759923/Urban_skyline_planning_strategy_analysis_Case_study_London, accessed on June 7, 2020.
Lynch, K. 1982. *The Image of the City*. Moscow: Stroyizdat,.
Nasar, J.L. (1998). The Evaluative Image of the City, Thousand Oaks, CA: Sage Publications.
Savarenskaya, T.F., Shvidkovsky, D.O., Kiryushina, L.N. 2004. *Urban culture of France in the 17th-18th centuries*. Moscow: Editorial, URSS.
Stepanov, A.V. 2016. *Phenomenology of architecture of Saint Petersburg*. Saint Petersburg: Arka.
The UNESCO Recommendation on the Historic Urban Landscape. The Resolution adopted by the UNESCO General Conference on the report of the CLT Commission at the 17th plenary meeting on November 10, 2011.
The Venice Charter for the Conservation and Restoration of Monuments and Sites, 1994. Material base of the cultural sphere: domestic and foreign experience in solving managerial scientific and technical problems. Information Collection, Moscow.
The Washington Charter: Charter on the Conservation of Historic Towns and Urban Areas, 1994. Material base of the cultural sphere: domestic and foreign experience in solving managerial scientific and technical problems. Information Collection, Moscow.
Vedenin, Yu.A. 2019. Cultural landscape as a keeper of the memory of oecumene. *Human Being: Image and Essence. Humanitarian aspects* 1 (36). Access mode: https://cyberleninka.ru/article/n/kulturnyy-landshaft-kak-hranitel-pamyati-oykumeny, accessed on February 2, 2019.
Vienna Memorandum on "World Heritage and Contemporary Architecture - Managing the Historic Urban Landscape", 2005.
World Heritage: Defining and Protecting Important Views. Montreal, March 18-20, 2008. https://www.patrimoinebati.umontreal.ca/en/activities/round-tables/.

Classification of ordinary development of Saint Petersburg in system of urban development regulation

I.L. Pasechnik
Saint Petersburg State University of Architecture and Civil Engineering, Saint Petersburg, Russia
NIiPISpecrestavratsia LLC, Saint Petersburg, Russia

ABSTRACT: Preserving the identity of cities can be attributed to the number of priority tasks of our time, given the speed and scale of globalization. The characteristics of the human environment and their sustainability are the basis for maintaining local cultural identity. The dynamic nature of cities recognized by experts in the field of heritage protection and the fundamental admissibility of including new objects in the historical urban fabric require special attention to the establishment of legislative restrictions on monuments, urban planning and other activities carried out on the territory of historical settlements. Taking into account the status of Saint Petersburg as a World Heritage Site, the conducted research summarized the principles proposed by methodological documents and international experts for determining the value of both individual objects and a holistic historical, architectural and urban planning environment. The author examined the quarters of historical buildings located in different zones from the point of view of the integrity of the environment, identified the categories of historical buildings in relation to different urban territories, proposed criteria for determining the value of ordinary historical buildings.

1   INTRODUCTION

Development is one of the most important material elements of the historical and urban environment. Fixing the urban planning framework, which is based on natural features and layout of the territory, development forms a recognizable silhouette of the city and its composition. Relying on people's ability to perceive and the memories they keep, development also participates in creating a visual and mental image of the city.

Ordinary development occupies a position subordinate to valuable elements of the environment. Nevertheless, it turns out to be that component that becomes the basis for maintaining and developing stable environmental characteristics, and its loss "sometimes leads to a distortion of the city's image, which is comparable with the damage caused by the destruction of the monuments and ensembles" (Regame et al. 1989). It is ordinary construction that is, as a rule, much less protected than monuments, and can undergo significant transformations and "suffers most of all from the current changes - demolitions, restructures, replacement of authentic parts, etc." (Ivanov 2000).

In domestic scientific literature and practice there are no approved assessment criteria that could be applied to ordinary buildings (Pasechnik & Marushina 2019). At the same time, an analysis of existing legislative restrictions aimed at ensuring the preservation of the historical and urban planning environment shows the need to introduce additional requirements that would create the conditions for maintaining the integrity of the environment and at the same time could become the basis for sustainable development of urban areas.

The possibility of changing the historically formed urban planning fabric, on the one hand, is considered one of the main components of the urban development process and is recognized by experts (Bandarin & Van Oers 2012, Gutnov 1984), on the other hand, it requires a careful approach to determining the permissible limits of transformation. Development of any legislative restrictions affecting the territory of a historic city is based on the analysis of its characteristics, stages and principles of its development. At the same time, the researchers noted that the information accumulated to date is mainly concentrated on the "artistic-figurative features" of buildings, and "specific information about mass development appears only fragmentarily" (Molotkova 2019).

Nevertheless, a large number of studies are devoted to architectural and urban planning features of Saint Petersburg, which focus on various aspects of the conservation and development of the urban territory (Golovina 2019, Lavrov et al. 2019, Mangushev et al. 2019, Sementsov 2012, Voznyak 2017).

The integrity of the environment of the historical center of Saint Petersburg allows us to talk about the general principles of its formation, but cannot lead to the absolute unification of the requirements for urban planning activities within the boundaries of environmental zones that differ from each other in the nature of their development.

2 METHODS

In order to identify the valuable characteristics of the ordinary development of Saint Petersburg, a typological and structural analysis of the historical and urban planning environment was carried out using graphic analytic methods and the information of the geographic information system. The data obtained on the basis of the analysis of historical cartography and iconography was checked during the on-site examination and photo fixation. The research results are presented in the form of analytical diagrams and tables in Excel and MapInfo format. Text tables contain information on the category of architectural and urban planning value of development objects, graphic materials clearly capture the results of the analysis.

3 RESULTS AND DISCUSSION

The historical and urban planning environment of Saint Petersburg is characterized by the integrity and authenticity of its key components: a stable planning structure; the architectural environment formed by both monuments and ensembles, and ordinary buildings; preserved visual connections of various elements of the historical urban landscape and the perception of prevailing views and panoramas. These features served as the basis for the inclusion of Saint Petersburg in the World Heritage List. Today, Saint Petersburg agglomeration forms a single organism, the development of which is determined by the originally established principles of environmental organization.

One of the threats to the integrity of the historical urban environment of Saint Petersburg is the possibility of its spatial fragmentation. In this regard, the preservation of the historical urban landscape of Saint Petersburg requires a systematic approach to regulation, which should take into account all its constituent elements, including objects of cultural heritage, ordinary development and new construction projects, which are supposed to be included in the historical environment.

Currently, the only effective instrument for protecting the historical environment in Saint Petersburg is protection zoning. Understanding of the impossibility of maintaining the integrity of the environment during fragmented protection led to the formation of a system of protection zones, the effect of which is ensured by compliance with the developed land use regimes and the requirements for urban planning regulations.

Already in 2009, after the entry into force of a regional law that established the boundaries of cultural heritage protection zones in Saint Petersburg, researchers noted that "the lack of specific information about what can and should be and what should not be allowed during the transformation is felt almost every time when there is an investment interest in a particular territory, site or structure in the historical center" (Slavina 2002). Despite this, to date, the situation with changes relating to ordinary development has not received significant progress.

In the absence of the practice of complex reconstruction of neighborhoods, activities related to regeneration and reconstruction are undertaken at the initiative of users of individual sites. This approach requires constant expert support and limitation of the spontaneity of the process of building transformations. This raises the question of the need to include special requirements in the current legislation.

The implementation of the principle of detailed regulation of urban development, taking into account the specific qualities and characteristics of the historical environment, will determine the parameters of acceptable interference, ensuring the preservation of key elements of the historical and urban environment and the continuity of development of urban areas.

Ordinary historical development plays a paramount role in the formation and maintenance of valuable characteristics of the urban environment. Thus, consolidation by an ordinary building of a historical layout, which does not have a special historical and artistic value, allows saving such important layout features as the module, block structure, etc. (UNESCO 2011).

The position and role in the formation of the historical and urban planning environment by a significant number of specialists is recognized as the main criterion in determining the value of the object of historical development. In this regard, it seems appropriate to consider separately objects located in the structure of a holistic environment and beyond, extending this approach to the territory of the new building. In this context, the transmission of valuable environmental qualities to newly developed urban areas is becoming a key principle in the sustainable development of Saint Petersburg. This approach is based, in particular, on recommendations developed by UNESCO experts on the preservation of historical urban landscapes (UNESCO 2011).

A separate building, being an element of the urban environment, can play a different role in shaping the image of the city: it can be active or neutral, determine or disrupt the nature of the environment, while the degree of its impact is determined not only by location (urban value), but also by the external appearance and architectural solution (architectural value). Moreover, today the value of the object is not considered in isolation from authenticity and integrity.

In accordance with the provisions of international methodological documents, the authenticity of any object can be expressed, in particular, through the authenticity of form and design, materials and substances, location and environment, use and functions, spiritual and physical perception, etc. (World Heritage Center 2019). The authenticity of the location and environment describes the role of the object in the formation of the urban environment.

Authenticity of the form and design, materials and substances, as well as spiritual and physical perceptions are directly related to the architectural features of the building.

Integrity in the interpretation proposed by UNESCO means a measure of the wholeness and intactness of the natural and/or cultural heritage and its attributes (World Heritage Center 2019). With regard to historical development, it can be understood as follows: an object is recognized as having remained integrity if all elements that are an expression of its value are preserved unchanged.

Thus, given the status of the historical center of Saint Petersburg as the main component of the World Heritage Site, it seems logical to consider the objects of historical development in terms of their urban planning and architectural value, which should be supported by the integrity and authenticity of both the environment in which the object is located and the components of the object.

The architectural value of an ordinary building object is determined by the degree of preservation (integrity) of the true appearance of the building, structure, construction, which is ensured by the preservation of the three-dimensional structure, architectural design of the front facades, and the configuration of the front roof slopes.

In this case, it is necessary to distinguish between buildings that form the front, and buildings that are elements of intra-quarter development. Each group requires a special approach to determining value.

For historical buildings forming a street front, 3 categories of architectural value can be distinguished:

- a valuable element of development (complete preservation or predominance of fully preserved elements of true appearance);
- an ordinary element of development (partial distortion of true appearance);
- an element that has lost authenticity (significant (i.e., affecting two of the three components of external appearance) or complete distortion of the elements of true external appearance).

In relation to the objects of intra-quarter development, it is advisable to limit ourselves to two categories:

- an ordinary element of development (full or partial preservation of true appearance);
- an element of development that has lost authenticity (significant (i.e. affecting two of the three components of external appearance) or a complete distortion of the elements of true external appearance).

The presence of preserved (completely or with slight distortion or loss) valuable interior elements even with a significant or complete distortion of true appearance is the basis for classifying the property as ordinary building elements.

The urban planning value of the object is due to its role in the formation of the environment (Table 2).

At the same time, the integrity and authenticity of the urban planning environment is determined by the integrity of the totality of its key historically formed elements - the planning structure, development and features of the street front (including the historical lines of building blocks, the type of development, the type of organization of the street front, the role of elements of historical landscaping, the type of organization of courtyard spaces, the presence or absence of discordant objects). The conclusion about the integrity of the environment is made according to the results of the analysis of the quarterly development when correlating them with the prevailing architectural and urban planning features within the environment zone or its sections.

The role of the development object in the structure of the urban environment is determined by participation in the formation of the street front, intra-quarter development, or inclusion in disperse development.

For the street front, the following categories of urban importance can be distinguished:

- the building plays an accent role (it closes the perspective of a street, avenue, lane, is located at the corner at the intersection of streets, lanes, avenues or forms a street front of embankments, main highways (avenues), squares that are key elements of the historical planning structure of Saint Petersburg);
- the building forms the street front of the street or lane.

For intra-quarter development:

- the building is an element of genuine intra-quarter development;
- the building is an element of unformed intra-quarter development;

Urban development value is not established for disperse development.

Table 1 presents the architectural and urban planning matrix containing the main elements of the methodology for determining the value of the elements of ordinary development, designed to determine the conservation parameters and allowable transformations of the existing historical environment.

Analysis of the characteristics of the historical and urban planning environment of a particular quarter performed using the matrix, allows detailing the requirements of land use regimes.

Figure 1 Shows an example of the analysis of the development of one of the quarters located within the boundaries of the development and economic regulation zone (Zone 1). It is a visual diagram with the designation of cultural heritage objects, historical buildings (historical buildings are the ones built before 1957, inclusive), dissonant objects and an indication of

Table 1. Architectural and urban planning matrix for determining the value of ordinary development objects.

Architectural value Urban importance		A valuable element of development	An ordinary element of development	An element of development that lost authenticity
Street front	building plays an accent role	conservation	conservation	conservation
	the building forms a street front	conservation	conservation	possible transformation
Intra-quarter development	an element of genuine/ prevailing intra-quarter development	–	possible transformation	possible transformation
	an element of unformed intra- quarter development	–	possible transformation	possible transformation
Disperse development		conservation	possible transformation	possible transformation

Figure 1. Analysis of the development of the quarter (base quarter 2044, Vasileostrovsky district).

their role in the formation of the historical and urban planning environment.

The result of the analysis performed using the proposed architectural and urban planning matrix, are recommendations for each individual building site (with the exception of cultural heritage sites and non-historical buildings not having architectural and urban value).

Table 2. Fragment of the analytical table (base quarter 2044, Vasileostrovsky district).

№	Address	Year of construction	Building category	Role in the formation of the urban environment	Recommendations
1	Saint Petersburg, Maly prospect of Vasilevsky Island, 34, building A	1956	Valuable element of development	Urban accent	Conservation
2	Saint Petersburg, Maly prospect of Vasilevsky Island, 30-32, building A	before 1917	Valuable element of development	Urban accent	Conservation
3	Saint Petersburg, Maly prospect of Vasilevsky Island, 30-32, building B	before 1917	Ordinary element of development	Example of genuine/existing intra-quarter development	Transformations are possible
4	Saint Petersburg, Maly prospect of Vasilevsky Island, 30-32, building C	before 1917	Ordinary element of development	Example of genuine/existing intra-quarter development	Transformations are possible
5	Saint Petersburg, Maly prospect of Vasilevsky Island, 30-32, building D	1959	Non-historical building	Element of the existing intra-quarter development	–
6	Saint Petersburg, 11th line of Vasilievsky Island, 60, building A	before 1917	Valuable element of development	The building forms a street front	Conservation
7	Saint Petersburg, 11th line of Vasilievsky Island, 56, building A	before 1917	Ordinary element of development	Example of genuine/existing intra-quarter development	Transformations are possible
8	Saint Petersburg, 11th line of Vasilievsky Island, 54, building A	before 1917	Valuable element of development	The building forms a street front	Conservation
9	Saint Petersburg, 11th line of Vasilievsky Island, 54, building C	before 1917	Ordinary element of development	Example of genuine/existing intra-quarter development	Transformations are possible
10	Saint Petersburg, 11th line of Vasilievsky Island, 52, building A	before 1917	Valuable element of development	The building forms a street front	Conservation

4 CONCLUSIONS

Studies related to the determination of areas of the environment with similar parameters were carried out in Saint Petersburg, starting in the second half of the 20th century. However, to date, a differentiated approach to determining the parameters of acceptable interference, based on the consideration of the stable characteristics inherent in a particular environmental zone (or local environmental areas), has not been fully implemented.

Preserving the urban planning fabric of Saint Petersburg requires an approach based on the principle of translating the stable patterns of the formation of the historical environment into newly developed or regenerated territories and reproducing its basic parameters on a scale corresponding to a particular environmental zone.

Implementation of land use regimes should be carried out with the introduction of specific methods of rehabilitation of the urban environment such as restoration, regeneration and renovation, applied taking into account the historical and modern patterns of land use, as well as the level of historical and cultural potential of various fragments of urban fabric.

Thus, the urban planning significance, the degree of preservation, and the features of the transformation of the valuable parameters of the territory can be taken as the main criteria in determining the ways of urban planning development, as well as establishing restoration regimes (preserved valuable elements of the environment), regeneration (with a partial loss of integrity), and renovation (taking into account the traditional typological characteristics of the urban planning environment and the formation of an urban environment that is adequate to the historical one, including the methods of co-scale and taking into account the context of new construction) and their various combinations.

REFERENCES

Bandarin, F. & Van Oers, R. 2012. *The historic urban landscape: managing heritage in an urban century.* Oxford: Wiley-Blackwell.

Golovina, S.G. 2019. Architectural and design features of the historical residential development stages in

Saint Petersburg of the XVIII – early XX centuries. *Bulletin of Civil Engineers* 6 (77): 36–43.

Gutnov, A.E. 1984. *The evolution of urban development.* Moscow: Stroyizdat.

Ivanov, A. 2000. Danish methodology for assessing the historical SAVE development: possibilities for use in Russia. *Arhitekturny Vestnik* 2: 10–15.

Lavrov, L.P. et al. 2019. Morphotypes of quarters of the historical center of Saint Petersburg. *Academia. Architecture and Construction* 4: 52–59.

Mangushev, R.A. et al. 2019. Saint Petersburg «genetic code». The XVIII century and the XXI century. *Bulletin of Civil Engineers* 5 (76): 33–40.

Molotkova, E.G. 2019. About the morphotypes of city quarters in the historical center of Saint Petersburg. Transparency and connectivity. *Bulletin of Civil Engineers* 5 (76): 41–54.

Pasechnik, I.L. & Marushina, N.V. 2019. Value category in theory and practice of conservation of historical urban environment. *Vestnik Tomskogo gosudarstvennogo arkhitekturno-stroitel'nogo universiteta. JOURNAL of Construction and Architecture* 3: 9–19.

Regame, S.K. et al. 1989. *The combination of new and existing buildings during the reconstruction of cities.* Moscow: Stroyizdat.

Sementsov, S.V. 2012. Saint-Petersburg historic agglomeration as an urban planning world-wide object. *Internet-Vestnik VolgGASU. Series: Multi-Topic* 1 (20).

Slavina, T.A. 2002. Transformation of historical cities in the 21st century (Saint Petersburg). In *Toward 21st Century Architecture. A potiori: collection of scientific articles.* Moscow: Russian Academy of Architecture and Building Sciences: 83–88.

UNESCO 2011. *Recommendation on the historic urban landscape.* Accessed June 6, 2020. https://whc.unesco.org/uploads/activities/documents/activity-638-98.pdf.

Voznyak, E.R. 2017. Research methodology the detail of the facades of historic buildings based on the theory of architectural forms. *Modern High Technologies* 1: 22–26.

World Heritage Center 2019. *The operational guidelines for the implementation of the World Heritage Convention.* https://whc.unesco.org/en/guidelines/.

Reconstruction and Restoration of Architectural Heritage – Sementsov et al (eds)
© 2020 Taylor & Francis Group, London, ISBN 978-0-367-65357-6

Renovation of coastal industrial areas. European experience of 2000s

O. Pastukh & A.G. Vaitens
Saint Petersburg State University of Architecture and Civil Engineering, Saint Petersburg, Russia

ABSTRACT: As part of the modern development of the urban environment in many countries of the world, comprehensive programs are being developed and implemented that cover the issues of urban revitalization and improvement. In this article, the authors will review the most significant and large-scale projects (being developed, implemented and currently being implemented) for the renovation, development and reconstruction of industrial coastal territories in the largest cities of Europe, in order to identify current trends and ways to work with adjacent territories and their modernization. Today, a special interest is not only the economic feasibility of renovation of coastal areas, but also the issues of architectural appearance and functional purpose. Both practitioners and theorists of architecture, engineers and designers expect to use the latest achievements of science and technology for this purpose. The use of large-span shell structures, including geodesic domes, made of wood and polymer materials in the construction of buildings and structures for various purposes during the renovation of industrial coastal areas will not only create a unique silhouette of the coastline of large cities, but also take care of the historical value of cultural heritage objects, taking into account the environmental aspect.

1 INTRODUCTION

Renovation of coastal territories in megacities is one of the most important problems of city authorities today. The transformation and development of coastal industrial areas will improve the environmental situation of megacities, as well as solve transport problems, not forgetting the increase in tourist attractiveness and economic component of the issue. The congestion of large cities with cars forces the use of coastal territories for the construction of multi-tiered transport overpasses, a large network of bridges, various industrial structures and numerous cargo berths is created. An important factor is the aesthetic perception of the silhouette of the coastline, the so-called "sea facade" of the city. Architects and engineers pay special attention to creating a unique architectural appearance from buildings of various functional purposes in the spaces of the coastal urban strip.

The presence of historical and cultural heritage sites that are protected by UNESCO in industrial areas requires a professional individual approach to integrate and adapt buildings to the modern needs of society (State Duma of the Russian Federation 2002).

2 METHODS

2.1 *The construction of residential blocks in the creation of quays at the port areas in Stockholm, Sweden*

1. On the North-Eastern outskirts of Stockholm, a program for the development of former port and industrial areas is being implemented (the project is called Norra Djurgordstaden). Mass construction of new residential areas began in 2011 on the site of former warehouses and other industrial parks. By 2025, 12,000 new homes are planned to be commissioned in the Swedish capital (Figure 1).

Mixed use of space will eliminate dead and dark areas at night and increase the security of your stay. Some of the old buildings will be restored and integrated into the new architecture. Each building has a unique architectural appearance. About 30% of the energy will be generated by renewable sources (Ilicheva 2016).

2. Industrial district in the Royal seaport – more than 2 thousand hectares of land 3.5 km from the city center. In 2008, a program began to transform this area into an environmentally oriented, self-sufficient area with full infrastructure. In total, it is planned to build at least 12 thousand new homes and create 35 thousand jobs, 600 thousand square meters will be allocated for commercial space. The city authorities paid special attention to the preservation of cultural heritage.

3. One of the latest and most successful examples of architectural and planning rethinking of coastal areas is The Hornsbergs Strandpark embankment. This project won the Swedish Sienapriset award in 2012. The Park is more than 700 m long and consists of four parts. In this Park, water and land were combined in a modern design of a winding Bank, smooth organic forms and clean lines. The Park itself faces West, and the embankment has three floating piers, which gives visitors the

Figure 1. Construction of residential buildings on port territories, Stockholm, Sweden. (photo Pastukh O. A., 2014).

opportunity to get as close to the water as possible. The Park also has swimming areas, showers, bike paths, and grilling areas. In the free areas between groups of plants, additional public spaces and facilities have been created for various events. When developing the concept of the Park, the main goal was to create an atmosphere of a cozy suburb that would contrast sharply with the busy urban environment in the heart of Stockholm (JDS 2020).

1.1.4. There are more than a dozen UNESCO world heritage sites located on the territory of Sweden, some of which are located on coastal territories and it is possible to adapt them to the needs of modern society.

2.2 *Transformation and development of the coastal territories of the Netherlands on the example of the port territories of Amsterdam and Rotterdam*

1. At the beginning of the XXI century, the UNESCO World Heritage List in the Kingdom of the Netherlands contains about ten names. The use of large-span wood and polymer materials in the renovation and development of coastal areas will not only ensure the preservation of cultural heritage sites, but also to repurpose and add modern functioning of these territories, taking into account the needs of society and the deteriorating environmental situation (Zhivotov & Pastukh 2020).
2. The Coastal territories of Amsterdam were formed from the end of the XIX century. before the beginning of the XX century – in the 1870s-1920s. These were the Eastern docks-artificial (alluvial) Islands on the IJ river. In the 1920s and 1970s, these artificial Islands maintained their

port and warehouse functions. Since the 1970s. due to the fact that sea passenger transport was replaced by passenger transport, and large-capacity vessels could not enter this part of the river, port and logistics functions were moved downstream of the IJ river, closer to the North sea. The vacated hulls and peninsulas began to fall into disrepair. Until the 1980s, the coastal territories of the left Bank of the IJ river served as berths. In the 1990s and later, housing construction began to develop here. The main type of development in these territories is blocked houses that are directly exposed to water. In the 2000s, a complex of the Ministry of Justice of the Netherlands was completed here. At the same time, the reconstruction of Amsterdam station was completed. An interesting example of adaptation for the modern use of former flour mills (silos) in housing (Figure 2). The authors of the project have preserved the main historical load-bearing structures of buildings, giving them a new life.

3. Rotterdam is the second most important city in the Netherlands and is currently the largest port in Europe. In the 1980s, several tenders were held for housing construction on the territories of the inner harbors (Old harbors), as a result of which reconstruction began. A typical example is

Figure 2. A - city cinema center; b - residential buildings in former flour silos, Amsterdam, the Netherlands (photo by A. G. Vaitens, 2015).

The Blaak residential building (1988 arch. Piet Blom), the Maritime Museum of Rotterdam. In the 1990s, the next stage of reconstruction of the coastal territories of Rotterdam began. On the initiative of the city municipality of Rotterdam in 1991 a competition was announced for the reconstruction of the left-Bank part of the center of Rotterdam-the former cargo harbor of Kop van Zuid. After the construction of the bridge (Erasmus bridge) – one of the symbols of Rotterdam, which connected the right Bank (the historical center) with the territories of former commercial harbors, active design and construction of new buildings began on the territories of this harbor (Figure 3).

Next to the former passenger terminal in the mid-1990s, an office building of the Netherlands Lloyd's was built according to the project of N. foster, and in the early 2000s, a high-rise residential building of Montevideo was built. A little earlier, the building of the Rotterdam Telecom (arch. R. Piano, late 1990s) and the complex of the Supreme court of the Netherlands (Wilhelminahof) were built. All these buildings are built on former wharf territories, which were called Queen Wilhelmina wharves, hence the name. The former Entrepot warehouse complex has been converted into a shopping complex, with

cranes and lifting devices left as "place memory". In 2014, the Market Hall complex was completed, with housing located in the curved part. The construction of this complex revived the social activity of this part of the old harbors. Currently under active reconstruction of one of the former cargo harbor of Rotterdam – Katendrecht. The former warehouses house the city's market for environmental products.

2.3 Reconstruction and development of coastal industrial territories in Germany on the example of the HafenCity district in Hamburg

Hamburg is the second largest city in Germany after Berlin and the third largest cargo port in Europe after Rotterdam and Antwerp.

In the southern part of the center of Hamburg, opened to the North Elbe, from the middle of the XIX century began to form a logistics center - HafenCity. This area was a seacoast, unsettled suburb of Hamburg, built up with warehouse brick buildings built along artificially dug canals. As such, HafenCity existed for the first half of the twentieth century. (Figure 4).

In the late 1950s, these territories began to decline. Since the mid-1990s, planning for the reconstruction of HafenCity has begun. The total area of HafenCity is 155 ha, half of which is occupied by artificial channels and bays. The master plan was approved in 2000. By 2010, only the Western part of HafenCity was built up - 30% of the entire territory. Full

Figure 3. A-Erasmus bridge (1993-1996), a view of the embankment development; b-buildings of the Netherlands Lloyd's (1990s, F. L. Wright) and Montevideo (2000), Rotterdam, the Netherlands (photo by A. G. Vaitens, 2015).

Figure 4. A- HafenCity development project, b - view of the port from the water, Hamburg, Germany (photo by A. G. Vaitens, 2017).

implementation of the reconstruction and development of HafenCity is scheduled for 2025. By this time, about 700,000 m^2 of housing, more than 1 million m^2 of office space, and more than 500,000 m^2 of social infrastructure and public spaces, as well as transport infrastructure, including metro lines and stations, are expected to be built on existing and alluvial territories.

3 RESULTS

Considering the issue of transformation, renovation, reconstruction and adaptation of coastal territories, a whole range of tasks is solved: aesthetic, technical, transport, economic, and environmental. From a technical point of view, we can offer a large number of innovative design solutions for buildings that meet current trends. Of course, to this day, the main emphasis in the design issue is on protecting the water area from water, such as an improved design with effective wave-damping properties, which solves the problems of stability and durability of berths exposed to the intense action of waves and climate. Examples of such structures can serve as hydraulic structures of the past centuries, created by outstanding engineers (Wu & Takatsuka 2006).

But it is also important to use modern structures for both residential and public buildings and structures for various purposes in the formation of the architectural silhouette of the coastline: business and office centers, concert and exhibition halls, museums, sports complexes, warehouses, engineering facilities, etc. A way to solve this problem can be the use of large-span spatial structures in the formation of the overall appearance of the urban environment and the silhouette of the coastline, the so-called "sea facade" of the city.

4 DISCUSSION

The authors of the article suggest using large-span shells and other engineering spatial structures made of modern materials in the reconstruction and development of coastal industrial territories. The main purpose of such buildings is considered not only industrial enterprises and factories, warehouses, logistics centers, but also public buildings for various purposes: cultural (exhibition and Museum centers), entertainment (concert and cinema halls, water parks), sports (stadiums and swimming pools).

The use of these structures will give a unique architectural appearance to the structures (Globen arena in Stockholm, Aquarium in Genoa) that form the city's coastline, demonstrate modern achievements in construction technology and engineering, and enable a wide range of various processes in buildings and structures of various functional purposes. The use of

this type of construction not only allows you to cover large spaces (especially important, given the size of the territories of former industrial areas), but also the use of modern building materials (glued wooden structures and polymers) will ensure high energy efficiency and environmental friendliness of buildings, in accordance with international quality standards and green construction (BREAM, LEED, GREEN ZOOM).

5 CONCLUSION

1. The Review of foreign experience in reconstruction, adaptation of existing industrial port areas and development of the coastal strip in the urban area, conducted in the article, demonstrates a constant growing interest in spaces near the water. Especially important is the issue of preserving cultural heritage in the context of modern globalization and the deteriorating economic and environmental situation around the world.
2. In many countries, territories near water, together with the surface of the water itself, have long been one of the most priority spaces in terms of placing the main urban functions on them. Thus, these spaces near the water, which have a great potential for natural and urban resources, become additional reserves of urban territories.
3. The issue raised by the authors of this article has profound development potential. Not only theoretical discussions, but also practical implementation of the conceived ideas of progressive engineering and architectural thought will have a positive impact on the economic and environmental situation, developing the infrastructure of the coastal territories of the world's largest cities.

REFERENCES

Ilicheva, D. 2016. An international experience in coastal areas usage. Architecture and Modern Information Technologies 3 (36).

JDS 2020. KAL/Kalvebod waves. http://jdsa.eu/kal/.

State Duma of the Russian Federation 2002. Federal Law No. 73-FZ dd. June 25, 2002 "On cultural heritage sites (historical and cultural monuments) of the peoples of the Russian Federation" (updated on July 18, 2019). http://www.consultant.ru/document/cons_doc_LAW_37318/.

Zhivotov, D. & Pastukh, O. 2020. Construction of geodesic domes made of wood and composite materials during restoration and conservation of cultural heritage objects. E3S Web of Conferences 164: 02020.

Wu, Y. & Takatsuka, M. 2006. Spherical self-organizing map using efficient indexed geodesic data structure. *Neural Networks* 19: 900–910.

Historical and architectural heritage of railway complexes in Belgorod region

M.V. Perkova, L.I. Kolesnikova & Y.A. Nemtseva
Belgorod State Technological University named after V.G. Shukhov, Russia

ABSTRACT: The problem of adaptation of historically formed small railway station complexes and their adaptation to modern conditions is currently most acute in small and medium-sized cities of Russia. The study examines the historical and architectural heritage of railway complexes on the territory of the Belgorod region from the second quarter of the XIX century to the beginning of the XX century. Preconditions of formation of the station complexes in the Belgorod region and the periodization of the stages of development of the station complex as an important communications hub in the city. The prerequisites for the architectural modernization of railway station complexes at the present stage are formulated: socio-economic, technological, environmental, urban planning, functional planning and architectural art. The directions of development of historically formed railway station complexes as multifunctional communication nodes of the city and tourism objects are proposed. a theoretical model of modernization of railway station complexes in modern socio-economic conditions is Developed, taking into account the investment attractiveness by choosing the type of architectural modernization and tested on the example of project proposals for the modernization of railway station complexes in cities: Valuyki, Alekseevka and Proletarsky.

1 INTRODUCTION

Currently, there is a tendency to form railway station complexes as multifunctional urban structures. In these conditions, the problems of architectural modernization of historically established railway station complexes become urgent, since they do not meet the modern requirements imposed on them. This is due to a whole set of problems that arise in the course of their operation, in particular: mismatch in capacity with increased passenger traffic, technological backwardness, lack of service network and cultural sphere, low level of comfort, inefficient use of the territory, low environmental and aesthetic qualities.

The problem of the research is the urgent need to adapt the historically formed small railway station complexes of the Belgorod region and adapt them to modern conditions by creating a well-founded conceptual project proposal that will preserve the cultural heritage object and will help attract investment to the territory of the settlement.

This problem is most acute in small and medium-sized cities in Russia, where existing historical buildings and structures of railway station complexes are not always included in the register of historical and cultural heritage.

The research hypothesis suggests that the modernization of railway station complexes with adaptation for modern use of cultural heritage objects is based on the criteria of the identity of the environment for accounting for the investment attractiveness of existing objects. This will allow us to form an effective concept for the strategic development of railway station complexes as city-forming objects.

2 MATERIALS AND METHODS

The methodology and methods of scientific research are based on a comprehensive study of the processes of adaptation of objects of historical railway station complexes in the conditions of changing socio-cultural priorities. The study uses archival sources and cartographic materials, materials from Museum collections, and examines existing regulatory and Advisory documents in the field of evaluating architectural monuments and regulating their use, the most fundamental of which is: Federal law No. 73FZ of 25.06.2002 (ed. from 13.07.2015) "on objects of cultural heritage (historical and cultural monuments) of the peoples of the Russian Federation". The research is based on a full-scale survey of historical architectural objects and analysis of implemented projects of foreign and domestic experience in adapting railway station complexes to modern conditions. When solving these tasks, the method of retrospective analysis and comprehensive assessment of the territory was used; cartographic analysis and landscape-visual analysis of the territory; environmental approach in the formation of a conceptual project proposal and quantitative methods for evaluating conceptual design solutions, adapted by the author for use in the practice of adapting architectural heritage.

3   RESULTS AND DISCUSSION

The history of architectural and urban development of railway station complexes is considered and the periodization of stages of development of housing and COMMUNAL services from the second quarter of the XIX century to the beginning of the XX century is revealed. The main stages of the emergence of the railway station complex as an important communication hub in the life of the city are highlighted (Petukhova, N. M. 2010):

1) The Stage of development of railway station complexes from the second quarter of the XIX century – the stage of construction of the first Railways and the emergence of stations (Gulidova Yu. I. 2011, Kraskovsky E. Ya., M. M. Uzdi, 1994).
2) The Stage of development of railway station complexes since the second half of the XIX century, the stage of typological formation of railway station complexes (Gulidova Yu. I. 2011, Kraskovsky E. Ya., M. M. Uzdi, 1994).
3) The Stage of development of railway station complexes from the end of the XIX – beginning of the XX century – the stage of flourishing of the architecture of railway station complexes, the stage of construction of the largest railway buildings complexes (Gulidova Yu. I. 2011, Kraskovsky E. Ya., M. M. Uzdi, 1994).

The role of the station complex in the urban environment is defined. The station is an element of the railway transport infrastructure and plays an important role as an intermediary between the urban environment and the railway (Kolesnikova L. I., Nemtseva Ya. A., 2019). Historically established railway station complexes have not only a transport and historical and cultural function, but also a city-forming and socio-economic function, being a kind of reference points for the further development of territories according to the concept of sustainable development (Perkova, M. V. 2015). As a result of consideration of modern problems of adaptation of historically formed railway station complexes, it is determined that further development of historically formed railway station complexes is necessary by developing the main ways of modernization (reconstruction, or modern adaptation). Without timely modernization, such complexes do not sufficiently use their potential as urban infrastructure objects in the current economic situation (Perkova M. V. 2018).

The analysis of foreign experience in the reconstruction of railway station complexes allowed us to determine the main trends for further development of this direction:

1) Conversion of railway stations into terminals that coordinate the work of transport corridors for passenger and train services;

2) Ensuring high quality of service through the use of various types of transports;
3) Combining projects for the development of railway stations and forecourts, as well as adjacent blocks with mixed-use construction;
4) Increasing the share of private operators of railway stations, optimizing the management of railway stations, organizing the internal space of railway stations.

It is established that the main direction of development of foreign railway stations is the transformation into multifunctional, compact transport and communication hubs that take into account the needs of not only passengers and visitors, but also the surrounding urban areas.

The railway station complex is a part of the urban transport system and includes the railway station and its adjacent territories, objects, buildings and structures that are physically, technologically or otherwise connected to the railway station and subject to a single legal regime of operation and development.

The questions of typology of historically formed railway station complexes are considered: the existing classification of domestic railway stations; the placement of railway station buildings, passenger platforms relative to the receiving and departure routes; the location of the station building relative to the city line. The historical aspects of the formation of the management structure of the construction and operation of Railways from the Formation of the Main society of Russian Railways with the participation of foreign banking capital to the establishment of the Ministry of Railways are considered.

The historical background for the formation of railway station complexes in the Belgorod region in the second quarter of the XIX century is revealed:

1) Creation and development of infrastructure of the South Eastern railway;
2) Creation of joint-stock companies for the construction of Railways on the territory of the Central Chernozem region;
3) Economic and industrial development of the Central black earth region;
4) Economic and industrial development of Belgorod region;
5) The development of agriculture.

The article considers the development of the structure and architectural planning aspects of railway station complexes on the territory of Belgorod and highlights the main trends in the functional and urban development of railway station complexes:

1) defining the station as a "center of attraction" in the urban environment, increasing its transit function (historically, in most cases, the station is located in the Central part of the city) and reducing the functions of accumulation and waiting for passengers; development of social and

commercial functions (previously unusual for a railway station);

2) Development and regulation of the infrastructure of the station complex, pedestrian communications (development of above-ground, underground space), including linking the station with the forecourt, etc;

3) Development of recreational areas on the territory of the railway station and environmental protection (recreation areas, exhibition halls; greening of design solutions) (Krushelnitskaya E. I. 2015).

Thus, railway stations become multifunctional objects of railway transport and the urban environment, connecting the external and internal transport systems of the city.

A universal algorithm for forming a conceptual project proposal for the architectural modernization of historically formed railway station complexes is proposed, based on the example of station complexes of small cities in the Belgorod region.

1) The pre-project stage involves identifying individual characteristics of the territory (analysis of the existing situation);

2) The conceptual stage includes setting the territory's problems according to criteria (resources, transport, business, society), forming concepts for adapting historically developed railway station complexes based on solving these problems;

3) The project stage involves the development of project proposals for objects within the framework of the accepted concept.

The developed algorithm was tested on the example of Gotnya, Valuiki and Alekseyev ka stations in the Belgorod region. At the first stage, a pre-project analysis was carried out, which revealed common problems for the three stations:

1) Unused historical and cultural potential of the historical territory;

2) Insufficiently developed transport and pedestrian network;

3) unfavorable investment climate for investors in terms of large investments in the modernization of complexes for the purpose of preserving cultural heritage objects in accordance with the Federal law "on cultural heritage objects of the peoples of the Russian Federation";

4) insufficient knowledge of citizens 'information about historical railway station complexes in Valuyki, Alekseyevka and Gotnya, an unformed brand of a significant historical and cultural object of regional and Federal significance.

At the second stage, a concept was formulated for the modernization of railway station complexes at Gotnya, Valuiki and Alekseyevka stations according to the criteria. According to the "Resources" criterion, it is assumed that the existing buildings and structures of the railway station complex will be preserved and strengthened with native resources. It is revealed that the objects of the railway station complex are objects of cultural heritage. It is proposed to recreate the identity of the place by carrying out repair and restoration works. Preserving the original appearance of historical buildings will allow the railway complex to become one large Museum of the history of the railway boom of the XIX century on the territory of the Belgorod region. Strengthening the natural framework by improving and greening the forecourt and adjacent blocks of mixed-use development, as well as creating recreational areas.

According to the "Transport" criterion, it is planned to introduce new transport systems and improve the transport and pedestrian infrastructure.

According to the "Business" criterion for Valuyki and Alekseyevka stations, it is proposed to divide the station forecourts into functional zones for building investment sites with a fixed type of activity: trade territories, cafe territories and open terraces, tourist and recreational areas, Parking areas and public and private transport stops. Objects of historical and cultural heritage of regional significance should be considered as potential centers of tourism, including the prospective development of the territories where they are located (Perkova, M.V., Krushelnitskaya, E.I. 2016, Perkova, M.V. 2016). The solution for the Gotnya station according to the "Business" criterion will be to create an open-air Museum of railway equipment with recreation areas and cafes. It is proposed to apply the same approach to the territory of Gotnya station as to the building of the "insiti" open-air Museum, which has a value as a historical exhibit of railway heritage. The Museum function will also ensure the social orientation of the business according to the criterion "Society".

According to the "Society" criterion, it is proposed to create a "place brand", create an urban environment that is attractive to all types of population, implement social projects, create new jobs and, as a result, get a place of attraction at the local and regional levels.

4 CONCLUSIONS

. The conducted research has shown that the modern period can be designated as a qualitatively new stage in the development of railway station complexes. In this regard, a number of prerequisites for the architectural modernization of historically established railway station complexes have emerged, which is of crucial importance at this stage of development of small and medium-sized cities. This study solves the important task of determining the direction of modernization of railway station complexes and allows us to make the following conclusions and recommendations:

1. In the formation and development of the architecture of railway station complexes (the second quarter of the XIX century and the beginning of the XX century), three historical periods were identified. The historical and architectural potential of railway

station complexes is determined in the context of the overall process of socio-economic development of the country.

2. The prerequisites for the architectural modernization of railway station complexes at the present stage are Identified: socio-economic, technological, environmental, urban planning, functional planning, and architectural and artistic.

3. The directions of development of historically formed railway station complexes as multifunctional communication nodes of the city and tourism objects are Proposed.

4. Developed a theoretical model of modernization of railway station complexes in modern socio-economic conditions, taking into account investment attractiveness by selecting the type of architectural modernization and tested on the example of project proposals for the modernization of railway station complexes in the cities of Valuyki, Alekseevka and Proletarsky:

– Type of communication and transport node direction of modernization of transport infrastructure based on the station complex Valuyki;

– Type of public transport hub - formation of a multi-functional structure

– The station complex and adjacent quarts of mixed development on the basis of the station complex Alekseevka;

– Type of tourist and recreational transport hub-the formation of a multi-functional structure of the Gotnya and Proletarsky railway station complex and adjacent blocks of mixed-use development.

REFERENCES

Petukhova, N. M. 2010. Urban planning role of railway stations in Russia and the evolution of their architecture: 1830-1910s. - Saint Petersburg.

Gulidova Yu. I. 2011. Railway revolution in Russia in the second half of the XIX early XX century. Economic journal. 2: 135142.

Kraskovsky E. Ya., M. M. Uzdi, 1994. History of railway transport in Russia, JSC "Ivan Fedorov".

Kolesnikova L. I., Nemtseva Ya. A., 2019, Historical and architectural heritage of objects of the Southeast railway on the territory of the Belgorod region, BSTU.

Perkova, M.V., Blagovidova, N.G., Tribuntseva, K.M. 2015, Fea tures of design of ecovillages in depressed areas in the city, Research Journal of Applied Sciences.

Perkova M. V. 2018, Methods of identification and resolution of civil conflicts and contradictions of development at the municipal level, Academia. Architecture and construction, 4: 61–70.

Krushelnitskaya E. I., Perkova M. V. 2015, Formation and development of architectural planning organizations tourist recreation territories, Belgorod, BSTU.

Perkova, M.V., Krushelnitskaya, E.I. 2016, Recreation and tour ism organization issues in modern urbanization conditions, International Journal of Pharmacy and Technology.

Perkova, M.V. 2016, Regional settlement system, International Journal of Pharmacy and Technology.

Reconstruction and Restoration of Architectural Heritage – Sementsov et al (eds)
© *2020 Taylor & Francis Group, London, ISBN 978-0-367-65357-6*

Principles of urban environment development based on identification features of historical city

F. Perov, O. Kokorina, U. Puharenko & Yu.N. Lobanov
Saint Petersburg State University of Architecture and Civil Engineering, Saint Petersburg, Russia

ABSTRACT: Identification characteristics of the urban environment play a crucial role in the existence and development of society, enable people to establish their place in the world, discover their belonging to a certain culture. Architecture is a milestone that makes it possible to preserve identification. The article suggests principles of urban environment development considering the three identification levels of the urban environment of a historical city.

1 INTRODUCTION

In terms of architectural decisions, the current practice of urban environment development in many respects looks random, small-scale, and in many ways tasteless. In many cities, so-called "modern" architecture appears which fails to create a comfortable living environment, is extravagant, and does not evoke the sensation of "birthplace" in people. It becomes apparent that attractive and authentic architecture cannot be created in a city without a single architectural and planning concept.

Upon the comprehension of the general tendency of construction, it becomes obvious that systematic analysis is needed to understand issues related to the identification of the urban environment for the people who are living in it.

The conservatism of people to changes in the urban environment plays a positive role (Glazychev 1984). Urban renewal is forced to occur evolutionarily, making it possible to choose the best solutions for renewal and reconstruction. New buildings in the center seem alien to older generations of residents because they saw the city without these buildings. They are quite organic and familiar for young people who grew up in the changed environment. Urban renewal strategy, considering the identification signs should be adopted based on the tasks set for its development. However, the obtained results often contradict with the preservation of the city's historical appearance.

2 DEVELOPMENT OF HISTORICAL CITIES AS OBJECTS OF IDENTIFICATION

Adult residents living in the city for many years, young people, and those who have come to a new place of residence, businessmen developing their business, tourists — they all have different points of view regarding the issues of identifying the urban environment of a historical city (Kachanov & Shmatko 1993). Historical cultural centers are largely oriented towards development while preserving the established architectural appearance. Without its preservation, they can lose their qualities, the source of income, with tourists as the main customers.

A city is not a petrified organism. It is developing to meet the needs of its residents. The facades of the historic houses in Florence demonstrate the efforts of residents to rearrange their habitat. However, the traces of reconstruction left on the facades have become a monument to the city's history for modern residents. Residents are often dissatisfied with its insufficient renewal level, which results in unresolved issues related to transportation, development of social infrastructure, and poor condition of buildings. Modern people hardly would be able to live in the city's historical center 100–200 years ago. In that case, the historical city would instantly turn into an unsuitable living environment, in their understanding.

An example of Venice, a true museum city, shows that city development is impossible without reconstruction. Venice's source of income is its history and the originality of its architectural environment. This helps to strictly limit and regulate new construction, preserve the architectural appearance of the city. The entire level of the pedestrian zone in the city has been completely reconstructed and no longer conforms to the city's historical appearance. It is unlikely that in the Middle Ages the city would shine with so many shop windows (Figure 1): all the first floors have been built from scratch rather than rebuilt.

In the city center, new buildings of hotels and museums have been built along the Grand Canal. Sometimes, these are modern architecture buildings, although they have been built following the strict height restrictions and in accordance with the

Figure 1. Venice. Reconstructed pedestrian zone.

structure of the historical center. Observation of the urban development rules does not make them extravagantly modern. It is clear that a city could not serve millions of tourists without these reconstruction measures.

The level of the identification of modern architecture with historical architectural surroundings is extremely high. In addition to the height restrictions, the space planning structure, principles of site development, materials of walls and architectural details are fully preserved (Figure 2). Modern architecture "blends" in the historical environment. The development rules are aimed at preserving the historical development of a city as a tourist center.

A city is characterized not only by its aesthetic qualities. It is a space of business activity of people.

If private investors get complete freedom of action, historical cities may be reconstructed until they are completely destroyed.

The space of a city is a public space. Only laws, regulations, and building rules can create a business activity area that will consider both the interests of preserving the urban environment and the possibility of its development and modernization.

Figure 2. Venice. Modern buildings in the city center.

3 REQUIREMENTS FOR ARCHITECTURE OF NEW BUILDINGS CONSIDERING THE IDENTIFICATION FEATURES OF HISTORICAL CITIES

The Great Encyclopedic Dictionary defines identification as establishing the identity between two objects based on the coincidence of their features. Based on this definition, let us suppose that identification can exist at several levels:

– level 1 — urban-planning level;
– level 2 — object level;
– level 3 — element level.

In our case, to ensure the level 1 of identification, it is necessary to find out what architectural and urban-planning features are the most typical for historical cities.

Many architects of Saint Petersburg studied the identification features of the city. In particular, architect Mamoshin M.A. determined the following identification features of the architectural environment of the city (Mamoshin 2016). He highlighted the following identification characteristics at the level of the urban-planning approach:

1. Skyline (according to D.S. Likhachev).
2. Red line rule (valid from the moment the city was founded).
3. Ensemble nature of buildings (city — a group of ensembles).
4. Contextuality of new buildings (interaction with the surrounding buildings).
5. Public spaces of embankments (the main distinguishing feature of Saint Petersburg embankments).
6. Perimetric nature (the basis for the development of Saint Petersburg quarters).
7. Firewall rule denoting the boundaries of intra-quarter demarcation.
8. Height restrictions of buildings (historical restrictions and modern zoning by height in the historical center).
9. Curvature of regular geometry in Saint Petersburg (Saint Petersburg is characterized by the fact that everything is slightly curved in plan, even within the framework of rectilinear geometry).
10. Contrast of the live water line and the correctness of development (contrast of near-water development is especially pronounced on the Neva and other rivers and canals of Saint Petersburg).
11. Perpendicular location of buildings with regard to water (priority orientation).

A comparative study of the features of two historical cities — Saint Petersburg (Russia) and Florence (Italy) — was carried out at the Department of Architecture in the Saint Petersburg State University of Architecture and Civil Engineering as part of an international workshop (Figure 3).

A comparison of two historical cities revealed the characteristic features of Saint Petersburg. As

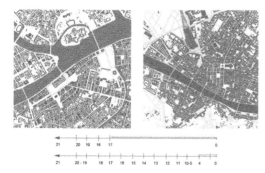

Figure 3. Comparative analysis of the urban texture of Saint Petersburg and Florence.

Figure 4. Analysis of riverbed width influence on the network of city highways.

Figure 5. Comparative analysis of main street profiles.

a result of the comparative analysis, the following conclusions can be drawn:

1. Saint Petersburg is distinguished by the scale of the urban structure elements. The modularity of quarter development in Saint Petersburg is much larger than in Florence. Noticeable differences are due to many reasons, most of which are determined by the historical development of each city. The cities of Saint Petersburg and Florence emerged in different historical eras. The medieval city of Italy was created inside a city fortress, which determined its dense urban structure and fine road network. Saint Petersburg, the city of 18th – 19th centuries, developed freely, without an external perimeter binding it.

2. The study showed the impact of natural objects on the planning structure of the cities — the ground profile and the space of rivers flowing through the cities. It was much easier to build bridges across the narrow riverbed of the Arno River in Florence than over the wide and deep Neva. This directly affected the network of city highways connecting city parts located on opposite banks (Figure 4).

3. The level of transport development also played an important role in determining the need for the planning structure to correspond to vehicle and pedestrian traffic.

This is clearly visible in the street profiles (Figure 5). The profiles of the main thoroughfares and main streets in Saint Petersburg as well as the distance between them are much larger and meet the requirements of both the 18th century and modern reality. It is worth noting that in Florence, after World War II and the bombings, reconstruction works were performed and the transport structure of the city center was developed.

The object level of identification (level 2) is directly related to the urban-planning level (level 1). Each city has its own features related to the characteristics of buildings being erected here.

Mamoshin M.A. determined the following identification characteristics of buildings in Saint Petersburg:

1. Facade length (along the building line of 25–50 running meters, historically derived from the merger of land plots).

2. Axial structure of facades forming the street development (mono-axis symmetry, multi-axis structure of the facade composition).

3. Odd number of windows on a facade (as a result of axis designation).

4. Diagonal axial construction of compositions of corner buildings and their elements (in the areas of street intersections).

5. Decrease in the building volume in the upward direction (following traditional tectonics); traditional three-part tectonic structure of the facade composition (base, middle part, attic).

6. Mandatory presence of a base (traditionally, a Putilov slab).

7. Traditional geometry of regular shaping (a combination of shapes derived from a square, circle, or trapezoid).

8. Vertical window openings (there are no horizontal windows in the historical Saint Petersburg).

9. Mandatory presence of quarters, minimum 1/2–2 bricks, with the construction of rainwater drainage.

10. Window shape (T-shaped wooden windows (with ventilation windows)).

101

11. Facade microplasticity (availability of horizontal cornices and belt courses, head moldings, caissons, medallions, decor on the facades.

4 MAIN CITY IDENTIFICATION FEATURES FOR THE DESIGN OF MODERN BUILDINGS

The study of all identification signs of level 2 enabled us to identify the main ones to be integrated into the city identification code, including the design of new buildings.

1. The most important criterion for identity is the scale and nature of development (Figure 6). The concept of scale should include height, the division of building facades into separate modules or sections, the ratio of the height of a building to its length. These parameters are not constant and depend on the historical and natural context.
2. Construction of quarters instead of neighborhoods and creation of a pronounced street-front make it possible to obtain separate yard areas. Yard areas are a clear space for arranging the life of tenants of several buildings, which can be controlled and maintained by them. It is crucial to remove parking lots from the yards. Access roads to houses must be provided, however, it is not reasonable to park cars in the yards. Parking should be allocated outside the yards, including under the ground.

However, the development of an endless line of facade fronts has a certain danger (Figure 7). The development of historical centers provides an answer to this question by shifting the development network and arranging visual dominants.

3. The division of houses into three parts in the historical center is not only an architectural technique.

Building in a straight line creates an endless monotonous outlook; shifting the development network and arranging visual dominants enables to split the path into segments and create landmarks

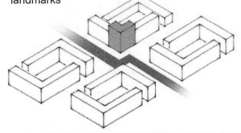

Figure 7. Principle of forming the building line.

For historical reasons, the first floors were provided along the main streets to accommodate service facilities or for technical purposes. The best apartments were located on the second or third floors. The upper floors and attics, due to their difficult and inconvenient access by stairs, were intended for cheap living and had low ceilings.

Currently, the functional meaning of the three-part division of facades remains. The provision of social and commercial functions on the first floor is both a continuation of historical traditions and a widespread international practice making it possible to create a full life in the area of street public spaces. Floor-by-floor zoning of a house with the allocation of apartments of various classes on different floors should remain in modern practice as well (Figure 8). The appearance of elevators has fundamentally changed the significance of the last floors (in comparison with the historical buildings of Saint Petersburg). Penthouses with apartments with views, galleries, non-standard layouts have appeared.

Incorporation of existing buildings into the environment
Consideration of the historical and natural context
Compliance with the scale and nature of development

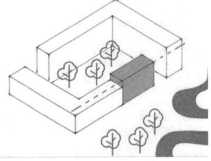

Figure 6. Principle of scale and nature of development.

Three-part division of a facade with a strongly pronounced base

Vertically elongated window and facade plastic using glazed projecting and recessed balconies

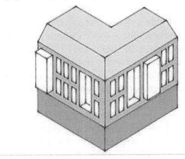

Figure 8. Principle of horizontal division of facades.

4. The element level (level 3) is aimed at solving the issue of giving buildings their own appearance, however, formed according to certain rules.

The traditions of the historical city show that the fundamental solution is to provide each house with its own facade. People naturally want to recognize their home, associate themselves with it. Times of standard projects with the same facades failed to destroy this natural human desire. The traditional and clear way to solve this issue is to use color and material (Figure 9). Color plays a crucial role in the architecture of Saint Petersburg. In the city with a lot of gray and rainy days, colors should be used extensively. However, color is also a method of identification, and researches determine the dominant color range that has established in Saint Petersburg.

The above methods ensuring the use of three levels of identification should be adopted from the historic development zone since it is obvious that the distinctive features of the city managed to form fully and establish themselves most clearly in this zone.

Currently, the majority of modern quarters conform only to the first two principles of historical development: usually, they are absolutely horizontal, their structure of building facades is quite consistent with that observed in the historical center. The difference lies in the principles of development, the scale of the street network, in the absence of wall surface texture of prefabricated apartment blocks, which results in a conspicuous monotony.

Breakdown of a house into sections by combining textures, colors and decorative elements: house as a puzzle of unique pieces connected together

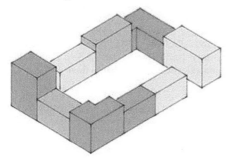

Figure 9. Principle of combining textures, colors, decorative elements.

REFERENCES

Glazychev, V.L. 1984. *Socio-ecological interpretation of the urban environment*. Moscow: Nauka.

Kachanov, Yu. L. & Shmatko, N.A. 1993. Semantic spaces of social identity. In V.A. Yadov et al. (eds), *Social personal identity*: 47–68. Moscow: Institute of Sociology of the Russian Academy of Sciences.

Mamoshin, M.A. 2016. Architectural and urban identity of Saint Petersburg. *Arkhitecturny Peterburg* 1 (38): 12–13.

"Garden cities" along railway lines in Saint Petersburg

N. Petukhova

Saint Petersburg State University of Architecture and Civil Engineering, Saint Petersburg, Russia

ABSTRACT: The article addresses the formation of a "garden city" system along railways of Saint Petersburg in the late 19th – early 20th centuries. Three types of "garden settlements" forming this system are considered: stations that appeared in historical "garden cities"; stations that appeared in populated places that had not had any "garden city" criteria before their appearance; stations that appeared in unpopulated areas.

1 INTRODUCTION

Following the evolution of the "ideal city" concepts, deurbanistic "agglomeration models" started to appear since the end of the 19th century. In fact, the emergence of railways contributed to their formation in no small way (perhaps, even to a great extent). The most well-known of these models, which was widely spread around the world, is the concept of a "garden city" by Ebenezer Howard, formulated in his book published in 1898 "To-Morrow: A Peaceful Path to Real Reform" (better known by the title of the 1902 edition — "Garden Cities of To-Morrow") (Howard 1965). The model of a "garden city" proposed by him formed an entire city-planning trend that developed during the 20th century and continues to evolve today.

A "garden city" was supposed to combine the physical and social advantages of the village and the city, and only a rational layout, which would provide a peaceful way to change society, was necessary for its creation. Following anarchist P.A. Kropotkin, he believed that "on a small scale, society can be made more individualistic and more socialistic" (Ikonnikov 2001).

Several similar cities are connected in a single complex, each of which is different from the other and has, among other features, the main city in the center. According to his concept, the number of inhabitants in standard garden cities should not exceed 32 thousand people, and in the central city — 58 thousand people.

The opportunity of implementing such settlement models was, first of all, due to the creation of a developed railway network by the end of the 19th century, able to unite separate settlements into a single system.

The construction of railways in Russia began in Saint Petersburg (the capital of that time) with the following railways stretching from it in the 1830–1850s: 1836 — Tsarskoye Selo Railway (since 1900 — a part of the Moscow–Windau–Rybinsk Railway), 1851 — Saint Petersburg–Moscow (Nikolaevskaya) Railway, 1851 — Varshavskaya (Warsaw) Railway, 1853–1857 — Baltiyskaya (Peterhof) Railway. In the 1970s, Finlyandskaya (Finland) Railway was added to them.

2 INITIAL "SUPPORTING NODES" OF THE RAILWAY SYSTEM

2.1 *Historical "garden cities"*

Settlements included in the historical Saint Petersburg agglomeration became the initial "supporting nodes" of the railway system. Starting from the 18th century, the agglomeration formed based on principles that anticipated the Howard's "garden city" model. According to S.V. Sementsov, "a 'garden city' is a native architectural and urban-planning phenomenon for Saint Petersburg agglomeration". Its model "existed and continues to exist in Saint Petersburg agglomeration throughout all (or almost all) decades of its development" (Sementsov 2019).

The system of suburban imperial residences around Saint Petersburg served as a core for the formation of a belt of settlements, which by Imperial order started to form around them already in the 1810s and 1820s (slobodas in Tsarskoye Selo, Peterhof, Pavlovsk, etc.). South of Peterhof, a system of regular settlements was established, named after the wife and children of Nicholas I of Russia: Vladimirovo, Kostino, Luizino, Maryino, Mishino, Olino, Sanino, Sashino. In 1831–1841, model area planning designs for settlements were developed for Saint Petersburg.

New settlements and cities formed a clear centric urban structure with the metropolis in the center, the composition of which strongly resembles the "garden city" model developed by E. Howard at the end of the 19th century. It was also multifunctional: in addition to the imperial residences, it included fortress cities and factory cities.

The first railway in Russia, Tsarskoye Selo Railway, connected Saint Petersburg with the historical "garden cities" — Pavlovsk and Tsarskoye Selo, where railway stations were built in 1837. The Peterhof Railway laid to the classic "ideal city" of Peter the Great — Peterhof, "strang" on its axis a number of settlements in the "Petrine" agglomeration: Ligovo (1857), Sergiyevo (1857), Strelna (1857). In 1864, this railway was extended to Oranienbaum, where another palace and park complex was located, which previously had belonged to Menshikov, an associate of Peter the Great. The connection with Gatchina — another element of the historical agglomeration model of the "ideal city" — a "garden city" — was consolidated with the appearance of two railway lines at once: Varshavskaya (Warsaw) Railway in 1853, Baltiyskaya (Peterhof) in 1872. Primorskaya branch of Finlyandskaya (Finland) railway extended to Sestroretsk, a town at a factory of the Petrine time, passing through the historical residences of Peter the Great — Dalniye Dubki and Blizhniye Dubki.

Thus, by connecting the historical "garden cities" in a single system, the railways almost duplicated existing traffic arteries, consolidating and developing historical agglomeration connections. Further development of this system took the way of "compaction", due to the expansion of previously existing settlements and the appearance of new ones, with railway stations as their cores.

2.2 Types of stations based on "garden city" criteria

We can distinguish three main types of stations that affected the formation of "garden cities" along the railways: stations that appeared in historical "garden cities"; stations that appeared in populated areas that had not had any "garden city" criteria before their appearance; stations that appeared in unpopulated areas.

For settlements of the first type — historical "garden cities" — the railways were an impulse for the formation of new suburban zones, which were located both around historical settlements and in their territory, partially changing their functional orientation. Thus, the suburban function begins to develop after the appearance of the railway in Tsarskoye Selo, Sofia and Freudental settlement.

The development of suburban function (after Tsarskoye Selo Railway became a part of Moscow–Windau–Rybinsk Railway in 1900) initiated the appearance of the Pavlovsk II station (nowadays — Pavlovsk), designed for summer residents, at the place of maximum approaching to Tsarskoye Selo. Suburban zones were formed near Tyarlevo platform and Pavlovsk station: Tyarlevo – Glazovo – Zverinets – Matrosskaya Sloboda – Lipitsy (Sementsov 2019).

All the mentioned suburb settlements had a regular planning structure. Zverinets and Glazovo were built-up according to the "ideal" projects of the best architects of the classical period: Cameron and Rossi, based on a radial-ring layout. Zverinets is a historical district of Pavlovsk. In accordance with the project of Cameron, it had a strict geometric shape with a circular pond in the middle, with eight rays extending from it. Glazovo was designed in the shape of a circle, with a pond in the center, where farmhouses were located in sectors extending from the pond.

These historical "garden cities" "set the tone" for newly emerging suburban settlements surrounding them, in accordance with certain "ideal" characteristics: regular layout, architectural appearance, etc. Due to the rapid communication with the metropolis, they soon became places of permanent residence.

At the same time, at the regional urban-planning level, mini-models of the Howard's "garden city" were formed, consisting of a historical "garden city" with a station that served as a metropolis, and suburban areas located nearby.

At the urban-planning level of individual settlements of this type, evolutionary development was characterized by the stability of the urban structure of a "metropolis" — a historical "garden city", usually based on a regular planning structure with a clearly defined main center (palace and park ensemble, factory, fortress). A railway station created a new compositional center with the direction of planning axes connected not with the general planning structure of the city but with the direction of the railway, which led to the formation of a poly-centric urban-planning model. In this case, two regular planning (usually) structures overlapped. As a rule, the railway did not cause any fundamental changes in the social and organizational structure of historical settlements. The first historical "garden cities", where railway stations appeared — Tsarskoye Selo and Pavlovsk — serve as an example of such development (Figure 1).

Figure 1. Tsarskoye Selo layout, 1838. Russian State Historical Archive, Fund 1293, List 167, Case 139, Sheet 1.

3 APPEARANCE OF NEW STATIONS AND FORMATION OF "SUBURBAN ZONES" AROUND SAINT PETERSBURG

3.1 *Principles of railway stations' arrangement*

The railways created prerequisites for the formation of a new settlement system by "compacting" the historically formed structure of "garden cities" with new stations, which in turn became the cores of new settlements. The further development of the railway system with the construction of new lines extending from existing railways was the reason for their appearance. The arrangement of stations along such lines was due to both the functional requirements of the railway, and the need to transport people and goods from certain places of residence or production.

The functional arrangement of stations on the railway was determined by the requirements of operation, which included two key conditions: the possibility to run a locomotive without replacement, which at that time was determined by its technical characteristics of 75 versts (262,500 feet), and the need to supply a locomotive with water and fuel. Based on these functional requirements, railway stations were divided into four classes. In 1843, P.P. Melnikov, when designing Saint Petersburg–Moscow Railway, developed "Regular Designs for Intermediate Stations of Four Classes", where he defined the principles of dividing stations into classes and their location along the main track. This classification became regular and started to be applied everywhere on the railways of Russia.

3.2 *Arrangement of stations near populated areas*

Vyritsa station with Posyolok can be given as an example of a "garden settlement" of the second type, developed in terms of urban planning and social characteristics, which did not have ideal city criteria before the appearance of the railway station. The station was built in 1904 in accordance with the functional requirements of the railway, near the small village of Vyritsa at Tsarskoye Selo Railway, which by that time was a part of Moscow–Windau–Rybinsk Railway.

The construction of the line caused a "suburban boom". A project of a conceptual "garden city" was developed, and an active advertising campaign was conducted to sell garden plots. By 1915, already more than 1.5 thousand cottages were built in Vyritsa, where up to 13 thousand residents of the city lived in the summer. Churches, educational and commercial institutions, theaters, shops, a cinema, a post office, and pharmacies were built in Vyritsa; the village had a railway and telephone connection with the city, 18 versts (63,000 feet) of "macadamized roads", three churches, eight versts (28,000 feet) of horse-drawn railway. An Improvement Society and a voluntary Fire Society were established. To the south of Knyazheskaya valley (one of the areas in the settlement), a new settlement appeared, which was created as a future "garden city" and at that point was called Posyolok (which means "settlement") (it still has this very name). From a private railway line built by one of the wealthiest residents of Vyritsa, Antip Efremov (father of a famous science fiction writer) to his sawmill on the Oredezh River, a railway line was laid to Posyolok, where in 1910, a station with the same name was opened, which provided direct communication first with Tsarskoye Selo, and then with Saint Petersburg.

3.3 *Formation of suburban zones along railways*

The suburban area near Vyritsa station united the following settlements: Vyritsa –Mina – Zarechye – Novoye Petrovskoye – Nikolskoye – "Knyazheskaya Dolina", from Ontsa stream (in the south-west) up to Pizhma–Vvedenskoye highway (in the north-east), with a branch line from Vyritsa railway station to Posyolok railway station with 5 platforms (Sementsov 2019).

With the appearance of Krasnoselskaya (Krasnoye Selo) branch of Baltiyskaya (Peterhof) Railway, built in 1859 and extended to Gatchina in 1872, the following stations were constructed: Gorelovo (1898), Skachki (1898), Mozhayskaya (1877-1879), Taytsy (1876), Pudost (1877), Marienburg (1885).

The construction of Primorskaya branch of Finlyanskaya (Finland) Railway gave a new impetus to the development of territories in this direction, where settlements began to arise and grow actively near the railway stations: Novaya Derevnya (1869), Staraya Derevnya (1869), Yakhtennaya (1894), Lakhta (1894), Olgino (1911), Lisiy Nos (1894), Aleksandrovskaya (1894), Tarkhovka (1895), Razliv (1894).

Some of these stations appeared precisely in connection with the development of suburban construction along the railway, to ensure comfortable communication with the city (sometimes at the special request of "Improvement Societies" organized there — e.g. Olgino station).

Suburban settlements were quite independent in organizational and managerial terms. They appeared as a result of decisions of ministries and departments, as well as on the initiative of city councils and governorate authorities.

Many suburban settlements with railway stations or towns as their cores organized Improvement Societies with their own Charters. These Societies not only supported the existing utilitarian urban-planning criteria of a "garden city" but also introduced the missing social component, which is one of the basic criteria of the Howard's "garden city" (Von Troil 1908).

Suburban zones of various densities and sizes were formed along railways. With an increase in the travel speed provided by the railways, the suburban areas expanded almost to the borders of Saint

Petersburg Governorate. Under those circumstances, natural differentiation of suburban settlements was observed. It can be explained by their price depending on their location relative to the historical "garden cities" as well as climatic and topographical features.

Engineer Glazyrin determined three types of settlements in accordance with their location relative to the base city: "garden outskirts" — directly adjacent to the city, "garden suburbs" — located at some distance from the city, and "garden quarter" — a small fragment of housing development (Glazyrin 1928).

By using the same three-part zoning structure (garden outskirts, garden suburbs, garden quarter), S.V. Sementsov (2019) significantly enlarges the scale of these elements to zones and belts and expands their geography to remote suburbs located near the borders of Leningrad region up to Luga in the south-east and Ladoga Lake in the north. This geographical distribution is due to the specifics of settling in suburbs of Saint Petersburg Governorate, where on some railway directions (Finlyandskaya (Finland), Varshavskaya (Warsaw)), suburban settlements formed almost continuous belts and extensive zones up to the borders of Saint Petersburg Governorate (Figure 2).

3.4 Appearance of stations in unpopulated areas

Some of the railway stations forming suburban areas appeared in areas unpopulated previously (stations of the third type).

The appearance of railways and railway stations in areas unpopulated previously made it possible to create new urban-planning structures meeting the "ideal city" criteria, based on a single plan and regularity, without account for the existing development,

which could interfere with the style and planning unity.

The construction of railway stations from scratch predetermined their formation as integral localities, consisting of not only buildings and structures included in the railway process but also residential buildings with accompanying infrastructure (locomotive depots, workshops, water houses and water-lifting towers, services, etc.). Local complexes of stations of each class served as an analog of the historical centers of cities, around which new residential and industrial areas appeared afterward.

The construction of stations along railways based on the "garden city" project became global in the late 19th – early 20th centuries with the expansion of railway construction geography, aimed at the development of remote territories: construction of Great Siberian Railway, Circum-Baikal Railway (Dalny city, Garden city, etc.). A linear ensemble of stations with a length of 600 km, built by Savva Mamontov in 1895–1900 according to the project of architects Kekushev and Ivanov-Shitsa when laying a railway line from Vologda to Arkhangelsk, is less known.

However, even in the central governorates, the concept of a "garden city" was very popular during the construction of houses for railway employees, which was given great importance at that stage. Such an approach reflected both the general interest in the housing issue, typical for the social consciousness of the turn of the century, and the features of railway construction requiring the resettling of employees in new places of residence.

As stated in Zheleznodorozhnoye Delo journal (1895), "without a doubt, possible improvement of employees' life should be one of the main concerns of any railway society or railway administration; since cheap and hygienic residential premises is a prime necessity, its provision to employees, especially those having a low salary, serves not only the interests of such employees but also the interests of railway companies" (Zheleznodorozhnoye Delo 1895).

In social and organizational terms, attempts to introduce cooperation in accordance with the doctrine of "freedom and cooperation", declared by Howard were made. One of the diagrams in the first edition, which has not been reproduced later, describes how an association of citizens paying moderate rent is able to ensure, first, the payment of interest on borrowed capital, then the gradual payment of the main loan, and finally the formation of their own pension fund and a fund for the development of the education and health system (Howard 1965).

3.5 "Garden city" projects for railway employees

The "garden city" concept was very popular in Russia. At the end of 1913, the "Society of Garden Cities" was organized in Saint Petersburg. One of its founders, A.Yu. Blokh published a Russian translation of the Howard's book entitled "Cities of

Figure 2. Layout of suburban areas along the railways of Saint Petersburg.

To-Morrow" in 1911. This Society was also re-established in the post-revolutionary period in 1922. Theoretical concepts developed as a part of the social experiments of Soviet architects in the 1920s, which set themselves a task of overcoming the difference between the city and the countryside, reflected this approach. Mikhail Okhitovich is the author of the urban-planning concept of deurbanism (Khan-Magomedov 2009). The most detailed argument in favor of the "garden city" concept was contained in reports and articles by famous urban planner V. Semenov (1922). The concept of a "garden city" as a "social city" was supported by Nikolay Milyutin in the book "Socialist city. Problem of building socialist cities" (Milyutin1930). Most of these concepts, in one way or another, took into account the importance and influence of railways for their implementation, following the stormy urban-planning discussions of that period.

In the 1910s, civil engineer V.A. Glazyrin developed an all-Russian system of garden cities for railway employees, which provided for their settling near capitals, major cities, and junction stations, as well as near factories (Glazyrin 1928).

He offered general schemes for placing settlements along railways, in accordance with their location relative to the "capital" city: "garden suburbs" — located at some distance from the city, "garden outskirts" — adjacent to the city, and "garden quarter" — a small fragment of housing development, planned as a garden city at stations with low traffic.

In each of these types of garden settlements, designers should ensure some mandatory types of industrial and public buildings: schools, hospitals, migration centers, halls for public meetings, clubs, kindergartens, nurseries, markets, and libraries. Several types of housing development were recommended as examples of planning organization: open; semi-open; semi-open and linear; group; ribbon; etc.

Glazyrin developed various planning schemes for the formation of different types of garden settlements near railway stations. Settlements by railways usually had a poly-centric structure due to the presence of two stations: a passenger station and a marshaling yard, as well as public centers in the area of housing development.

Programs for creating garden settlements for railway employees were also developed by railway engineers N.V. Kunitsyn and V.I. Boshko (Yensh 1910).

In 1916, the Ministry of Railways decided to build new garden settlements to provide about two million employees of Nikolayevskaya, Omskaya, Tomskaya and Moskovsko-Kazanskaya railways with houses.

To develop projects of "garden cities" along railways in Petrograd in 1916, competitions were announced: the Society of Architectural Knowledge announced a competition for the design of layout for Elytsy Park garden settlement near Suyda station

and Pribytkovskaya platform on Varshavskaya (Warsaw) line of North-Western Railway, and the Society of Garden Cities announced a competition for the design of projects for three garden settlements along Rybinskaya Railway (13–22 versts (45,500–77,000 feet) away from Petrograd).

In 1916, architect S.S. Nekrasov developed a "garden city" project for Suyda station of North-Western Railway. A project of a "garden settlement" near Srednyaya platform was developed for Detsko-selskaya line (former Tsarskoye Selo line) of North-Western Railway. A project of "garden outskirts" was developed for Novo-Farforovsky post of Octyabrskaya Railway (Glazyrin 1928).

Projects of "garden cities" at railway stations were created on the basis of picturesque urban-planning models laid out in the schemes by Howard and Fitch: a centric layout with several centers connected with curved roads, green public areas, etc. These projects considered only the planning and spatial organization of the territory, without developing a social and organizational component. World War I and the revolution prevented the implementation of these projects.

In addition to the important but nevertheless "auxiliary" participation in the implementation of anti-urban concepts of the "ideal city", the railways themselves caused the creation of new settlements based on the "garden city" concept, representing an ideal new ground for their implementation — a clean sheet that makes it possible to uncompromisingly follow the idea that Plato called for.

The formation of suburban zones in Saint Petersburg Governorate, based on the "garden city" concept, was promoted with a number of factors of temporal and spatial character — they emerged at the right time in the right place: the period of their formation in the late 19th – early 20th centuries coincided with the period of the greatest popularity of this concept, and the traditions of regularity, typical for Saint Petersburg, and many examples of historical "garden cities" created the necessary spatial frame.

REFERENCES

Glazyrin, V.A. 1928. Garden towns and villages at transport routes. Leningrad Institute of Transport Engineers. Leningrad: s.n.

Howard, E. 1965. *Garden cities of to-morrow*. Cambridge: MIT Press.

Ikonnikov, A.V. 2001. *Architecture of the 20th century. Utopia and reality. Vol. 1*. Moscow: Progress-Traditsiya.

Khan-Magomedov, S. O. 2009. *Mikhail Okhitovich. Creators of the avant-garde*. Moscow: Fond "Russky Avangard".

Milyutin, N.A. 1930. *Socialist city. Problem of building socialist cities*. Moscow, Leningrad: State Publishing House.

Semenov, V. 1922. About garden cities. *Kommunalnoye Khozyaystvo* 8–9: 7–8.

Sementsov, S.V. 2019. Creating garden cities and garden town around Saint Petersburg — a new high-quality level in the spatial development of Saint Petersburg agglomeration. In *Gardens and parks. Encyclopedia of style. Proceedings of the 25th Tsarskoselskaya Scientific Conference*. Part 2: 199–208. Saint Petersburg: Serebryany Vek.

Von Troil, B. 1908. *Charter of the Society for the Assistance with the Improvement of Kellomyaki country place*. Saint Petersburg: Stroitel.

Yensh, A. 1910. Garden cities. *Zodchiy* 17: 307–310, 319–323, 327–330, 335–338, 345–348, 353–354. 1895. Revisiting the construction of residential buildings for railway employees. *Zheleznodorozhnoye Delo* 13–14: 124.

Revitalization of architectural landscapes of Alexander and Babolovo parks of "Tsarskoe Selo" Museum-Reserve in 2014 - 2019

M.N. Ryadova
Saint Petersburg State University of Architecture and Civil Engineering, Saint Petersburg, Russia

ABSTRACT: The article analyzes the architectural landscape of the Alexander and Babolovo parks of the palace and park ensemble of Tsarskoe Selo. The special feature of the studied architectural landscape is that the style-forming objects of the extensional composition consist of palaces and pavilions, and of the environ-ment-forming – objects of the landscape gardening art. Work on preserving the planning structure and volume-spatial composition can be carried out by different methods, depending on a number of criteria. The selection and definition of a method is not always obvious even if there is a large block of regulatory docu-mentation. Objective: to determine the main directions of preserving the architectural landscape of the palace and park ensemble of Tsarskoye Selo in a contemporary museum. In the course of the study, the following methods were used: archival and bibliographic, cartographic, analytical, field studies.

1 INTRODUCTION

The Alexander and Babolovo parks are monuments of landscape gardening art of the XVIII – early XX centuries, objects of cultural heritage of federal significance, part of the "Tsarskoe Selo" State Museum-Reserve. The theoretical issues of preserv-ing the cultural heritage are dealt with by Russian and foreign scientists: I. A. Bondarenko, Y. A. Vedenin, G. V. Esaulov, A. M. Razgon, B. M. Kirikov, A. S. Schenkov, S.V. Sementsov, Ch. Brandi, B. Szmygin, V.V. Lavrov.

The palace and park ensemble is now not only an architectural landscape, an object of cultural heri-tage, but also a separate type of expositional space "insitu", when historically valuable monuments and objects are located in their natural, historical or arch-aeological environment (Maystrovskaya 2016). Before conservation work, it is necessary to consider not only the material but also the intangible compo-nent of the architectural landscape (Vecco 2010).

2 MATHEMATICAL MODEL

The combination of water spaces (the Kuz`minka river, artificial canals and ponds) with open and half-open spaces of oak groves, deciduous and coniferous alleys is characteristic of the Alexander park. At the beginning of the XIX century, the borders of the park were expanded to the north, which significantly increased the territory.

The ensemble of the park consists of separate regular and landscape areas, combined by a planning structure and a three-dimensional composition. Landscape zoning within the boundaries of the "Tsarskoe Selo" Museum-Reserve was developed in 1986 by Natalia E. Tumanova as part of the restor-ation project of the Alexander park. On the presented scheme, zoning is proposed within the historical boundaries of the park with the addition of territories that are not part of the museum, but are an integral part of the ensemble. I. Regular garden; II. Chinese village; III. Palace area; IV. Area of the White Tower; V. Meadow area; VI. Chapelle area; VII. Area of Lamsky ponds; VIII. North-east area (Pen-sioners' stable); IX. Arsenal area; X. Forest; XI. Oak area; XII. Lamsky pasture (Zoological reserve); XIII. Area of the Kuz`minki river; XIV. Green-houses; XV. Imperial Farm area; XVI. Area of the Martial Chamber; XVII. Fedorovsky town; XVIII. Area of the Convoy Barracks of His Own Imperial Majesty.

The areas vary in size, configuration and purpose. The center of the landscape composition of the Alex-ander park is the woodland of the Arsenal area (former Menagerie) with a predominance of conifer-ous trees (Figure 1). From the south-east, the area of meadow landscapes adjoins to the forest of the Arsenal district, which is a natural transition from the dark forest area to the artificial Regular garden. The meadow area is a system of glades elongated in the shape, with open meadows in the north-east sep-arated by a narrow strip of plantings from the land-scapes of the Palace area, connected with the Alexander palace and three ponds. In the south-west, the Meadow area transits into the Chapelle landscape area, which is a system of open and half-open

Figure 1. The Alexander park. Landscape zoning scheme.

landscapes with monumental groups of oaks. The area of Lamsky ponds adjoins the forest massif from the west with open water spaces, meadows and glades directly passing into the Big meadow of the North-Eastern landscape area (Kishchuk 2016).

In 2010, the transformation of the Alexander park entered an active phase. Elements that were subject to mandatory preservation and restoration in the course of work are: combination of open spaces (meadows, fields, reservoirs, roads, playgrounds), half-open (plots with groups and single trees), enclosed spaces (massifs), with historical buildings and structures, visual connections between cultural heritage sites that make up the ensemble, location, spatial, structural, planning and architectural and artistic characteristics of buildings and structures, as well as landscape objects located in the park, volume-spatial composition, relief, planning system, road network.

Suggestions have been developed to increase the park's accessibility for residents and tourists, parking spaces have been organized, and routes —pedestrian, eco-trails, bicycle and ski, horse riding — have also been distributed (Ryadova 2020).

Regular garden. The Regular (New) garden, bounded by the Krestovy canal, is divided by four wide cross-located avenues into four square, which are diverse in their planning decision. It has preserved its historical regular layout and consists of four bosquets: "Parnassus", "Skarpir (Mushroom)", "Ozerki", and "Theater". The territory is in a difficult technical condition, but it is a unique monument and serves as the central core for the ensemble of Alexander Park.

Currently, bosquet Mushroom has a regular planning. The trellis was restored in 1970 on old roots. On the contrary is located the curtain "Parnassus" with an artificial hill. The multi-trunk linden trees that have grown from the old trellis have preserved

the layout of the composition of bosquets at the foot of the mountain. The project provides for the restoration of the historical space-planning solution of the curtain, with the conservation of old-growth trees in bosquets (Rjadova 2019).

The Chinese theater, destroyed during the Great Patriotic War, is a ruin. The roofing structures, the designs of the stage space, the stalls and the side suites are completely lost.

Because of the special historical, artistic and cultural value of the Chinese theater, it is proposed to restore the historical spacing in its pre-war form: the central volume with side 2-storey outbuildings.

To adapt the building, an air conditioning and ventilation system is needed, but taking into account the preservation of the historical volume of space, the placing of the refrigeration machines in the building was prohibited. It was decided to re-equip the nonhistoric building, located next to the theater, to a transformer substation and installation of the refrigeration machines.

Palace area. Since 2011, work has been underway on the restoration and adaptation of the Alexander palace, the architecture of the facades is restored to its original period, without the complements of the early XX century. The restoration of the lost historical layout, volume and spatial composition, including its own garden and a system of outdoor lighting and hydraulic structures is planned after the completion of the restoration of the Alexander palace.

Area of the White tower. The reconstruction of the "White Tower" complex, which began in the 1990s, was completed in 2012. The planning of the territory within the boundaries of the bastion and the fence is made on the basis of historical plans.

Inside the bastion children's equipment and activities for sports games of emperor's children were located on specially planned sites. Not far from the bastion, an earthen fort was preserved (restored in the 1990s), which was used to teach the fortification of the children of Nicholas I.

The planning of the territory inside the bastion was recreated in 2012.

Chapelle area. Chapelle pavilion was built under the project of Adam Menelaws in 1825-1828. During the World War II, the pavilion received numerous damage. In 2016-2018, specialists worked not only on the restoration of the structure itself, but also on the adjacent territory on a large area. The pavilion is the compositional center of a large landscape area. It is made in the form of a romantic ruin and, apart from its surroundings from the corresponding type of environment, cannot fulfill its meaning. Therefore, restoration in isolation from the landscape for such objects is not acceptable. In addition to melioration work, restoration of the road-path network in this area of the park, the restoration of the stone bridge over the Vittolovsky canal and the reconstruction of the lilac alley were completed. Simultaneously with the construction of the Chapelle pavilion in the XIX century, a lilac alley was planted. It stretches from the Krestovy canal to the Pink Guardhouse, it is a link between the Alexander and Babolovo parks.

Restoration of the melioration system, including the demolition of self-sowing trees and shrubs, the installation of a closed drainage system, including the laying of ring drainage for organizing drainage around the Chapelle pavilion, restoration of historical drainage collectors, deepening and strengthening of ditches to historical marks, cobblestone lining of the bottom according to a historical pattern.

North-east area (Pensioners' stable). Pensioners' stables are a small two-story stone pavilion, which housed the horses of His Own Majesty Imperial saddle. The two-storey building with a round stair-tower cum belvedere, two three-sided bay windows and a single-storey extension was made of brick in the English Gothic style. Wooden summer stalls, a well, and latrine have not preserved to our time. Part of the clover meadow to the west of the stable building was turned into pasture for the horses. Behind the stable, opposite the tower, there is the cemetery of the royal horses.

This complex is unique. After its restoration, there will be a museum dedicated to horses and a children's environmental center in the building.

Large-scale work is being carried out in the cemetery. The melioration system of the entire district was restored in order to drain it. Planned work on the restoration of historical tombstones and the reconstruction of the lost.

Arsenal area. The main building of the district is the Arsenal pavilion, located in the geometric center of the park at the intersection of eight vistas. The building was destroyed by fire and for a long time had no roof. The restoration and reconstruction of the pavilion entailed the conservation work of the park. The road-path network was repaired, and lost walking routes were restored. But the most important event for the architectural landscape was the discovery of the main axis between the Alexander and Catherine parks, and the formation of a specific perspective on the Arsenal from the Catherine palace (Figure 2). This perspective existed in the XVIII century, when this area of the park was a hunting ground – Menagerie. In the XIX century, during the forming a landscape park, it was hidden, planted, and instead of a direct clearing, a bend road was constructed - that is clearly visible on the plan of the 1850, made by Grekhnev surveyor. In the middle of the XX century, the prospect was restored, but due to the thickened crowns and branches, the visual corridor was not visible. Work in 2016 was carried out taking into account the biological and physiological characteristics of the tree species of this site. It was important to reveal the perspective and at the same time preserve the natural decorativeness of the trees. In this case, the hierarchy of architectural interconnections prevailed over the decision of landscape

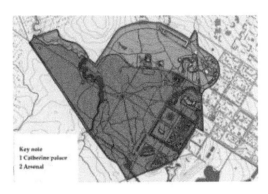

Figure 2. View perspective between the Catherine Palace and the Arsenal Pavilion.

composition. From the opposite side of the Catherine Palace, this prospect continues into the Catherine Park to the Hermitage pavilion. It is the axis on which the central areas of the two parks seem to be strung.

Oak area. The restoration of the Spruce alley, which is one of the main axes of the Alexander park, was completed. It leads to the Golden gate of the Catherine palace. The historian Ilya Yakovkin wrote: "This road was laid for passage directly through the former menagerie by the command of Elizabeth I and arranged in its present form by the command of Alexander I, as well as the most lattice gates, for the passage of persons of the Imperial court" (Yakovkin 2008). Plots of land on both sides of the alley were cleared of weed species. As a result, picturesque groups of old-age trees, including oaks and larch trees, were opened - they underwent a course of treatment.

After self-sowing vegetation, shrubs and trees are cut down, many visitors to the park ask why the landscape that has become familiar to them has changed so much. The loss of transparency of landscapes is perceived by them as something right. Tourists do not realize that landscape parks are not the creation of nature, that every tree, bend of the road was carefully thought out by an architect, gardener and an artist working with them during the construction of architectural landscapes. Transparency creates open spaces and in the process of the existence of large park territories it is very difficult to maintain, but it must be done, otherwise the spatial-spatial composition will inevitably begin to degrade visually (Eremeeva & Lavrov 2016).

The spruces in the alley were cured — the root system, which literally hung in the air in some trees was strengthened in the ground. A packed gravel road of 1 kilometer length was restored. In the future, it is planned to place interactive zones for families along the Spruce alley.

In order to develop infrastructure beyond the borders of the territory at the entrance to the Spruce alley from the Volkhonskoye highway, an open parking lot was arranged for visitors.

Imperial Farm area. The work of restoration and adaptation of the Farm complex began in 2012. It consists of 9 buildings, some of which were completely destroyed. The historical layout of the complex and the historical decisions of the buildings, internal layouts and architectural solutions of the facades of historical buildings will be restored. The drainage system was repaired and restored, and a sewage system was organized. Reconstruction of the external and internal engineering networks of power supply, electric lighting, water supply and sewerage complex. The complex of the Imperial Farm will be available for disabled visitors. The estimated completion date is 2022.

Area of the Martial Chamber. The Martial Chamber is a complex of structures: the building of the Chamber with a corner octagonal tower with a tented top is surrounded by a fence and galleries with three towers, forming an irregular pentagon in plan. Building received this form in connection with the outlines of the part allocated for the construction of the park. After Second World war, the badly damaged building was used as housing, later it was adapted for the restoration workshops of "Lenoblrestavratsiya".

In 2009-2010, a project was developed to adapt the Military Chamber to the museum function. Its implementation continued until 2014. In the course of the work, the historical spatial design of the building with tops acting as park dominants was restored. Late dissonant buildings in the territory of the complex were demolished. The only Russian museum of the First World War was inaugurated in the building. In the surrounding area were constructed a bus parking and parking for staff cars.

Especially in order to ensure the accessibility of the new museum object under the support of the Government of Saint Petersburg, the Farm road was reconstructed. From a broken dirt road, it turned into a citywide road. The road runs right through the park. Red asphalt was chosen as a coating, which imitates a stuffed coating, but at the same time has high wear resistance.

Over the past three years, the Alexander park for safety reasons has been surrounded by a transparent metal fence on the border of the land plot which belongs to the museum, made by the model of the fence designed by Silvio Danini.

Babolovo park is not filled with architectural structures to such an extent as Catherine and Alexander parks. The main attraction is in its volumetric and spatial composition and old trees.

The landscape zoning scheme includes: I. English garden; II. Peat field; III. Bolshaya Polyana; IV. Area of Silver willows; V. Kuz`minki river area; VI. Area of the Baur canal; VII. Area of the Krainaya Road; VIII. Severnaya meadow; IX. Hay massif; X. Forest; XI. House of charity of maimed warriors; XII. Area of the School Garden.

During the Second World war, the park suffered great damage. Many old trees were cut down, buildings and hydraulic structures were damaged or completely destroyed. Over the past decades, the park has largely lost its compositional structure. Indigenous species of wood plants displace ornamental species, meadows overgrow with weeds and are flooded, the appearance of alternating open and closed spaces has changed. The landscape park with carefully thoughtout planning, groves and alleys has turned into an almost untidy forest.

In 2014, Babolovo park became part of the museum. The main task of the museum is to gradually recreate the lost composition of the landscape park (Figure 3).

Volume and spatial structure: formed by natural elements (II, III, IV, V, VII, VIII, IX, X), architectural and hydraulic elements (I, VI, XI, XII). Planning structure: tracing of historical roads, alleys and paths of the New garden and the park. Landscape elements: elevation of elevations at the Babolovo palace, Kuz`minka river, Babolovo pond, Vittolovsky aqueduct; historical green spaces of the park, park compositions.

A project was developed to maintain old trees and an approach was developed to preserve them. A huge complex of work on sanitary felling in the English garden was done. The garden was laid out by the personal direction of Catherine II at the Babolovo palace and is a picturesque meadow framed by groups of old-age trees. Thanks to sanitary felling, the park has acquired yet another miraculous attraction – picturesque meadows, painted with bright colors of meadow and forest herbs.

The riverbeds were cleared, old dry and tumbled down trees were removed, measures were taken against the propagation of pests.

In 2015, upon the request of the museum, a project was developed to preserve old trees on the Willow

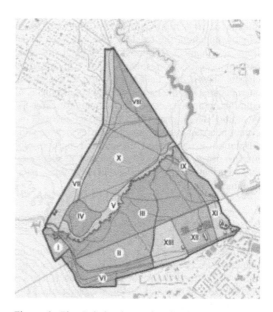

Figure 3. The Babolovsky park. The landscape zoning scheme.

alley. Silver willows were planted at the beginning of the XX century. Sixty years later, their crowns were severely cut off, but due to the lack of systematic care, the trees survived to our days in a deplorable state. Around the year, specialists formed crowns on the Willow alley, and, in fact gave the trees a second life. Every year, tree rehabilitation measures are carried out in accordance with the developed program. The alley is gradually returning to its former appearance, becoming picturesquely silver.

In 2016, conservation of the Babolovo palace was completed. Project work is underway to restore the space-planning solution and adapt the building to the museum pavilion, equip it with engineering systems (heating, electricity, water, wastewater, low-current networks). The Krainaya alley, a landscape road leading from the Alexander park to the Babolovo palace, forming the eastern border of the park, was repaired.

The urgent issue is the need for watering of the park system. The main artery supplying Babolovo, Catherine and Pavlovsky parks is the Taitsky aqueduct, which has been completely destroyed. This unique hydraulic structure, 16 kilometers long, originating from the Taitsky aqueduct, is interrupted along the highway by new roads, has collapsed underground parts (mine gallery), and the structures are in disrepair. Since the 1980s, options have been offered to restore the monument, but so far no solution has been found.

In 2005, the territory of the former school garden (XII), and the buildings and structures located on it, became the property of "Steelmar Scandinavia". The concept of adapting the territory to modern use implies its development with 70 residential cottages. Such a decision met with great resistance from the local community, town protection movements, the monument protection authority; in the course of lengthy litigation, the territory has now been saved from development (Filippova 2019). But the issue has not been finally resolved so far. This situation clearly shows that under the current legislation, attempts are possible barbaric treatment of objects of cultural heritage. According to many researchers, including Berdiugina Y.M., in Russia there is no clear understanding of the boundaries between the adaptation of an object of cultural heritage for modern use as a way to preserve it and an adaptation that contributes to the loss of value of an object (Berdiugina 2018).

3 CONCLUSION

The need for developing new heritage protection scenarios has been identified. This is precisely what many theoretical studies of conservation processes do (Dietze-Schirdewahn 2010, Sesana et al. 2020, Yankovskaya et al. 2020).

From practical point of view, the traditional heritage protection paradigm is out of date since it neither describes nor standardizes the contemporary reality. In spite of a number of theoretical and practical actions, however, no new paradigm has been developed and adopted yet. For this reason, the traditional, twentieth-century historic preservation theory is still widely used as a source of reference (Szmygin 2016).

The results of the preservation and development of architectural landscapes of Tsarskoye Selo are presented. Its will not only improve the quality of preservation of cultural heritage objects and the natural landscape environment, but will also contribute to the development of the museum territory, ensuring the sustainable use of the full range of cultural heritage objects, and will allow preserving the ensemble the continuity of the formation of the planning structure and volume-spatial composition.

REFERENCES

Berdiugina, Y.M. 2018. Criteria for the adaptation of cultural heritage sites for modern use. *Academicheskij Vestnik UralNIIProekt RAASN* 1 (36): 47–51.

Dietze-Schirdewahn, A. 2010. The dynamic relationship between culture and nature: the management of historic parks and gardens. In M. Mälkki & K. Schmidt-Thomé (eds), *Integrating aims — built heritage in social and economic development*: 83–100. Espoo: Centre for Urban and Regional Studies Publications.

Eremeeva, A.F. & Lavrov L.P. 2016. Lost transparency of Saint-Petersburg historic open spaces. *Bulletin of Civil Engineers* 6 (59): 35–48.

Filippova, O.A. 2019. School garden institution — the cradle of Tsarskoye Selo parks. In *Gardens and parks. Encyclopedia of style*: 304–314. Saint Petersburg: Serebryany Vek.

Kishchuk, A.A. 2016. Act on the results of the historical and cultural examination of project documentation for the conservation of the cultural property of federal significance "Alexander Park" developed by OOO PG RIEDER KGIOP Archive.

Maystrovskaya, M.T. 2016. *Museum as an object of culture. The art of exposition ensemble*. Moscow: Progress-Traditsiya.

Rjadova, M.N. 2019. Application of GIS technologies for conservation of landscape architecture objects. *Vestnik Tomskogo gosudarstvennogo arkhitekturno-stroitel'nogo universiteta. JOURNAL of Construction and Architecture* 21 (6):70–78.

Ryadova, M.N. 2020. The stages of formation (1710-1917) and transformation (1918-present) of the architectural landscape of the Tsarskoye Selo palaces and parks ensemble. *Construction Materials and Products* 3 (1): 95–103.

Sesana, E. et al. 2020. An integrated approach for assessing the vulnerability of World Heritage Sites to climate change impacts. *Journal of Cultural Heritage* 41: 211–224.

Szmygin, B. 2016. Protection of historic monuments and sites — achievements, problems, perspectives. In B. Szmygin (ed.), *Heritage in transformation: cultural heritage protection in XXI century — problems, challenges, predictions. Heritage for Future* 1 (3): 191–199.

Vecco, M. 2010. A definition of cultural heritage: From the tangible to the intangible. *Journal of Cultural Heritage* 11 (3): 321–324.

Yakovkin, I.F. 2008. *Description of Tsarskoye Selo or a guide for those visiting it. With an opening letter and commentaries by G.V. Semenova.* Saint Petersburg: Kolo.

Yankovskaya, Y. et al. 2020. Structure, adaptability and security of an architectural object. *E3S Web of Conferences* 164: 05007.

Reconstruction and Restoration of Architectural Heritage – Sementsov et al (eds)
© 2020 Taylor & Francis Group, London, ISBN 978-0-367-65357-6

Real historical skyline of metropolitan Saint Petersburg by 1917

S.V. Sementsov

Saint Petersburg State University of Architecture and Civil Engineering, Saint Petersburg, Russia

ABSTRACT: The article presents some results of long-term comprehensive historical, archive, cartographic, bibliographic, and morphological studies on the specifics of Saint Petersburg's historical urban-planning skyline before 1917. Systemic specifics of the city historical skyline are identified combining six typological classes of different-scale facilities (that have taken up certain spatial and role tasks in the unified skyline organization of city's spaces during development) and forming the multi-layer, hierarchically arranged skyline of the city. Interrelated urban-planning compositional roles are studied in terms of the historical Saint Petersburg skyline formation in the structures of the semantic and symbolical environment (temples of various confessions), residential and public buildings and structures, production facilities (industrial facilities and complexes), structures of urban-planning and monumental design (memorials and monuments). Some common factors of man-made formation of skyline perception surfaces are taken into account: from the main city's skyline to the contact perception at the level of the man's height.

1 INTRODUCTION

Skyline is one of the most important and most keenly felt and visually perceived characteristics of any historical and modern city that expresses its uniqueness, authenticity and novelty to the maximum extent. Therefore, for dozens of years, the study of city skylines and specifics of their perception has been one of the most relevant challenges in the history of urban planning, theory and psychology of spatial perception. The issues of studying the skyline of urban spaces manifest as problems related to the urban planning of man-made landscapes (problems of city planners) (Belyayeva 1977, Bostanci & Oral 2017, Lynch 1982, Nazarova 2019) intertwined with problems related to spatial perception (problems of physiologists and psychologists) with account for the specifics of spatial perception — simultaneous and successive syntheses (Gillam 2000, Rubinstein 2010, Sechenov 2014).

For Saint Petersburg formed in 1703–1917 across vast lowland near-Neva areas as the Russian capital with specific urban-planning and compositional ambitions (with collective coordinated efforts on the part of city planners and architects, engineers, city officials, and state authority in general, including all Russian emperors), to establish a unified, well-considered, clearly modeled "skyline of Saint Petersburg", as D.S. Likhachev called it, has always been one of the major tasks. It has also been one of the tasks of Leningrad and Saint Petersburg researchers (Baranov 1980, Krylova 2014, Stepanov 2016, Zakharov 1984). Any elements forming the skyline solution were strictly controlled within the unified system of urban-planning and architectural legislation, and almost every visual dominant (first of all,

related to public and residential facilities) was unambiguously and clearly enunciated within the city's space. All such dominants created a unified, mathematically precise spatial system of vertical and visual, semantic and symbolical dominants. This can be seen in some specimens of survived city-planning layout documents (Plan of the city of Saint Petersburg 1821). This unique skyline pattern of the historical Saint Petersburg (as it was called by D.S. Likhachev) was included in important documents as well as in the Petersburg Strategy of Cultural Heritage Preservation (Government of Saint Petersburg 2005, Likhachev 1989).

A particular spatial understanding of Saint Petersburg formed during the 20th century is captured in the social consciousness and documents. The city is perceived as lowland, with the historical background development "below the unified cornice" (not exceeding the height of the Winter Palace cornice — 11 fathoms from the daylight surface level = 23.43 m), with a unified uniform horizontal skyline that stretches along the near-Neva lowland and the water area of the Gulf of Finland over dozens of square kilometers, with individual sparsely placed high dominants (temples) looming over the endless areas.

This rather approximate and apparently inaccurate ("in bold strokes") opinion must be specified through detailed researches. The rule of building below the unified cornice started to form as early as in the 1710s during the reign of Peter the Great, at first for the first-floor level and the first floor over the basement, and then for the second floor, and after the 1820s — for the third and fourth floors. This is when a life-changing decree was signed under Alexander I and Nicholas I setting out that all buildings in Saint

Petersburg must not be higher than the Winter Palace cornice. Compliance with this decree with simultaneous leveling of all urban buildings continued until the 1960s or even early 1970s, being privately preserved during the eras of avant-garde, Stalin's classicism, and even at the first stages of the five-story industrial house construction. These requirements for Leningrad (the former name of Saint Petersburg) were tacitly preserved only in the historical development zones while the periphery was dominated by the requirements of the construction industry transitioned to large-scale construction of 9–18-story buildings. The overall picture is much more complicated if we consider the system of vertical dominants of the city in general. Unfortunately, it is not studied (especially taking into account that various types of vertical dominants were systematically destroyed in the 20th century). Therefore, today we all see an absolutely different skyline pattern of the historical city as compared to the period until 1917. We have accustomed to and unconsciously absorbed (individually and socially) the outcomes of the terrible destruction of the Saint Petersburg historical skyline.

2 METHODS

The research is based on the author's results of undertaken historical, archive, cartographic, morphological studies on Saint Petersburg facilities with an increased number of stories (or height dimensions) that played a specific role in forming the city's skyline. To address this task, several questions must be answered: A) What elements of the urban planning and architectural fabric of Saint Petersburg took part in the formation of the city's historical skyline and to what extent? B). How was the spatial system of the Saint Petersburg skyline crystallized? C). To what extent was it destroyed after 1917 and what is left now?

3 RESULTS AND DISCUSSION

A). What elements of the urban planning and architectural fabric of Saint Petersburg took part in the formation of the city's historical skyline and to what extent?

The skyline of any building, group of buildings, block, group of blocks (man-made zone or area), the city in general, is formed by all elements providing vertical completion of these various-scale facilities. This also includes the skyline of the historical Saint Petersburg. Therefore, paying attention to some group of elements while abandoning others would be incorrect. It should be remembered though that each group of the same type has its own specifics (morphologic, semantic, symbolical, scale, dimensional, stylistic, etc.), so their role and relevance in the skyline formation differ.

It appeared that the perception and evaluation of the city's skyline (or its fragments) have significantly different spatial scales of perception and specific levels of perception horizons. At least two such important planes of skylines can be identified: city's skyline (formed by large-scale background development and vertical accents dominating over it); contact skyline at the level of the human height (formed by skyline specifics of structures and street furniture close to people).

In general (very approximately), elements taking part in the formation of the historical Saint Petersburg skyline could be divided into the following groups.

– Those carrying visual and symbolic-semantic information for the city in general and its spatial zones: temples of any confessions.
– Those carrying visual and applied information for the city in general and its spatial zones: stacks of industrial facilities, chimneys of buildings and structures.
– Those carrying visual and symbolic-semantic information for spaces close to people: first of all, chapels, monuments, street furniture.

B). How was the spatial system of the Saint Petersburg skyline crystallized?

The skyline field of the historical Saint Petersburg started to form as early as in the 1710s as a hierarchical system including six classes of structures inter-related in size, mostly temples. Each class of high structures differed in specific dimensions (special structures, radii of visibility) and structures of a specific class compositionally drew specific spaces to themselves. The papers by O.N. Zakharov and S.V. Krylova (Kurakina) specify heights and skyline specifics of Saint Petersburg temples (Krylova 2014, Zakharov 1984).

Class 1. Dominants of governorate significance. Buildings and structures 80–100 m high and more, dominating over the surrounding buildings by 6–10 times at the time of construction. There are few such main dominating temples: the bell tower spire of the Peter and Paul Cathedral in the Peter and Paul Fortress (112 m high in 1717–1857; 122.5 m high after bell tower reconstruction in 1857); the Resurrection Cathedral (Figure 1) of the Smolny Convent (with the central dome 98.0 m high) and its incomplete bell tower (with the design height of 144 m); the St. Isaac's Cathedral (with the cross above the dome 101.5 m high). It was under Peter the Great when the major temple of Russia, Saint Andrew's Cathedral, was planned to be built at the Spit of Vasilyevsky Island (130 m high, the project was not completed). Among non-temple structures, the construction of the Kronstadt Lighthouse was started in the 1720s (with the design height of 250 m; not completed). All these dominants provided (or must have provided) a vertical arrangement and spatial perception at distances up to 30–40–50 km and more (Sementsov 2020).

Figure 1. Panorama of the Smolny convent with the resurrection Cathedral (picture of the 1900s).

Class 2. Dominants of regional significance. Structures 70–90 m high, dominating over the background buildings by 4–5 times at the time of construction. Just as in class 1, they included both temples and industrial buildings: the standalone bell tower of the St. Nicholas Naval Cathedral in Saint Petersburg (74.5 m high) while the height of the cathedral itself is much lower (48 m high); the tower of the Admiralty (Figure 2) (with the spire of 74.0 m high); the Kazan Cathedral (with the dome top at the level of 72.0 m); the standalone bell tower of the Saint Vladimir Cathedral in the Vladimir Square (72.0 m high); the Church of the Savior on Spilled Blood (81.0 m high); the Trinity Cathedral (80.0 m high); the Naval Cathedral of St. Nicholas in Kronshtadt (70.6 m high), the Cathedral of Saints Peter and Paul in Peterhof (70.4 m high). These vertical dominants provided a vertical arrangement of areas and spatial perception at distances up to 10–20 km (Sementsov 2020).

Class 3. Dominants of city significance. Dominants of this class (50–60 m high) can be divided into symbolic-semantic (temples) and practical ones (industrial stacks at factories). They were constructed both in the city and in vast suburban areas. These include but not

limited to: the Holy Trinity Cathedral of the Alexander Nevsky Lavra (62.0 m high) (Figure 3); Simeon and Anna Church (with the bell tower spire 51.0 m high) (Figure 4); Saint Andrew's Cathedral on Vasilyevsky Island, and many others. For example, for dominant temples, this was one of the most popular classes in Saint Petersburg. During construction, they dominated over the background buildings of 1–2 stories high by 3–4 times. But after the number of stories and the height of buildings increased (from 1–2–3 to 3–4), their domination over the new level reduced to 2–3-fold in many places. Such height dimensions of

Figure 3. Cathedral of the presentation of the Holy Virgin of the Semenovsky Regiment. Lithograph based on the drawing by I.I. Charlemagne (1830s–1840s).

Figure 2. Admiralty (picture of the 1880s).

Figure 4. Simeone and Anna Church (1731–1734). Photo by the author.

dominant buildings and structures provided a visual arrangement of areas and spaces with a perception radius of 6–10 km (Sementsov 2020).

Class 4. Dominants of district significance. Low vertical dominants, including large-scale construction temples with the height of domes and spires, high towers of residential and public buildings, top outlines of industrial stacks, all 40–50 m high (Figure 5). They include palace churches of suburban imperial palaces: Peter and Paul Church in Peterhof (Figure 6), the Resurrection Church in the Catherine Palace of Tsarskoe Selo, the Grand Ducal Shrine in the Peter and Paul Fortress. The designed and constructed height of the buildings of this type during their construction exceeded the average height of the background buildings by 2.5–3 times providing a visual arrangement of surrounding spaces with a radius of 4–7 km (Sementsov 2020).

Class 5. Dominants of local significance. This is also one of the most popular types of historical development with high buildings including structures of various functions: temples, residential, public and industrial buildings, and their parts having an increased architectural and compositional height, up to 25–40 m high (Figure 7). At the initial stages of city development when the background buildings had 1–3 stories on average (up to 10–15 m with roofing), such "tall" buildings were 25–30 m high. But in the 1820–1910s, when the background buildings were constructed up to 23.4 m high (to the top line of cornices), such dominants (temples, towers on buildings, stacks of industrial facilities) were constructed a little bit higher — up to 40 m high. Thus, such a height 1.5–2-fold exceeded the height of the background development. The location of temples and towers complied with the urban-planning requirements: they were constructed in squares, on the corners of buildings, in the areas of emphasis. These dominants were well read within the visual field having perception radii up to 2–5 km (Sementsov 2020).

Class 6. Local dominants. Such dominants can be divided into two groups: 1) for the upper (skyline) belt of the city in general, large urban-planning zones and blocks; 2) for the "contact",

Figure 6. Peter and Paul Church of the Peterhof Palace (the 1750s). Photo by the author.

Figure 7. Moskovsky (Nikolayevsky) railway station, Saint Petersburg (1842–1851). Photo by the author.

basement zone, located close to people. For the upper (urban-planning and skyline) belt of buildings, dominants include small towers, spires, domes, special lucarnes, etc. (with the total height of 20–22m above the ground level by the end of the 18th century, and 30–35 m by the end of the 19th century) and with the height above the line of cornices in historical buildings of 2–5–8m, so that they only 1.1–1.2-fold exceeded the height of surrounding buildings. This group includes various towers, domes, pediments, attics,

Figure 5. Armory (photo of the 1900s).

finials of various shapes, and also hundreds of crosses, bell towers, bell cots of residential chapels arranged above the roofing level. Most residential chapels in Saint Petersburg were located on upper stories and their crosses and bell cots protruded above the roofing level. Such architectural and sacral temple tops of buildings and structures were vastly spread in the city and suburban development providing a visual arrangement of spaces with a practical minimal radius of perception of 0.5–2.0 km.

There also was the second layer of skyline pattern of this class for all people in the city space. These are widely spread buildings of the near-ground, contact, basement zone (at the level of pedestrians or people using various types of ground transport). First of all, this zone includes sacral and symbolical buildings such as chapels, chantries, silhouettes of gratings, gates, etc., and monumental structures: numerous sculptures and monuments (5–10 m high each) (Figure 8). At the level of the man's height and in the areas of human contact perception, they formed a very important emphasizing urban-planning and symbolical configuration of the environment. The characteristics of the skyline pattern also played a special role here. It is not by chance that all chapels and monuments were constructed so that they could be seen against the sky background (Sementsov 2020).

Figure 8. Chapel of the Saint Andrew's Cathedral (1878–1879). Photo by the author.

C). To what extent was the Saint Petersburg skyline as a spatial system destroyed after 1917 and what is left now?

As we saw in the previous paper, most vertical dominants (temples of various height) were destroyed in the 1920–1960s. Most finials of residential and public buildings suffered significantly during the last 100 years of the city's life during wars, renovations, etc. Many house chimneys were demolished during the reconstruction of industrial facilities and the elimination of production facilities. Thus, a very dense field of hierarchically interacting high elements existing before 1917 is now represented by several randomly survived buildings, structures and fragments.

4 DISCUSSION

This article is the first to describe the formation of the urban-planning skyline of the historical Saint Petersburg as a multi-layer hierarchically arranged system combining structures of various nature and elements of visual, sacral, symbolic, semantic and monumental nature. This includes temple construction, buildings and structures of residential, public and industrial purposes. This system started to form since the first years of Saint Petersburg's existence.

5 CONCLUSIONS

As a result, the multi-level system for the formation of the unified urban-planning skyline of Saint Petersburg has been identified, which combines buildings, structures and fragments of six classes of visual, sacral, symbolic and semantic dominants forming the unified spatial and visual field inside Saint Petersburg (from several largest temples of governorate significance to mass local temples). The parallel existence of the urban-planning skyline pattern (city's skyline) with the contact silhouette pattern (at the level of man's height) has been highlighted. The extent of destruction and losses have also been considered in the system of vertical dominants of Saint Petersburg in general.

6 RECOMMENDATIONS

The article can be recommended to historians of urban planning and architecture, modern city planners and urbanists modeling urban spaces and accounting the specifics of their perception, specialists in protection, preservation and restoration of historical and cultural heritage, urbanists and designers engaged in the matters of the development of modern cities and agglomerations as well as the preservation of historical cities and landscapes.

ACKNOWLEDGMENTS

The author would like to express his gratitude to his compatriots who introduced him to the papers on city space perception — M.P. Berezina, Kh. Steinbakh. The author would also like to thank his colleagues who were patient and reverent to long-term time-consuming researches in this area, including S.V. Krylova (Kurakina), N.A. Akulova, Ye.R. Voznyak, K.A. Romanchuk.

REFERENCES

Baranov, N.N. 1980. *City skyline*. Leningrad: Stroyizdat, Leningrad Department.

Belyayeva, Ye.V. 1977. *Urban architectural and spatial environment as an object of visual perception*. Moscow: Stroyizdat.

Bostanci, S.H. & Oral, M. 2017. Experimental approach on the cognitive perception of historical urban skyline. ICONARP. *International Journal of Architecture and Planning* 5 (Special Issue): 45–59.

Gillam, B. 2000. Perceptual constancies. In A.E. Kazdin (ed.), *Encyclopedia of psychology*. Vol. 6: 89–93. Washington, DC: American Psychological Association.

Government of Saint Petersburg 2005. Decree of the Government of Saint Petersburg No. 1681 "Concerning Saint Petersburg strategy on preservation of cultural heritage" dated November 01, 2005.

Krylova, S.V. 2014. Architectural, spatial, and composite features of the temples location in Saint-Petersburg and the surrounding counties in the 18th – 1st half of 19th centuries. *Vestnik MGSU* 3: 27–35.

Likhachev, D.S. 1989. Skyline of the city on the Neva. *Our Heritage* 1: 8–13.

Lynch, K. 1982. *The image of the city*. Translated by V. L. Glazychev. Moscow: Stroyizdat.

Nazarova, A.Yu. 2019. Theoretical aspects of research and preservation of historic city skyline. *Vestnik Tomskogo gosudarstvennogo arkhitekturno-stroitel'nogo universiteta. Journal of Construction and Architecture* 21(3): 77–85.

Plan of the city of Saint Petersburg compiled in the General Headquarters of His Imperial Majesty by Major General I. Vitztum and engraved in the military topographical depot. 1821. Saint Petersburg.

Rubinstein, S.L. 2010. *Basics of general psychology*. Saint Petersburg: Piter.

Sechenov, I.M. 2014. *Elements of thought*. Saint Petersburg: Lenizdat.

Sementsov, S. 2020. Spatial development of Orthodox temple construction in Petrograd and its environs by 1917. *E3S Web of Conferences* 164: 04023.

Stepanov, A.V. 2016. *Phenomenology of Leningrad architecture*. Saint Petersburg: Arka.

Zakharov, O.N. 1984. *Architectural panoramas of Neva banks*. Leningrad: Stroyizdat, Leningrad Department.

Reconstruction and Restoration of Architectural Heritage – Sementsov et al (eds)
© 2020 Taylor & Francis Group, London, ISBN 978-0-367-65357-6

Saint Petersburg style of 21st century

M.A. Shapchenko
Saint Petersburg State University of Architecture and Civil Engineering, Saint Petersburg, Russia

ABSTRACT: The article describes the basic methods of urban planning, architectural and design research to determine the genetic code of Saint Petersburg, sets out the territorial boundaries of the object of study - the city of Saint Petersburg; draws general conclusions of research and design work. The paper proposes a methodology for the study of a historic city with addition of modern buildings. Carriers of the style of Saint Petersburg - the parameters of the environment that underlie the identity of the urban environment – have been identified.

1 INTRODUCTION

The urban planning fabric of Saint Petersburg has been formed for over 300 years. Over the past 70 years there has been a change in urban planning techniques (Kamensky & Naumov 1966). Due to the widespread use of typical, mass housing, impersonalized micro districts appeared. New construction does not preserve the genetic code of the city, which has been created for centuries (Hayman 2012). The genetic code of the city is the prevailing patterns, techniques, features of urban development decisions of buildings, ensembles and squares that form the identity of the place (Sementsov 2003).

It is impossible to create cities by following a short list of rules that take into account functional zoning, utility, symmetry axes and regular structures, because the image of the city is a much more complex and deep matter. City layers are connected by time and appear before us with all their positive and negative sides. It is necessary to accept the city as it is (Lynch 1982).

Saint Petersburg style is a combination of signs of the urban environment of Saint Petersburg, which forms its urban planning and architectural identity.

The aim of the work is to identify, systematize and use Saint Petersburg style carriers for different types of urban development.

2 METHODS

To determine the signs of the genetic code of Saint Petersburg, the stages of urban development of the city with the search for patterns of development at each stage of development were analyzed (Geddes 1915). In this case, the pattern refers to the typological unit of the urban environment, united by common features, interpreted in a single time period in different territories (Alexander et al. 1997). Studying the typology of

the development of Saint Petersburg allows us to analyze not only the formation of a single point, but also to compare them with each other, to trace the evolution of the urban environment and to understand whether the development of the 21st century corresponds to the development principles of historical Saint Petersburg. By grouping these patterns, urban fabric can be divided into the following types: the city center of the 18th century, the city center of the 19th century, the gray belt, the city center of the 20th century, typical mass micro districts, modern micro districts.

Also, in order to search for the uniqueness of the Saint Petersburg environment, a comparison was made of the development in Saint Petersburg with the largest cities in Europe: London (Ikonnikov 1958), Paris and Rome (Internet Archive 2020, Topuridze 1970). Conclusions about the unique structure of the city as an ideal city, without the addition of medieval structures and about the increased dimension of the city blocks and streets (Likhachev 1989) were drawn. In the framework of this comparison, the natural framework, administrative and territorial division, historical development, urban planning framework were analyzed; central and peripheral patterns were compared (Bunin & Savarenskaya 1979).

In the framework of a more detailed analysis of different types of Saint Petersburg environment, the boundary of the study was determined, which is a beam 2 km wide, from the historical center of Saint Petersburg to the south to the periphery. This beam includes all types of urban development.

For the analysis of the beam, three main levels were studied and proposed as a research method: urban planning, architectural and design code analysis.

As part of the urban development analysis, the urban framework, urban fabric, ensembles and urban interiors were studied (Gutnov 1984). Architectural analysis consisted of an analysis of the structure of a single quarter in each type of development, the systematization and typology of buildings of the background development, an analysis of the architecture of building facades (Hillier et al. 2010). Analysis of the design code included color schemes, materials, urban furniture and elements of visual culture.

Systematizing the revealed signs of each type of environment, the matrix of all existing parameters of the urban environment of Saint Petersburg was compiled. This matrix displays all the elements and characteristics of all types of urban development, which has become the basis for identifying the carriers of the Petersburg style. The practical application of these parameters is the final link in the work on the Petersburg style.

3 RESULTS

One of the tasks of the work is to design new quarters while preserving the features of the genetic code of Saint Petersburg. In order to introduce more specific recommendations for the new development, plots with a high degree of incompleteness of the environment were selected. These territories included: the gray belt, typical mass housing of the middle of the 20th century and modern quarters.

The main parameters of the new development of the gray belt are:

- the creation of additional centers of attraction;
- the transition to the quarterly system (the city skeleton was reduced and additional routes were laid);
- the creation of ensembles of squares, the construction of architectural dominants as emphasis;
- the development of neighborhoods with the preservation of red lines, the permeability of yards, the preservation of the number of stories of the background buildings of the historical center;
- the fixed length of the facades, the mismatch of the cornices, the ratio of the plane of the wall to the plane of the openings.

The main parameters of the new development of the typical mass housing are:

- the reduction of the city framework due to the laying of additional highways. However, the scale of development remains within the scope of the city center of the 20th century for interaction and communication of the new framework with the historical one;
- the creation of ensembles of squares, the construction of architectural dominants as emphasis, the use of high-rise buildings as dominants;
- the development of neighborhoods with the preservation of red lines, large courtyards, preservation of the number of stories of the background buildings of the city center of the 20th century;

- the fixed length of facades, mismatch of cornices, the ratio of the plane of the wall to the plane of the openings, the creation of membranes.

The main parameters of the new development of the modern quarters are:

- the creation of a large center at the intersection of Leninsky Avenue and Novoizmailovsky Avenue with new high-rise buildings;
- dismantling of all typical quarters;
- the reduction of the city framework due to the laying of additional highways. However, the scale of development increases relative to the center due to an increase in the number of stories;
- the creation of ensembles of squares, the construction of architectural dominants as emphasis, the use of high-rise buildings as dominants;
- building blocks with the preservation of red lines, large yards;
- the fixed length of facades, mismatch of cornices.

Applying these principles, design decisions were made in these territories to demonstrate the architectural and urban planning environment. As part of the work, a master plan of the development beam from the center to the periphery was developed, where the design was carried out in areas with a low degree of completion of the environment, to best demonstrate the application of the developed methodology in practice. In addition to planning decisions, volumetric projects for streets, squares and neighborhoods were worked out.

4 DISCUSSION

The definition of the historical genetic code of the city was formulated. The article systematized knowledge in the field of urban planning and architectural identity. A methodology for the analysis of large historical cities on the example of Saint Petersburg, including the analysis of historical development and analysis of the modern environment, was proposed.

The morphology of urban planning, architectural and design environment of the city was revealed. Classifications of such concepts as an architectural ensemble, city highways, architectural dominant, urban interior, and typology of buildings were systematized.

The method for modeling the parameters of the historical genetic code of Saint Petersburg was proposed.

Identity parameters of various types of buildings in Saint Petersburg were revealed. The carriers of Saint Petersburg style that have an impact on the perception of the urban environment were identified.

Predictions were made on the further development of the urban fabric of the city. Concrete recommendations were given for the renovation, reconstruction and development of the neighborhoods and micro district of Saint Petersburg.

The solutions to the problem of the loss of historical identity of the urban environment of Saint Petersburg by restoring the historical principles for the development of new territories and areas that have undergone renovation were proposed.

Identity parameters of different types of the city environment were revealed and the Petersburg style was defined on the basis of these parameters.

5 CONCLUSION

The historical stages of the urban planning and territorial development of Saint Petersburg were analyzed. Based on this analysis, a typology of the historical and modern city development was revealed. Different types of buildings were grouped according to general characteristics. These groups are patterns of the environment of Saint Petersburg and are divided into 6 types: the city center of the 18th century, the city center of the 19th century, the gray belt area, the city center of the 20th century, typical mass micro districts of the 20th century, modern buildings

The method for identifying the parameters of the genetic code of the city on the basis of a joint study of urban planning, architectural and design qualities of the environment was proposed.

The parameters forming the urban planning and architectural identity of Saint Petersburg were determined.

The promising parameters of the new city development were modeled; draft plans for different types of environment were developed.

The historical genetic code of the city is the prevailing patterns, techniques, features of urban development decisions of buildings, ensembles and squares that form the urban, architectural and mental identity of the place, affect the perception of the urban environment by a person.

Saint Petersburg style is a combination of parameters of the urban environment of Saint Petersburg that form its historical genetic code.

6 RECOMMENDATIONS

For further work on this topic, it is necessary to systematize and refine the data on the urban planning situation in Saint Petersburg, to monitor new approaches to the study of identity.

Urban research lies mainly in the field of historical development, the study of legislative standards, or a comprehensive urban analysis, which, unfortunately, often does not reveal the parameters of identity.

In parallel, the analytical base of the architectural aspects of the city formation can be revealed. The environmental approach to the study of the urban environment underlies the methods of urban analysis, which in turn explains the mental processes of urban life. Thus, it is necessary to conduct comprehensive scientific research, combining different approaches to the study of the environment.

ACKNOWLEDGMENTS

Among the works on urban planning, architectural development of the urban environment, the works of N.V. Baranov, A.I. Naumov, A.V. Bunin, T.F. Svarenskaya, A.E. Gutnov, K. Lynch, E. Hyman should be named. The works exploring Saint Petersburg include the works of S.V. Sementsov, V.S. Antoshchenkov, V.A. Nefedov, S.S. Ozhegov, V.G. Lisovsky, E.I. Kirichenko.

REFERENCES

Alexander, C. et al. 1997. *A pattern language: towns, buildings, construction*. New York: Oxford University Press.

Bunin, A.V. & Savarenskaya, T.F. 1979. *History of urban art. Urban planning of the slave system and feudalism. Vol. 1*. Moscow: Stroyizdat.

Geddes, P. 1915. *Cities in evolution: an introduction to the town planning movement and to the study of civics*. London: Williams.

Gutnov, A.E. 1984. *The evolution of urban development*. Moscow: Stroyizdat.

Hayman, E. 2012. *New morphology of architecture. Why do buildings need genes?* https://archi.ru/russia/40448/novaya-morfologiya-arhitektury-zachem-geny-zdaniyam.

Hillier, B. et al. 2010. Metric and topo-geometric properties of urban street networks: Some convergencies, divergencies and new results. *Journal of Space Syntax* 1 (2): 258–279.

Ikonnikov, A.V. 1958. *Modern architecture in England. City planning and housing*. Leningrad: Gosstroyizdat.

Internet Archive 2020. Architects and Planners of Letchworth Garden City. https://web.archive.org/web/20080924201806/http://www.letchworthgardencity.net/heritage/index-4.htm.

Kamensky, V.A. & Naumov, A.I. 1966. *The general plan for the development of Leningrad*.

Likhachev, D.S. 1989. Skyline of the city on the Neva *Our Heritage* 1: 8–13.

Lynch, K. 1982. *The image of the city*. Translated by V. L. Glazychev. Moscow: Stroyizdat.

Sementsov, S. V. 2003. Urban planning protection of the historical heritage of Saint Petersburg. In B.M. Matveev (ed.), *Space of Saint Petersburg: Monuments of cultural heritage and the modern urban environment*: 56–76. Saint Petersburg: Philology Department of Saint Petersburg State University.

Topuridze, K.T. (ed.) 1970. *Le Corbusier. Architecture of the 20th century*. Moscow: Progress.

Reconstruction and Restoration of Architectural Heritage – Sementsov et al (eds)
© 2020 Taylor & Francis Group, London, ISBN 978-0-367-65357-6

Use of cork coating in architectural monuments of Saint Petersburg

P.G. Shchedrin
Saint Petersburg State University of Architecture and Civil Engineering, Saint Petersburg, Russia

ABSTRACT: The article discusses numerous examples of applications of cork and cork products (including cork sheets) as insulation, sound absorber and structural material during construction work in Saint Petersburg in the middle of the 19th - especially the beginning of the 20th century. This constructive and decorative material, new to Saint Petersburg, has found wide application. It was used both in ordinary buildings in the center of Saint Petersburg and iconic objects when performing works on the dome of St. Isaac's Cathedral and in the Yusupov Palace.

1 INTRODUCTION

In the old buildings of Saint Petersburg, built at the end of the 19th century or overhauls of the late 19th - early 20th centuries, cork materials are sometimes found in the form of heating or sound insulation and elements of enclosing structures, for example, as the facade walls of bay windows with lime mortar plastering.

With a high degree of certainty we can assume that cork began to be imported for construction work in Russia back in the 1830s.

In the second half of the 19th century, cork tree was imported in Russia annually in the amount of at least 100,000 rubles (Logginov & Gordienko 1976). Based on the world price for cork raw materials of that time, this was approximately 500 tons in volume and weight.

One cannot but admire the entrepreneurial spirit of our ancestors. Literally five years after the discovery of waste-free cork processing technology, a note was placed in the Brockhaus and Efron Encyclopedic Dictionary (Brockhaus & Efron 1991) in the economics section: "The production of cork products is gradually developing and becoming large, although it's based exclusively on the processing of foreign raw materials. The number of factories of this kind in 1896 reached 27, their productivity is 6,383 thousand rubles, the number of workers amounts to 4,947 people; all technical improvements have been applied here" [3].

It is further noted that, for example, in 1897, 912,508 poods worth 3,186 thousand rubles were imported from Portugal, France, Spain and Germany. In an advertisement placed in Zodchiy magazine, we read that for construction work, cork slabs 1.0x0.5 meters in size and 5-60 mm thick were offered [3]. According to the publication of E. Kern (1929), "for the partitions between the living quarters, etc. the thickness of cork plates of 2 cm (for sound insulation) can be accepted as quite sufficient". Fragments of such plates from interior partitions can be seen in many Saint Petersburg houses built in the 1900s.

2 METHODS

Here we can give, as examples, the addresses of some buildings where cork was used:

1. Leo Tolstoy Square. The house with towers (reconstructed in 1913-1915). Under the windows of the towers, walls with a thickness of 850 mm are made of cork bricks (Ludikov 2013).
2. Kamennoostrovsky avenue, 13, Mira street, 13, Divenskaya street, 2. K.Kh. Keldal's house (1902-1903). The overlap is insulated with cork sheets. (Figures 1–2).
3. Maly avenue, Petrograd side, 71, Barmaleeva street, 18, Podrezova street, 15. (1897-1908, 1910-1911). (Figures 3–4).
4. Podrezova street, 17. A.T. Must's tenement house. (1911). Cork walls of bay windows.
5. Nesterova street, 9. I.N. Borozdkin's tenement house. (1910-1911). Cork walls of bay windows.
6. Voskova street, 8-5. P.P. Chestnokov's tenement house. (1910). Cork walls of bay windows.
7. St. Isaac's Cathedral. Saint Petersburg, St. Isaac's Square, 4.
8. Yusupov Palace. Saint Petersburg, Moika Embankment, 97.
9. Lieutenant Schmidt Embankment, 31. F.F. Uteman's house. In F.F. Uteman's house cork is found in the form of (Archive of The Committee on State Control, Management and Protection of Historical and Cultural Monuments of Saint Petersburg) masonry bricks in the bay window of the main staircase (specific weight 180 kg/m^3) and sheet insulation in the floors of the 2nd floor (specific weight 144.8 kg/m^3). (Figure 5).

And many others.

Figure 1. K.Kh. Keldal's house. Insulation in the basement. Ceilings over the basement are on metal beams, filling: concrete vaults of crushed bricks. The cork was laid on a lime screed. Cork sheets of 490x1000x60 mm. Photo by the author.

Figure 4. Podrezova street, 15. The insulation in the form of cork bricks. Size 233x117x60 mm. Photo by the author.

Figure 2. K.Kh. Keldal's house. Insulation in the basement. Cut view of cork. Photo by the author.

Figure 5. Brickwork made of cork brick of the bay window wall of the F.F. Uteman's house. Photo by the author.

2.1 Saint isaac's cathedral

In 1913 (Russian State Historical Archive 1913b), the conical dome of St. Isaac's Cathedral was insulated (Figure 6).

The cork coating was made in two layers with a vapor barrier of roofing felt. The top layer of cork is plastered with lime mortar. Cork slabs were laid of clean hydraulically pressed cork with glue, 1000 x 500 mm in size (Russian State Historical Archive 1913a). Cork sheet thickness is 30 mm, with a density of 202.33 kg/m^3. The seams are smeared with ruberine (bituminous mastic). The work was completed in November 1913 (Figures 7–8).

By 2017, part of the protective layer of plaster was lost. The top layer of the cork sheet, in places of exposure, began to crumble; it was destroyed in some places. The total shrinkage of the cork sheet for 100 years was 1-1.5%. There were gaps between the insulation sheets. Thermotechnical calculation

Figure 3. Bay window of the house on Podrezova street, 15. Solid red bricks and cork bricks are visible. Photo by the author.

Figure 6. St. Isaac's Cathedral. General view. Photo by the author.

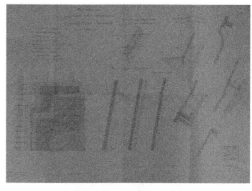

Figure 8. The drawing of the new insulation of the conical dome of St. Isaac's Cathedral. The design was developed by August Linde on September 7, 1913. Central Saint Petersburg State Archive of Literature and Art. Fund 598. List 1. Case 95. Sheet 59.

Figure 7. A detailed drawing on the arrangement of pipes in the dome for moisture removal. The archive of the Scientific and Technical Library of Emperor Alexander I Saint Petersburg State Transport University. Folder 8. Sheet 9. 1845. Size 638x980 mm. Signed by V. Schreiber.

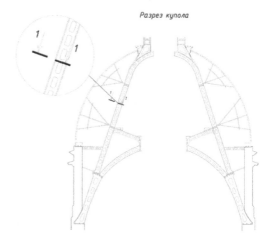

Figure 9. The project of the dome restoration of St. Isaac's Cathedral. Section of the dome. A cast iron frame is shown.

showed the need for laying an additional layer of insulation of at least 40 mm. For this, a cork sheet agglomerate 50 mm thick was chosen.

After approval of the restoration project of the dome of the cathedral (Archive of the Committee on State Control, Management and Protection of Historical and Cultural Monuments of Saint Petersburg) by the Committee for the Protection of Monuments,

restoration and repair work was carried out and the second layer of insulation in places of the destroyed plaster layer S = 300 m^2 and an additional layer of cork of 1100 m^2, with the thickness of 50 mm, density of 110 kg/m^3, was laid. The total weight of the newly installed cork made up 8 tons (Figures 9–12).

2.2 *The yusupov palace*

Cork coating is found in the form of (Archive of the Committee on State Control, Management and Protection of Historical and Cultural Monuments of Saint Petersburg) cork chips in the lime plaster of the outer wall of the Oak Dining Room; sound insulation in the floors of the Vigi Hall; bulk crumb in the ceiling above The Half of the Young

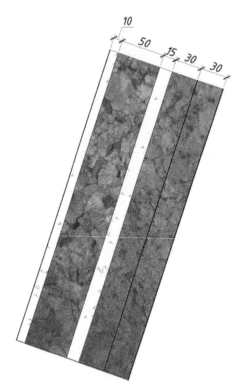

Figure 10. The section on cork slabs. Historic cork coverings are shown (right) and new elements during the restoration process (left).

Figure 11. The view of the historical cork plate of St. Isaac's Cathedral. Photo by the author.

basement in the silver boudoir room. The thickness of the backfill is 112-137 mm. Cork coating is in the form of cork brick in The Half of the Young bathroom (Figure 13).

Figure 12. Macro shooting of a cork plate sample. Photo by the author.

Figure 13. Cork filling in the ceilings of the Yusupov Palace in The Half of the Young. Photo by the author.

3 CONCLUSIONS

Studying the identified objects where the cork tree bark was used, one can come to the conclusion about the strong influence of industrial production growth and significant innovations in the designs and technologies of construction in Saint Petersburg in relation to historical buildings in the city, in Russia and abroad.

REFERENCES

Archive of the Committee on State Control, Management and Protection of Historical and Cultural Monuments of Saint Petersburg. P-59. The project for the restoration of the conical dome of St. Isaac's Cathedral in part of its insulation, at the address: Saint Petersburg, St. Isaac's Square, 4, building A. Section II. Design work. Vol. 1. Architectural and building solutions. No. 31603864274-02-12-AC-2. Saint Petersburg, 2017. Completed by Vozrozhdenie Peterburga LLC.

Archive of the Committee on State Control, Management and Protection of Historical and Cultural Monuments of

Saint Petersburg. P-61. The project for the restoration and repair of the interiors of the premises of the cultural heritage site of federal significance the Yusupov Palace located at the address: Saint Petersburg, Moika Embankment, 94, building A. The project was approved on December 14, 2018, No. 01- 26-6364/18-01. Performed by Vozrozhdenie Peterburga LLC.

Archive of The Committee on State Control, Management and Protection of Historical and Cultural Monuments of Saint Petersburg. P-516. The project of restoration and adaptation of the object of cultural heritage of regional significance F.F. Uteman's house. Main staircase and lobby. Located at: Saint Petersburg, Lieutenant Schmidt Embankment, 31. Performed by LLC SPbProektRestav ratsiya. Agreed on August 01, 2018.

Brockhaus, F.A. & Efron, I.A. 1991 *Russia: encyclopedic dictionary (reprint of 1898)*. Leningrad: Lenizdat.

Kern, E.E. 1929. Cork trees and cork problems. *Dostizheniya i Perspektivy*: 643–654.

Logginov, B.I. & Gordienko, M.I. 1976. *The experience of growing cultures of Amur corktree*. Moscow: Lesnaya Promyshlennost.

Ludikov, V.I. 2013. *The history of cork oak cultivation in Russia*.

Russian State Historical Archive 1913a. *Contracts for various works in the cathedral*. Fund 502, List 1, Case 957: 17–18; List 4, Case 48: 21.

Russian State Historical Archive 1913b. *On the insulation of the conical dome*. Fund 502, List 1, Case 957: 4.1900. *Zodchiy* 1.

Reconstruction and Restoration of Architectural Heritage – Sementsov et al (eds)
© 2020 Taylor & Francis Group, London, ISBN 978-0-367-65357-6

Specific features of manors of nobility in remote uyezds of Saint Petersburg Governorate (through example of Luzhsky Uyezd)

E.Yu. Shuvaeva
Saint Petersburg State University of Architecture and Civil Engineering, Saint Petersburg, Russia

ABSTRACT: The article addresses the prerequisites and specifics of the emergence of remote manors of the nobility through the example of Luzhsky Uyezd. Features of stately manorial estates located in remote uyezds of Saint Petersburg Governorate are described in comparison with standard manors and manor complexes of a similar status located in the uyezds near Saint Petersburg. Specifics of manors' location and pragmatic use of manorial territories, relations between the settlement structure and geographical characteristics, areas of manorial park and garden complexes inherent in such facilities are identified. Three manors of Luzhsky Uyezd that differ in location features — Nezhgostitsy, Perechitsy and Busany — are described as specific examples. The study is based on scholarly works of the past, extensive archival, bibliographical and carto-graphical materials, and on-site investigations.

1 INTRODUCTION

More than two thousand manors of the nobility existed in Saint Petersburg Governorate by the beginning of the 20th century. They started to actively appear in the third quarter of the 18th century during the reign of Elizabeth of Russia (Guseva 2008). Later, during the reign of Catherine the Great, manor house ensembles continued to be built along with the reconstruction of the existing complexes.

Many manors of the nobles (not only currently existing but also lost ones) located in the modern Leningrad and Novgorod Regions are described in papers by I.V. Barsova (1971), T.E. Isachenko (2003), O.V. Litvintseva (2006), and S.E. Guseva (2008). Publications by E.A. Kozyreva (Kozyreva 2016) also deal with studies of the nearest manors of the nobles.

This article describes the main features of the emergence of a certain type of manor — remote manors of the nobles in Luzhsky Uyezd. Highest officials ranked at least as the fourth class according to the Table of Ranks were owners of such country manors.

Remote manors of the nobles had characteristic functional, compositional and spatial peculiarities that distinguished them not only from usual manors of the nobility but also from similar manors located in the area adjacent to the capital — in the nearest uyezds (Petergofsky, Sankt-Peterburgsky, Tsarskoselsky, Shlisselburgsky Uyezds (Sementsov 2019). By now, 105 remote manors of the nobles, the majority of which (29 manors) are located in the former Luzhsky Uyezd have been studied (Figure 1).

2 MAIN PART

Luzhsky Uyezd was the largest of the uyezds forming the remote belt and located nearer to the borders of the governorate.

Landscape and climatic conditions of the territory were important initial conditions for the construction of manorial complexes there. Various hills and mountain chains that were primarily laced with lakes and swamps were characteristic of the southern part of Saint Petersburg Governorate (Guseva 2008). Manors in Luzhsky Uyezd were often built along rivers (Luga, Oredezh, Plyussa) and around lakes (Cheremenetskoe, Vrevo, Merevo, Pesno). Lesser number of populated localities and no manors of the nobility in the western part of the uyezd is explained by the marshland (Murashova & Myslina 2001). The geographic peculiarities affected the settlement structure: the majority of large manorial complexes were located in agriculture-friendly lands.

The Cheremenetsky Monastery located on the western shore of Cheremenetskoe Lake was the center of the largest group of manors in Luzhsky Uyezd (Chistyakova 2015). Around it (John the Apostle's Monastery founded in the 15th century) (Nabokina & Noskov 2015), lakeside manors of the nobles were located side by side: Nezhgostitsy, Bol-shoy Navolok, Rapti, Domkino, and Skreblovo. A road and rail network developed in the mid-19th century: the ancient route from Saint Petersburg through Gatchina and Luga to Pskov laid in the 1830s was duplicated by the Saint Petersburg – Warsaw Railway (Glinka 2017). Manor complexes were built along the highway owned by the Gover-norate Council (Zheltsy–Dubki, Zheltsy–Lilino, Zheltsy–Tosiki), along country routes (Shiltsovo,

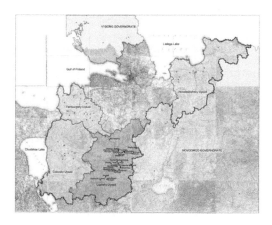

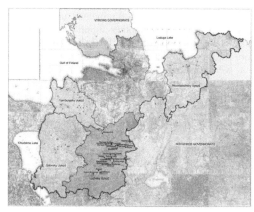

Figure 1. Layout of manors of the nobility in Luzhsky Uyezd.

Figure 2. Layout of main groups of manors of the nobility in Luzhsky Uyezd relative to water bodies and traffic arteries.

Ivanovskoe, Busany, Zapolye, Golubkovo, Yugostitsy, Pochap, Syrets), or at the end of a special bend (Nadbelye), near railway stations (Ilzho) (Table 1).

Thus, the studied manors of the nobility in Luzhsky Uyezd were located primarily in improved natural and landscape zones with consideration to the landscape and climatic conditions (near natural water bodies) and along inland traffic arteries (Figure 2).

In her PhD thesis "Park and garden complex of countryside manors of the Saint Petersburg Governorate nobility (typological aspect)", S.E. Guseva characterizes Luzhsky Uyezd as a territory "with the equal number of small and medium-sized manors

and with a small number of large manors" (Guseva 2008). Manors of the capital nobles located nearer to Saint Petersburg had a large area, but the more remote they were, the lesser was the territory of park and garden complexes. The average area did not exceed 10 ha (Rozov 2005) in the most remote uyezds (Gdovsky and Novoladozhsky), and in Yamburgsky Uyezd it was 18 ha. The landholdings in Luzhsky Uyezd were the largest among the remote uyezds (average area — 23 ha). Nikolskaya (21 ha), Ilzho (24 ha), Nadbelye (24 ha), Zapolye (25 ha), Syrets (25 ha) manors had a standard area. It is also characteristic that several manors (two to three) could be located in the same settlement. This is due

Table 1. Classification of manors of the nobility of Luzhsky Uyezd by location.

Lakeside			
Lake Cheremenetskoye	Lake Vrevo	Lake Merevo	Lake Pesno
- Rapti	- Domkino	- Merevo	- Lyubensk
- Bolshoy Navolok	- Skreblovo	- Kella	- Vechasha
- Nezhgostitsy			
Riverside			
Oredezh River	Luga River	Plyussa River	Gorodonka River
- Zatulenye	- Muraveyno	- Kureya (Butyrki)	- Lyushchik
- Kamenka			
- Perechitsy			
Along roads			
Highways owned by the Governorate Council	Railway	Roads of the uyezd council	Standalone manors along roads of the uyezd council
- Zheltsy–Dubki	- Ilzho	- Shiltsovo	- Pochap
- Zheltsy–Lidino		- Ivanovs- koye	- Nadbelye
- Zheltsy–Tosiki		- Busany	- Syrets
		- Zapolye	
		- Golubkovo	
		- Yugostitsy	

to the fact that the land was fragmented and inherited or sold in parts. Thus, another significant feature of manors of the nobility in Luzhsky Uyezd was formed: they were located not in a continuous circle, but were nested together. Smaller manors were located around the main, larger manors at a distance of 3–5 km. Such a system emerged due to the separation of large and the largest original manors, allocation of manors to heirs, which was accompanied by the fact that new manors were built near the main one. New manors consisted of an owner's house and a surrounding park, auxiliary buildings were less diverse. Their number varied from 2 (Muraveyno, Bolshoy Navolok) to 24 (Zatulenye). On average, 10 to 15 facilities with primarily household functions were situated in the territory of a manor complex (Shuvaeva 2020). This feature is preconditioned by the long-term or seasonal residence of the owners who required a large household area (Roosevelt 1995). In addition to parks and gardens, it included kitchen gardens and agricultural land (Isachenko 2003). The surrounding buildings and structures included utility facilities such as barns, cowsheds, stables, warehouses, milk houses, mills, shops, etc. There were almost no "stately" facilities characteristic of the nearest manors of the nobility (greenhouses with exotic plants, round pavilions, stables for pedigreed horses) (Kozyreva 2019). Since it was necessary to live there for a long time, additional residential space was required. This task was solved through designing a sufficient number of residential premises in the manor house and erection of houses for servants (the caretaker's house, the gardener's house, etc.). Household buildings constructed primarily of stone and bricks have been preserved the best.

Below, we describe three manors of Luzhsky Uyezd differing in their location: lakeside (Nezhgostitsy), riverside (Perechitsy), and roadside (Busany).

Nezhgostitzy. The largest manor of Luzhsky Uyezd, Nezhgostitsy, is located near the Cheremenets Monastery on a shore of Cheremenetskoe Lake (Figure 3). Its area, according to different data, is 85–120 ha.

The layout of the territory established in 1778 was thought out carefully, and all features of the relief were used most advantageously. A wooden two-story manor house with a mezzanine was situated at the edge of the table land, while other buildings, parks, and gardens were located at the upper terrace. The manor house had all the attributes and the scale of classical stone facilities of the capital, despite the simplified nature of wooden elements. A family chapel was situated in one of the side wings located on the same line as the manor house, and the other side wing had a kitchen (Nabokina & Noskov 2015). The household facilities included greenhouses, hothouses, warm houses, cellars, and ice cellars. The entrance road leading to the center of the ensemble, the manor house, altered with a regular garden merging into a landscape garden (Guseva 2007). The

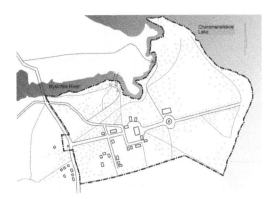

Figure 3. Layout of the Nezhgostitzy Manor.

whole territory was surrounded by meadows and water bodies. Such a large manor complex satisfied the requirements for the long-term comfortable seasonal residence of the owners to the fullest.

Perechitsy. Perechitsy village was situated on the high right bank of the Oredezh River, and a manor complex with the area of 17 ha was adjacent to it from the south-west. It was situated on an even terrace and on the southern slope (Figure 4).

In terms of the layout, the park and garden complex was divided into two parts — the regular and landscape ones. Observation decks were situated on the natural steps of the slope. The manor house was built on the edge of a steep (Murashova & Myslina 2001).

Busany. Busany is a manor medium in the area (21 ha). It is located near a road of the uyezd council on the south-western shore of the Vrevo Lake (Figure 5). The park and garden are situated on a plain with a step to the shore lowland (Murashova 2005).

The manor complex consisted of the upper and lower parks, the owner's yard, the area of kitchen gardens and other gardens, areas of household facilities located at both sides of the central axis. An entrance

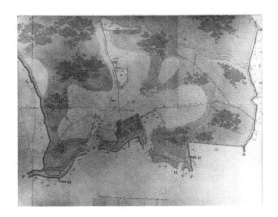

Figure 4. Perechitsy. 1782 (Saint Petersburg Central State Historical Archives, Fund 262, List 9, Case 2105).

132

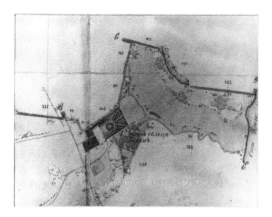

Figure 5. Busany country house. 1850 (Saint Petersburg Central State Historical Archives, Fund 262, List 101, Case 1035).

alley ran from the highway to the manor house. Orchards were set up along the alley. The manor house was rebuilt several times, but its location at the edge of the terrace with steep carefully designed slopes remained unchanged. A servants' side wing with an ice cellar was on the left of the house; stone stables, a coach house, a barnyard, a drying house, sheds, and granaries were located in the southern part. A regular park with a pond and three alleys in the form of a fan were set up behind the manor house. The precise layout of the landscape met the chaste architectural solution of the manor house.

3 CONCLUSIONS

Following the analysis of manors of the nobility built in Luzhsky Uyezd (one of the four uyezds of Saint Petersburg Governorate forming the remote belt) from the beginning of the 18th century till the beginning of the 20th century, we identified features that distinguish them from the manors of the nobility of the nearest belt (Petergofsky, Sankt-Peterburgsky, Tsarskoselsky, Shlisselburgsky Uyezds) and standard suburban manor estates. These manors were located in special large groups, so-called "nests" along the main highways or along the banks of rivers or lakes, taking into account the landscape and geographical specifics of the uyezd. The area of the largest manor complexes, as a rule, did not exceed 40 ha, manorial facilities were primarily of household nature and satisfied the requirements for long-term residence. As described through the example of Luzhsky Uyezd, all the above-listed features of the remote manors of the nobility that determine their specifics can be traced.

Therefore, we can distinguish a special group of stately manors in large territories of borderline uyezds of Saint Petersburg Governorate that had an impact on the development of the historic agglomeration.

REFERENCES

Barsova, I.V. 1971. *Manorial parks of the Leningrad Region and the principles of their use. PhD Thesis in Architecture.* Leningrad: Leningrad Institute of Civil Engineering.
Chistyakova, T.N. 2015. *Silver Belt of Russia.* Saint Petersburg: NP-Print.
Glinka, N.V. et al. 2017. *Leningrad Region. Historical sketches.* Saint Petersburg: Avrora.
Guseva, S.E. 2007. Alleys in the manors of Saint Petersburg Governorate. *Industrial and Civil Engineering* 7: 49–50.
Guseva, S.E. 2008. *Park and garden complex of countryside manors of the Saint Petersburg Governorate nobility (typological aspect). PhD Thesis in Architecture.* Saint Petersburg: Saint Petersburg State University of Architecture and Civil Engineering.
Isachenko, T.E. 2003a. Nature-cultural complexes of country estates in landscapes of Saint Petersburg region. *Izvestiya Russkogo geograficheskogo obshestva* 135 (2): 1–14.
Isachenko, T.E. 2003b. *Relationship between natural-and-cultural complexes of manors of the nobility and their landscapes. PhD Thesis in Geography.* Saint Petersburg: Saint Petersburg State University.
Kozyreva, E.A. 2016. The estates of the highest nobility located near Saint-Petersburg as a phenomenon of historical and town-planning life of the city (on the example of Ekaterinhof Park). *Bulletin of Civil Engineers* 1 (54): 5–11.
Kozyreva, E.A. 2019. Aristocratic society estates as phenomenon of historical and city planning life of Saint Petersburg (the Stroganov's estate case studies). *Vestnik Tomskogo gosudarstvennogo arkhitekturno-stroitel'nogo universiteta. Journal of Construction and Architecture* 21 (3): 67–76.
Litvintseva, O.V. 2006. *Formation of countryside manors of the nobility in the Novgorod Governorate at the end of the 18th– the 19th centuries. PhD Thesis in Architecture.* Saint Petersburg: Saint Petersburg State University of Architecture and Civil Engineering.
Murashova, N.V. 2005. *One hundred manorial estates of Saint Petersburg Governorate: a historical reference book.* Saint Petersburg: Vybor.
Murashova, N.V. & Myslina, L.P. 2001. *Manors of the nobility in Saint Petersburg Governorate. Luzhsky District.* Saint Petersburg: Blitz Russian-Baltic Information Center.
Nabokina, O.V. & Noskov, O.V. 2015. *Luga and its outskirts. An extract from the history of Luzhsky District.* Moscow: Tsentrpoligraf.
Roosevelt, P. 1995. *Life on the Russian country estate: a social and cultural history.* London: Yale University Press.
Rozov, N.G. 2005. *Belt of the Pskov Land. Manors of the nobility.* Pskov, Pushkinskie Gory: Sterkh Print Yard OOO.
Sementsov, S. & Akulova, N. 2019. Foundation and development of the regular Saint Petersburg agglomeration in the 1703 to 1910s. *Advances in Social Science, Education and Humanities Research* 324: 425–433.
Sementsov, S.V. et al. 2020. Estates of the highest nobility of the Saint Petersburg province as a special spatial structure of the historical Saint Petersburg agglomeration. *IOP Conference Series: Materials Science and Engineering* 775: 012073.
Shuvaeva, E. 2020. Developmental peculiarities of nobles' estates on the territories of distant uyezds in Saint Petersburg province. *E3S Web of Conferences. Topical Problems of Green Architecture, Civil and Environmental Engineering 2019 (TPACEE 2019)* 164: 04029.

Reconstruction and Restoration of Architectural Heritage – Sementsov et al (eds)
© 2020 Taylor & Francis Group, London, ISBN 978-0-367-65357-6

Influence of architectural fantasies of Yakov Chernikhov on projects of Zaha Hadid

A.A. Smirnov
Saint Petersburg State University of Architecture and Civil Engineering, Saint Petersburg, Russia

O.A. Kotlovaya
International Design School, Saint Petersburg, Russia

ABSTRACT: In recent years, they talk about the secondary nature of Russian architecture relative to Western counterparts. This myth is largely false. They forget that the Soviet avant-garde was at the forefront of world modernism of the twentieth century, and the German school of lightweight structures was largely based on the idea of hyperbolic paraboloids by Vladimir Shukhov. The purpose of this article is to show on specific projects the process of borrowing ideas realized in the 21st century by architect Zaha Hadid (1950-2016, Iraq, UK), from the work of Soviet avant-garde architect Yakov Chernikhov (1895-1952).

1 INTRODUCTION

This article reveals the similarities in the work at first glance of little similar architects of the USSR and Great Britain, whose work in recent years has caused considerable interest. It draws clear parallels, analysis of projects and reveals the similarity of architectural ideas. At first glance, the work of these architects is so varied that it seems impossible to compare them and draw any parallels. What could a native of Ukraine have in common during the years of Tsarist Russia, a Soviet architect who built little and was engaged almost exclusively in the genre of architectural fantasy, and the world-famous British architect of the 21st century, which has nothing to do with the USSR? This is only at first glance. This story demonstrates how the energy and talent of a person, put on paper in the past in difficult historical circumstances, can give powerful creative shoots after almost a century.

2 PROBLEM

The problem is the underestimation of the importance of Soviet architecture in the global context. Ignorance and misunderstanding of the historical interconnections of modern architecture with the experimental architecture of the past.

3 METHODS

Research methods are a comparative visual analysis of sketches and designs. For an objective illustration, we present several sketches of Yakov Chernikhov and projects implemented by Zaha Hadid.

4 EXAMPLES

Yakov Chernikhov (Figure 1) - Leningrad architect, born and studied in Ukraine in Pavlograd. In 1914 he entered the Academy of Arts at the Faculty of Architecture. He graduated in 1925. The leader was Leonty Benoit, a wonderful draftsman.

The Soviet avant-garde arose in Moscow in the 1920s, but the Saint Petersburg school remained neoclassical. Immediately after graduating from the Academy, he opened his workshop, where he began to experiment in the field of avant-garde architectural imagination.

In 1933 he published his central work, "Architectural Fantasies. 101 composition in paints" (Chernikov 2008), which becomes an anthology of forms of architecture of the Soviet avant-garde, not only in the USSR, but also in the West. The most striking of Chernikhov's projects is the Water Tower of factory factory "Red Gvozdil-shchik" (1931).

In the 1940s, Chernikhov, in the wake of interest in the theme of the "Palace of Soviets", he created a series of miniatures. This series of works "Palaces of Communism" was sent to TV. Stalin.

The phenomenon of the diversity of talent of Yakov Chernikhov is at a unique set of circumstances in his life: the shackles of the neoclassical school held back Chernikhov's energy for his breakthrough in the field of avant-garde architectural imagination. There was a motivation to create new and non-realization of ideas.

Zaha Mohammad Hadid (Figure 2) - was born on October 31, 1950 in the capital of Iraq, Baghdad. One day, Zaha told her parents that she wants to create cities similar to the dunes on which flowers bloom -she was only 10 years old. To receive primary

134

Figure 1. Yakov Chernikhov.

Figure 2. Zaha Hadid in her youth.

remained a paper underground. In the late 1980s, a crisis ensued in world architecture. The attitude towards deconstructivism changes after it is addressed by Philip Johnson, who in 1988 organized the exhibition "Deconstructivist Architecture" in New York (Tarkhanov 2019). They started talking and writing about deconstructivism. In the mid-1990s, the Frank Gehry Bilbao Museum project was implemented (Figure 3). Frank Gehry's innovation was in the fact that he adapted the NASA program for the design of self-aircraft (Kipina 2019) to create complex forms.

In the early 1990s, the grandson of Yakov Chernikhov, Andrei Chernikhov leaves the USSR and takes away with him the archive that he publishes in the west (Chernikhov 2009).

Zaha Hadid also fails in the 1990s, since in 1997 her first project was not implemented - a fire station in Weil am Rhein in Vitra (Figure 4). A fire station for 6 cars is aggressive, tough, crazy even by the standards of the late technogenic high-tech of the 1990s. Obviously, the idea of a triangle strung on straight rods, which we see in this construction, is similar to one of the sketches of Yakov Chernikhov (Figure 4).

One of the latest projects by Zaha Hadid was implemented in Moscow. At the Dominion Tower, we see respect for the architectural context. It resembles another sketch by Yakov Chernikhov (Figure 5).

Figure 3. Guggenheim Museum Bilbao.

education, parents assigned Zaha Hadid to a closed boarding house at the monastery of the French mission in Baghdad. Since childhood, Zaha Hadid knew several languages and was fluent in French, German and English. After graduation, Zaha enters the American University of Beirute at the Faculty of Mathematics. In 1972 Hadid leaves to study in Tondon, and enters the UK's most prestigious university of architecture Architectural Association School of Architecture. One of the professors called it "The Planet, which rotates in its own orbit." In 1978 she completed her education and received a diploma with honors/ Special Diploma Prize /. The first years she worked in the association of OMA architects together with Rem Koolhaas, and in 1979 founded her own workshop. All 1980s take place in architectural contests. In the 1980s, deconstructivism (Belogolovsky 2004)

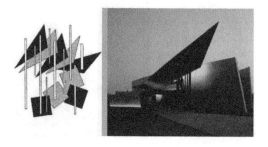

Figure 4. On the left is a composition from the personal archive of A. Chernikhov; on the right is the Vitra Complex, Weil am Rhein, Germany. 1994 (architect Z. Hadid).

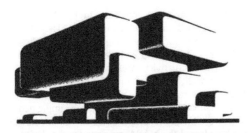

Figure 5. Above - a drawing of arch. Y. Chernikhova; below -Dominion Tower, Moscow (architect Z. Hadid).

Figure 7. Left - Water tower of the factory "Red Gvozdilshchik" (1931); right - Villa Tower - Barvikha Luxury Village, Moscow (2013).

Figure 8. On the left - sketch of arch. Y. Chernikhov; right - interior of the Roca London Gallery, UK 2011 (architect Z. Hadid).

The formal resemblance to Chernikhov's fantasy and the composition of the building of the Center for Contemporary Art in Cincinnati (2000) are interesting. The building expresses precisely the transition from brutalism to de-constructivism. Contrary to the fashion of 100% glazing in hi-tech architecture of the late 1990s - 2000s, the author interprets the building as heavy and opaque (Figure 6).

In 2004, Zaha Hadid came to Saint Petersburg and visited the Yakov Chernikhov Water Tower. In 2006, Zaha Hadid began work on a villa project in Barvikha Luxury Village in Moscow. This work led to the implementation of another idea of Yakov Chernikhov in cutting-edge architecture (Figure 7). Similarities are evident visually in other projects, such as the interior of the Roca London Gallery (Figure 8).

5 CONCLUSION

The fate of the two architects was, despite different eras, deeply similar. From obscurity, failure and self-seeking to recognition and applause around the world. But if Chernikhov anticipated the avant-guard, never having drunk a cup of self-actualization in stone, metal and glass and success, then Hadid might have found in his fantasies that inspiration and support, bridges to European architecture, which allowed her to mature in Russia his creativity to find himself and realize his ideas a century later, to translate into eternity. The influence of Yakov Chernikhov on Zaha Hadid is obvious. Perhaps this external influence is formal, but the Chernikhov -Hadid union is a unique, it is an architectural "message" in a century. Yakov Chernikhov, in the scale and futurism of his architectural fantasies, was ahead of his time and could not or did not have time to realize them in his lifetime. Zaha Hadid, on the contrary, has long spent its unique language and style, and meeting with his work gave her a fresh stream of inspiration. This suggests that the work of Zaha Hadid as one of the most influential

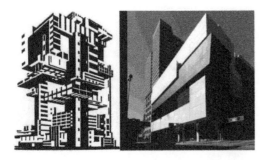

Figure 6. On the left is a sketch (architect Y. Chernikhov); right -Center for Contemporary Art in Cincinnati (architect Z. Hadid).

architects of the XXI century is largely inspired by the architectural fantasies of Yakov Chernikhov, as one of the most influential architects of the USSR.

REFERENCES

ArtFuture School 2019. The Bilbao effect in business. Accessed December 11, 2019. https://artfuture-school.ru/blog/effekt-bilbao-v-dele.html.

Belogolovsky, V. 2004. Pure architecture of Peter Eisenman. Accessed December 11, 2019. https://archi.ru/press/world/33635/chistaya-arhitektura-pitera-aizenmana.

Chernikhov, D.Ya. 2009. *Graphic masterpieces of Yakov Georgievich Chernikhov: The Collection of Dmitry Yakovlevich Chernikhov.* Berlin: DOM publishers.

Chernikhov, Ya.G. 2008. *Architectural fantasies. 1933. 101 compositions.* Moscow: Avatar.

Glancey, J. 2006. I don't do nice. Accessed December 12, 2019. https://www.theguardian.com/artanddesign/2006/oct/09/architecture.communities.

Hadid, Z. 2015. *Zaha Hadid at the Hermitage: Catalogue.* Saint Petersburg: Fontanka.

Horgan, J. 2001. *The end of science. Facing the limits of knowledge in the twilight of the scientific age.* Translated by M.V. Zhukova. Saint Petersburg: Amphora.

Khan-Magomedov, S.O. 1995. *Pioneers of Soviet design.* Moscow: Galart.

Kipina, J. 2019. Frank Gehry — expressive genius of deconstructivism. Accessed Decemver 11, 2019. https://losko.ru/frank-gehry/.

Tarkhanov, A. 2019. Portrait: architect Philip Johnson. Accessed December 11, 2019. https://www.admagazine.ru/architecture/portret-arhitektor-filip-dzhonson.

Reconstruction and Restoration of Architectural Heritage – Sementsov et al (eds)
© 2020 Taylor & Francis Group, London, ISBN 978-0-367-65357-6

Decline of "one-storied USSR" politics in cities of Russia

A.V. Surovenkov & M.V. Skopina
Saint Petersburg State University of Architecture and Civil Engineering, Saint Petersburg, Russia

G.V. Stukalov
Institute of Urban Planning "RosNIPI Urbanistiki" JSC, Saint Petersburg, Russia

Z. Tuhtareva
"Alfavill-Volga" JSC, Samara, Russia

ABSTRACT: The paper is based on retrospective research of development in dwelling zones in the cities of USSR. It is spoken in detail about reasons for beginnings of multistorey dwelling zones (microdistricts). As the basis for study it's taken the graphic analytic method of cartographic documentation analysis which is based on aerial and satellite views, including unclassified espionage views from U.S. Geological Survey's archive. In conclusion it is given recommendations for balanced development in dwelling zones in Russia by analogy with foreign experience.

1 INTRODUCTION

Nowadays it's typical for Russian cities almost not to have individual dwelling zones with developed civil engineering infrastructure. Although low-rise building quota is 40% of total input squares every year, this housing development is conducted fragmentary and far away from main urban parts. New house building in individual dwelling zone is not stylistic regulated, not equipped with engineering communications and it often looks chaotic. In low-rise building activity there are only a few big development companies in Russia.

In spite of pulling down and transformation of microdistricts in the cities of West Europe and USA, overwhelming territory of big cities is built up with dwelling formations of microdistrict type. Existing individual dwelling zones in the cities of Russia, as a formation of pre-microdistricts epoch, is endangered of settling. Because of the fact that city authorities are planning to build microdistricts almost everywhere, it's forbidden for living in individual dwelling zones people to rebuild houses and that leads to total territory decay like this one.

2 PROBLEM STATEMENT

Nowadays Russian town-planning is advanced on the way of making big dwelling microdistricts which consist of multistorey apartment building (9-25) and it doesn't correspond not only with town-planning traditions of majority countries of the world, but also to traditions of pre-revolutionary Russia (Stukalov 2015). In spite of some advantages of microdistricts, it makes monotonous and high-density human medium (about 20000 people for one square kilometer) and logically it generates problems with parking near the houses and on the streets (because of the little density of trunk road network in microdistricts it's about 8 kilometers for one square kilometer (Stukalov 2015)). Moreover there is not such a thing in microdistricts like closed private space for tennants, feeling of local community, on which developing it's paid so much attention by the program of sustainable developing UN-HABITAT (2020).

Not many in Russia know that before 1950s the town-planning USSR went on the way of developing of dwelling low-rise building. Owing to deep studies on this matter, the definition «One-storied USSR» is suggested by authors and examined in this paper

3 PROBLEM RESEARCH AND THE DEFINITION «ONE-STORIED USSR»

In the end of XIX century Russian cities mainly were minor dwelling zones with one or two-storied individual dwellings. Russia was one of the most urban weakest countries in Europe, there were a few big cities, but the most people were rural in contrast to Europe where urbanism was systematic. In many respects it's explained with late elimination of serfage.

But in the beginning of XX century, because of the increased urbanism, the main migration flow has become immigration from rural living area to the cities. Especially after the revolution, in the period of 1920-1990s, about 90 000 people in the USSR have moved to the cities (Novrosen 2020). It is explained by the fact of the industrialization in 1920-1940s and the postwar urban expansion with redistribution of population to the eastern USSR regions.

During the 1920s housebuilding question was a primary goal for Soviet architecture in the discussions of urban development. It has reflected in the discussion on the matter of socialistic resettlement 1920-1930s. But however, in light of the undeveloped constructional base, whilst these years urban people have built wooden off-the-grid houses in the uptown, according to the wooden architectural traditions (Figure 1b). Therefore «one-storied USSR» was formed, but already including an urban planning solutions such as an orthogonal grid of streets that was spread in Russian country towns since Catherine the Great. And alongside this it is also central squares management accounted for fests and parades.

Closer to the beginning of 1930s there were built the first big municipal buildings, multistorey apartment buildings for party bosses and also communal buildings for workers. These buildings were built in the central part of the soviet cities.

After the Great Patriotic War there were destroyed about 1710 cities and settlements, 70000 villages and 32000 industrial enterprises (TASS 2020). One of the major and primary targets has become a provision of housing for a great amount of people which had lost their houses and therefore lived in rough living conditions. With the advent of new industrial enterprises, energetic power improvement and transport development, there have begun to form workers settlements consisted of wooden barracks that were not corresponding with people's needs. Some of these buildings were made by prisoners of war (Meerovich 2009). Essentially it was a part of «One-storied USSR».

The task of quick and economical way of people's resettlement from the barracks and ramshackle buildings with no utility system and also development of constructional base in USSR in 1950-60s – it all helped the further extensive use of cheap panel housebuilding and it is allowed to build houses using the whole urban territory and providing housing for everyone who are in need.

What worked well with this idea was the Athens Charter concept (Modernist Architecture 2020), in which one of the main changing was a move from building urban quarters with individual dwellings and a little amount of multistorey houses to the building microdistricts only with multistorey houses. This method could be called hyper-quarter building. Quarters like this were understood as unique dwelling unit that is possible to use and spread about.

The idea of microdistrict allowed placing all social objects in the quarter's center by forming the core. Besides dwelling houses in the structure of dwelling unit were added day nurseries, kindergartens, schools, shops and other services for dwelling unit. But the main thing was a green garden disposing for communications of all who live in the hyper-quarter (Figure 1a). The role of microdistrict was understood as a creation of local dwelling formation in the social unity. The new conception was close to the ideology of socialistic country and that was a move for its quick spreading.

This method helped with solving different problems. For instance it's a great help in the question of children's safety, because they have to go to school by crossing busy streets. Taking into consideration the school capacity and is optimal accessibility for the children, it's possible to count that an average number of people in the microdistrict is 5000-8000 people.

By urban reconstruction the microdistrict's structure was based on the combination of a few existing quarters for the sake of placing new schools and kindergartens in a new-formed territory.

The deal is that the definition of «microdistrict» has two conceptual directions:

– Social-economical: as a basis for microdistrict construction it was placed family life sceneries and its functional intension.
– Territorial: economical dwelling unit that forms the unified the whole one (Fedchenko 2016).

As the problem research shows it's clear that main reasons of choosing the way of microdistrict building in 1950s are following:

– Big territorial formation (microdistrict) has minimized and made easier the building of transport and engineering infrastructure that was a good corresponding with economy policy in these years in the USSR.
– People has got relatively level playing field with no kind of social differences that inevitably appears by living in individual dwelling houses

4 COMPARATIVE ANALYSIS

But why does USSR not follow the American «suburbia» model by developing peripheries and providing every family with their own house, land property with developed engineering and transport infrastructure? There are some reasons for it which is antipodes for reasons chosen by USSR in building microdistricts that was described above.

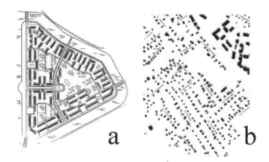

Figure 1. A) Microdistrict «Rusanovka», Kiev. Kindergardens, schools and public garden – in the center, dwelling houses are on the edges; b) The placing of individual houses of «Onestoried USSR» on the edge of the Ufa city.

During the cold war in 1946-1991s there was an expansion rise in the USA. Fear before the spreading of communistic ideas has lead to spreading anti-communistic idea. The country was ruled by authority and business on mutual agreed conditions (Khazin 2016). This idea of «suburbia» was a pure political move, because if a man has a land and his own house – he wouldn't be the communist, he just wouldn't have time for it (Khazin 2016).

Suburbia has got an opportunity to fulfill all people needs in dwelling zones. It was comfortable located for an automobilist and has a minimal necessary infrastructure. «One-storied America» approach has begun not with building houses, but with building infrastructure: highways, providing utility system (Khazin 2016). In the second half of XX century suburbia in the West Europe countries has also territorial increased.

It's typical almost for all American and European cities to have small-quarter town planning pattern with a lot of compact zones of dwelling zones which is connected with the historical ownership title and respect for private property. In the second half of XX century in many West European cities it was made small number of dwelling formation of microdistrict type in pure form.

Nowadays in foreign town planning processes, incredible metamorphoses are happening: during the last 20 years it's clear that there's a tendency of combination suburbia, historical town parts and microdistricts. In the town-planning it's used pragmatic approach in creation design models. For example, in West Europe cities new town-planning methods are developed such as «Compact building» and «New urbanism» (UN-HABITAT 2009). The «Compact building» conception is following: it's a creation of compact, but mainly mid-rise building by using traditional urban quarter network. The «New Urbanism» is a combination of specific land rise. That is why in some of West Europe countries and in the USA some definitions such as land rise, compact of building and number of stories are being criticized and watched over in some projects.

For example, the basis of planning solution by architect Jean Nouvel in the statement for the project of general layout of the Paris conturbation – it's an idea of double action which involves using «New urbanism» and «Compact building», so it's meant the renovation of existing quarters – in the uptaking, upperworks, impaction and development of urban functions (Nouvel et al. 2011). In this manner it'll be a combination of functions of trade, dwelling, office and ecological pure manufactures. The development of every place in the city will be examined in particular, in the context of existing situation.

One way or another, whether it is suburbia or «New urbanism», objective town planning parameters are the most important for further comfort living. So in this manner average number of density of population in new dwelling parts in the cities of West Europe – it's 5000 people for one square

kilometer, providing of primary highway system – till 15 kilometers of streets for one square kilometer (Demographia 2020). Moreover in the European town planning it's always suggested a reserved private space for inhabitants what creates a feeling of local community.

5 RESEARCH METHODOLOGY

Using the graphical analytic method of analysis of cartographic documentation on the basis of aero- and satellite views (including information from the U.S. Geological Survey's archive (USGS 2020)), it's possible to look at the results of building microdistricts by the example of the Ufa city (Figure 1).

Ufa was a center of the region in the USSR and one of the rapidly increasing cities in the XX century among the others industrial centers in the Volga region, it was a typical Soviet city that enlarged after the Great Patriotic War, because of the drift of the industrial assets from near-front zone. Before the war the midpart of Ufa was built up individual dwelling houses of «One-storied USSR» from 1920-1930s (Figure 2b). Multistorey buildings and microdistricts has begun to appear here already in 1950s, combining historical center and its wooden manors with northern industrial districts (Figure 2c)

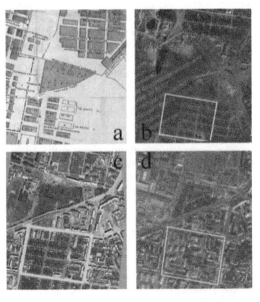

Figure 2. Evolution of functional-planning structure of the Ufa city's midpart. Yellow line is for the same territory. a) Ufa in 1925. This territory is not built up yet. b) German espionage aerophoto in 1941 (www.warfly.ru). The territory in 1941 consisted of 15 quarters of one-storied buildings of 1920-1930s c) American declassify espionage sattelite view in 1967 (USGS 2020). In 1967 the territory, within the Ayskaya, 8 March, Revolutionary and Mingazeva streets, was united into one hyper-quarter in preserved streets that became highways after. d) Satellite view in 2012 («Yandex maps»).

As the picture shows it was the process of «onestoried USSR» quarters unifying into united «hyperquarters» (microdistricts). This example illustrates how 15 quarters, which had been formed in 1920-1940s, were united into one microdistrict, old streets were transformed into passages in microdistrict or were destroyed. Only main streets as microdistrict's boundaries had been left without changing.

The first nine-storied building in Ufa was built in 1962 and became a reason to begin the building of multistorey dwelling houses. In 1970s new microdistricts were formed – Aysky, Telecenter, Green boscage, Enthusiasts etc. Nowadays almost the whole territory of Ufa is built up dwelling zones of microdistrict type with multistorey dwelling houses. Exception to this rule is only the historical center that kept within itself a few number of historical manors, small-storied houses and small garden squares.

6 SUMMARY

According to the text written above, the idea of microdistrict building had solved the following important problems in the urban reconstruction:

The boost of sanitary-hygienic standards by the rising the quality of existing dwelling houses, deconcentration, quarters ventingness, building parks, square gardens and artifical lakes, and also city lightning.

- Management of modern highways furthering safety in mechanized movement, management of mass transport and infrastructure that is necessary for the city.
- Equipment of dwelling and municipal buildings with plumbing, hot water supply, electricity, canalization, heat and gas supply system.
- Creation of necessary social infrastructure: cultural, educational, medical and trade objects.

This problem was possible to solve only be the whole housebreaking of «one-storied USSR», built in 1920-1940s, because of the fact the these years all the edges were built up wooden houses that hadn't any engineering communications. By this housebreaking it was easy to create a necessary infrastructure for the city. The fact that the land was the national property, this housebreaking of existing dwelling houses didn't make any troubles. We can conclude that existing inertial type of microdistrict building in Russia is an effect from two main reasons:

- Soviet building legacy – big existing microdistrict where all infrastructure of the periphery is made for a service of the formation like this, it leads to extensive tradition continuation in new dwelling placing.
- Low level of dwelling provision per capita (about 20 square kilometers per head that 2-3 times less

than in European countries) because of the late urbanism in Russia. Such dwelling provision creates a request for flats in dwelling groups of microdistrict types with 10/25-storied buildings, in spite of the building monotony and transport problems. It is also because of the fact that the major part of urban terrirtory in Russia is built up with microdistricts and these territories are not understood like ghetto among the people.

7 CONCLUSION

As a matter of American and European experience, it's possible to conclude that Russian town-planning needs to use a new approach, similar to «New urbanism» and «Compact building» - it's meant to build mid-storied buildings till 6 floors in historical centers and actively combine functions. It's also worth to do that in the case of small-quarter planning structure of the streets. In the periphery it's possible to build microdistricts (5/12-storied houses) and also individual dwelling groups.

As practicing city planners by research institution, authors mean that the problem of inertial microdistrict building and slow development of individual dwelling zones could be solved only by strong political decision to create conditions for development of these zones:

- Impact of land renting value for low-rise developers.
- National economic policy worthwhile for mass production of lite building elements for low-rise buildings.
- Changing the character of academic program in architectural universities moving up in designing dwelling zones of not-microdistrict type.
- Experience exchange with developed countries in the field of town planning and changeover project organizations in creation general layouts of the cities with mix building.
- Liberalization of national town planning policy to the existing individual dwelling zones in the cities of Russia – its housebreaking exception.
- Active awareness-building among the population, proving all the advantages of low quarter building.

For the replacement of some percent of developed individual dwelling zones it's necessary to have a complex economic approach. Among the others it's needed to name following actions:

- changeover to the new technologies in lowrise construction that makes its price minimum 30-40% lower and provides its flexibility in architectural and planning urban solutions;
- providing districts with necessary infrastructure: water and gas systems, canalization, electrification, communication lines;

– providing the territory with guaranteed transport infrastructure with objects in the areas of health service, education, culture and all of it according to the urban development master plan.

REFERENCES

Demographia 2020. Demographia world urban areas. 16th annual edition. http://www.demographia.com/db-worldua.pdf

Fedchenko, I.G. 2016. *Formation of residential planning units in the mid-20th – early 21st centuries. PhD Thesis in Architecture*. Moscow: Moscow Architectural Institute.

Khazin, M. 2016. Suburbia experiment — the end of the American dream. http://www.khazin.ru/articles/6-finansy-i-pravo/4896-%C2%ABeksperiment-suburbija%C2%BB-%E2%80%94-konets-amerikanskoi%C2%A0mechty

Meerovich, M.G. 2009. Ed. Howard's idea of garden-city and the Soviet working settlements-gardens. *Vestnik Tomskogo gosudarstvennogo arkhitekturno-stroitel'nogo universiteta. JOURNAL of Construction and Architecture* 4: 46–50.

Modernist Architecture 2020. CIAM's "The Athens Charter" (1933). https://modernistarchitecture.wordpress.com/2010/11/03/ciam's-"the-athens-charter"-1933/

Novrosen 2020. Population. Population migration in Russia. http://www.novrosen.ru/Russia/population/migration.htm.

Nouvel, J. et al. 2011. Grand Pari(s). Project International 29 (2): 108–135.

Stukalov G.V. 2015. *Metropolis urban planning*. Moscow: Sport i Kultura 2000.

TASS 2020. Unnecessary war. World War II in figures and facts. http://www.tass.ru/spec/wwii.

UN-HABITAT 2009. Planning sustainable cities: policy directions. London; Sterling, VA: Earthscan.

UN-HABITAT 2020. http://unhabitat.org/

USGS 2020. Official web archive of U.S. Department of the Interior. U.S. Geological Survey. http://www.earthexplorer.usgs.gov.

Reconstruction and Restoration of Architectural Heritage – Sementsov et al (eds)
© 2020 Taylor & Francis Group, London, ISBN 978-0-367-65357-6

Organizational and technological industrial area renovation systems: Basics of functioning

D.V. Topchiy & V.S. Chernigov
Moscow State University of Civil Engineering, Moscow, Russia

ABSTRACT: Any organizational activities should be structured to ensure a systemic approach to interconnections among various objects and modules. This approach is also valid with regard to creating organizational and technological systems of implementation of industrial facility conversion projects. However, efficiency of interaction among various management entities is currently at a low level due to lack of unified deterministic connections that would ensure immunity to internal and external impacts, but at the same time enable the manageability both of the entire system and its components.

1 INTRODUCTION

Industrial areas in Moscow occupy 20,000 hectares, or practically a fifth part of its territory. Moscow as the leader among the Russian cities implementing fast-paced renovation programs witnesses a twofold development trend: redeveloping an area into an innovative, science-intensive and environment-friendly enterprise while retaining the area's functional purpose, or changing the functional purpose of an area with inevitable demolition of buildings and soil recuperation.

By developing industrial areas municipal authorities create new jobs in close vicinity to residential areas at the periphery of the urban environment. This approach also offers solutions to some logistics problems, as the traffic flow will be directed from the central part of the city in the morning and from its periphery in the evening. As a result, municipal transportation infrastructure is greatly disburdened and more dynamic transport routes are formed enabling utilization of lands located far away from megalopolises.

2 MATERIALS AND METHODS

Redevelopment of industrial municipal areas creates the basis for implementing a large number of quarter-based residential housing projects as well as projects of social facilities, transport networks and "green" recreational territories. As an example, there are plans to redevelop some 15,000 hectares of land in Moscow with 5,000 hectares proposed for completed emollition of all facilities, soil recuperation and subsequent construction, while some 8,000 hectares will be partly repurposed with reconstruction and conversion of some buildings that will be subsequently used for social purposes of the residents.

The measure will provide, to a certain degree, the solution to acute municipal problems such as the lack of sports centers, parking lots, family entertainment centers, cinemas, etc. (Topchiy & Tokarskiy 2018).

However, the processes associated with industrial territory renovation and redevelopment projects are not systemic. This leads both to extensive production costs and reduced economic efficiency and technological safety of such projects.

Dating back to the middle of the last century, the principles of the "systemic approach" and the "system age" in scientific research in this country were formed in line with approaches to interpretation of scientific phenomena used by the great Russian scientists M.V. Lomonosov, D.I. Mendeleev, I.V. Michurin, S.I. Vavilov, K.E. Tsiolkovskiy, V.L. Ginsburg, L.D. Landau, to name but a few, and leading international researchers in various fields of science including Pierre Weiss, William Ashby, Ludwig von Bertalanffy, and Mihajlo D. Mesarovic.

One of the outstanding scientists, who proceeded from the systemic matrix principle in his research endeavors, was Niels Bohr. He based his quantum theory hydrogen-like atoms on a systemic approach, explained the fine structure of spectral lines, and formulated the Selection Rules, which he applied, in particular, to his harmonic oscillator. Lateron, the Correspondence Principle played a pivotal role in the development of sequential quantum mechanics, which in fact became the Interaction System. This approach to the structure of issues gives a wider perspective of the object of research and the possibility of recording the undergoing processes and arranging them in accordance with a unified principle.

On the contrary, Albert Einsteins ought to develop a system that, unlike the phenomenological approach proposed by NielsBohr, would describe the processes and changes within the system and would be dynamically adaptable to internal factors and the external environment, but at the same time remain sustainable. What Niels Bohr and Albert Einstein principally differed in was their approach identification of these sense of the system. The phenomenological approach provides insight into interrelations among system elements, while the method advocated by Albert Einstein identifies correspondences within the system empirically and aligns them in a logical necessity. However, both of these great scientists view structure development as the unified structure of an integrated approach to the unified whole (Lapidus & Abramov 2019).

"System" is an ancient term (it comes from ancient Greek σύστημα:"whole concept made of several parts, composition") used in the context of separate elements combined on unified principles, though frequently lacking unified requirements, criteria or functioning framework, as well as interaction structures. A characteristic feature of the systemic approach is analytical study of individual objects, in this case industrial territory scheduled for renovation, without precise formation of individual elements of this system, their constituent modules, or their interaction both within the system and with the external environment.

One of the first Russian scientists to study the processes of implementing construction projects, was a doctor of engineering science, professor Alexander Antonovich Gusakov, the author of Systemic Techniques in Construction, who not only defined the terms and principles for a systemic approach to various construction processes, but also developed the criteria and principles of interrelations and feedbacks from arising connections (Bolotova et al. 2019).

Development of unified selection criteria for the elements of the system of implementation of industrial facility renovation and redevelopment projects has resulted in identification of four main stages.

At the first stage one needs to identify the principal social and business-oriented elements of the system being shaped that reveal the general attitude to the redevelopment project both of the public and the state represented by various authorities.

At the second stage it is expedient to develop a mathematically interrelated network of connections, criteria and responses to internal and external drivers, to assess and shape the system's development capability, to provide a tool of user-friendly input of information and prompt receipt of responses, to assess the level of external impact, and to ensure sustainability and performance efficiency under various operating conditions (Klimina et al. 2019).

The third stage of development of a systemic approach to industrial facility conversion operations should ensure the applicability of this system to various projects associated with renovation of industrial territories.

And the final, fourth stage concerns the assessment of the system's performance, identification of critical ways of project implementation, and determination of levels and criteria of assessing the efficiency of the created system.

Therefore, it is obvious that formation of a systemic approach to the processes of renovation

and redevelopment of industrial territories requires identification and formulation of a systemic factor. This problem lies in the focus of the notion of the System to be applied in research on the development and formation of a unified system of industrial facility conversion.

The main characteristics of a system like this include a number of provisions as follows:

- A theoretically shaped system can become unified if it covers all participants, all isomorphic process regularities and interaction mechanisms of various system modules.
- The revealed isomorphic features of various system modules can be assessed if the isomorphism criteria are sufficiently weighty and significant, and the number of criteria lends itself to evaluation and mathematical analysis.
- Formation of the General System Theory requires identification and description of a systemic isomorphic factor.

Like any other system that describes inter connections among participants in construction processes, the organizational and technological system of industrial facility conversion is based on the mathematical system theory. The system itself is instrumental not only in the study, but also in the explanation of various processes involved in the project implementation. The idea of this mathematical model was described in detail in the works of Mihajlo Mesarovic, who formulated the sequence of applying a mathematical model in the study of characteristics of the model under review (deduction method) or by using computer modelling. As are result, the systemic approach methodology can be structurally presented as follows:

- Formalization. System formation on the basis of a statement of works.
- Deduction. Study of the formed system.
- Interpretation. Study of the obtained results of system operations.

Therefore, these quince of applying mathematical theory to the structure of the organizational and technological system of interconnections among parties involved in redevelopment operations can be formulated as follows: first, theoretical justification is undertaken followed by practical performance in a specific project and a subsequent assessment of the obtained results and other phenomena.

The obtained result as a critical element of the system performance should be thoroughly analyzed. To begin with, all system operations and various eventual variation series should be represented in various terminological results thus noting the significance of the mathematical model. This impact can be described by the following:

1) Kind and type of the expected result;
2) Stage-by-stage approach in obtaining the result;
3) System element(module) producing theresult;
4) Validation of the obtained result.

In fact, the theses described above are expressed by the main mechanism of the system, and at the same time they encompass the purpose of creation of the organizational and technological system.

It should be noted that the main elements of the organizational and technological system of industrial facility conversion, namely, information support and organizational structures, consist of various element-forming modules combined in accordance with unified attributes and internal ties. Notably, an important parameter under review is the "degree of freedom" both of the modules and the system elements in general, and, accordingly, the system itself. If the "degree of freedom", i.e. the boundaries of module operations, is noted entified, it is impossible to assess the module and correctness of its operations, because the information received from this module will be excessive and will not lend itself to processing or structuring. All degrees of freedom of a module that fail to produce any significant results should be deleted.

3 CONCLUSION

Accordingly, the obtained result is the main and indispensable element of the system operations, a tool ensuring a streamlined impact on its various elements and modules. If elements operate in accordance with this principle, the system is called functional.

To sum up the above assumptions, it can be argued that a "management system" is assumed to be located outside the managed object. Dialectically, the term management system means that such a system is fully fledged and fully sufficient.

REFERENCES

Abramov, I. et al. 2016. The analysis of the functionality of modern systems, methods and scheduling tools. *MATEC Web of Conferences* 86: 04063.

Bolotova, A.S. et al. 2019. Technical rationing of the construction technology of reinforced concrete floor slabs using nonremovable empitness-liners. *International Journal of Civil Engineering and Technology* 10 (2): 2160–2166.

Klimina, V. et al. 2019. Unified classification of defects detected by the technical examination. *E3S Web of Conferences* 110: 01086.

Lapidus, A. & Abramov, I. 2018. Studying the methods for determining and maintaining sustainability of a construction firm. *MATEC Web of Conferences* 251: 05017.

Lapidus, A.A. & Abramov I.L. 2019. Systemic integrated approach to evaluating the resource potential of a construction company as a bidder. *IOP Conference Series: Materials Science and Engineering* 603: 052079.

Topchiy, D. & Tokarskiy, A. 2018. Formation of the organizational-managerial model of renovation of urban territories. *MATEC Web of Conferences* 196: 04029.

Topchiy, D. & Lapidus, A. 2019. Construction supervision at the facilities renovation. *E3S Web of Conferences* 91: 08044.

Topchiy, D. et al. 2019. Construction supervision during capital construction, reconstruction and re-profiling. MATEC Web of Conferences 265: 07022.

Reconstruction and Restoration of Architectural Heritage – Sementsov et al (eds)
© 2020 Taylor & Francis Group, London, ISBN 978-0-367-65357-6

Principles of forming functional landscape for wooden architecture museum in southern part of Zaonezhye Peninsula

S. Zavarikhin, O. Kefala & T. Nesvitckaia
Saint Petersburg State University of Architecture and Civil Engineering, Saint Petersburg, Russia

ABSTRACT: The article determines conditions for placing historical chapels in the southern part of the Zaonezhye Peninsula as part of the Zaonezhsky tourism and recreational cluster.

1 INTRODUCTION

The Zaonezhsye Peninsula, located in the north-eastern part of Onega Lake, is a real reserve of wooden architecture of the Russian North. However, the continuing population decline leads to the disappearance of villages, churchyards, small settlements, slobodas, sketes, and changes in the typical functional landscapes. People are leaving the villages. Unique wooden churches and chapels are left without proper care, overgrown with trees and shrubs, historical buildings are decaying, and all this accelerates their destruction. Due to the inaccessibility of many of them, it is impossible to carry out the necessary repair and restoration works and organize tourist displays of these places. Therefore, the task of the concentration of architectural monuments in the territory of the wooden architecture museum is urgent. In many ways, this important task will be solved by a project of organizing the Zaonezhsky tourism and recreational cluster (TRC). However, a preliminary solution to another problem is required: the location of architectural monuments with account for their historical perception, which determines the fulfillment of certain conditions for their spatial organization according to centuries-old compositional techniques, and the formation of a functional landscape so that objects transported there or existing there could end up in their native landscape situation. The authors of the study believe this condition is crucial, therefore, by using cartographic, bibliographic, measurement, graphic, and photo materials, they classified landscape situations, compositional techniques, and architectural-and-spatial forms, using chapels (as one of the most common types of wooden architecture monuments in the southern part of the peninsula) as an example, and determined the main recommendations for their placement.

1.1 The territory of the Zaonezhye Peninsula was actively occupied and developed as a living environment by ancient hunters for wild deer since the Mesolithic era. In the 10th–13th centuries, the Sami clans were replaced by Finnic tribes (Veps, Karelians, Finns). In the 15th century, these lands became part of Muscovy. The "great salt road from Novgorod to Sumsky Posad" passed through Zaonezhye (Korablev et al. 2001).

Due to its remoteness from Central Russia and long-term (and almost complete) self-sustainability, Zaonezhye still has villages that are more than five thousand years old. Such self-sustainability contributed to the formation of strong traditions in all spheres of life, including architectural and construction activities. Traditional construction technologies, in turn, provided good preservation of wooden structures for centuries. They also contributed to the preservation of the unique landscapes of the peninsula, including a large number of lakes, water channels, hills and cliffs, forests and woodlands.

The patriarchal-natural way of life on the peninsula remained until the early 1930s when industrial technologies and new social "technologies of life" came to this historical "enclave". This led to the out-migration, which, in turn, resulted in the disappearance of traditional settlements. Unique wooden churches and chapels were left without people and transport communications. It is extremely difficult to carry out the necessary restoration work, preserving the functional landscape, and organizing tourist displays of these places due to their inaccessibility (Republican Center for State Protection of Cultural Heritage Sites 2020).

Therefore, the task of the concentration of wooden church architecture (in particular, chapels as the most common type) in the territory of the open-air wooden architecture museum is vital.

The representation of a typological series of chapel structures in the exposition of the Architectural Park on the historical and cultural landscape of the Kenozero National Park can be used as an example (Kenozero National Park 2016).

In 2014, the government of Karelia approved the "Concept of creating Zaonezhsky tourism and recreational cluster". In the course of its implementation, it is planned to establish a tourism and recreational area on the Zaonezhye Peninsula and adjacent islands of Onega Lake. This project will open the authentic

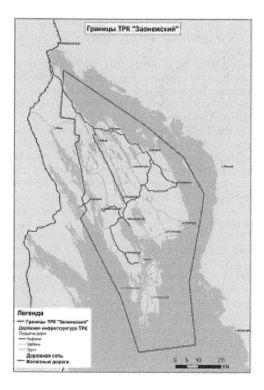

Figure 1. Borders of the Zaonezhsky TRC with a scheme of road infrastructure scheme (SPOK 2020).

Zaonezhye region to Russian and foreign travelers. In this territory, 126 architectural monuments, including historical chapels, will be presented as tourist resources (Maxpark 2020, SPOK 2020) (Figure 1).

The solution of this problem requires the prior solution of another task: identifying historical chapels located in remote places, determining peculiarities of their functioning in the cultural and historical landscape of this territory so that after their transportation, spatial and landscape characteristics of the objects could be similar to those that were observed in the area of their original location.

The purpose of the study is to substantiate the principles of wooden church architecture monuments' functioning, using chapels as an example, in the territory of the designed Zaonezhsky TRC.

The tasks of the study are as follows:

- to collect as much information as possible about the types of chapels in the territory of the Zaonezhye Peninsula;
- to identify features of the spatial perception of chapels;
- to collect as much information as possible about the types of landscape with regard to particular chapels in the territory of the Zaonezhye Peninsula;
- to analyze and summarize all the obtained characteristics of chapels, types of their spatial

perception in functional landscapes in final conclusions and recommendations for the representation of the objects in the Zaonezhsky TRC.

The study covers historical chapels located in the southern part of the Zaonezhye Peninsula.

In addition to the materials of a partial on-site survey, cartographic materials, literature sources, and photos were also used to solve the tasks. The classification of functional and cultural landscapes was the main method used when handling these materials.

Wooden architecture of the Russian North, in particular church architecture, is a clear manifestation of the authentic Russian national culture and a valuable architectural heritage. Constructive and decorative techniques of religious wooden construction were developed before the adoption of Christianity in Russia and were not affected by Catholicism, as evidenced by historian of Russian architecture A. Krasovsky in the second half of the 19th century (Musin 1997).

The uniqueness and value of wooden church architecture are determined by its functioning in a cultural landscape.

The concept of cultural landscape is defined as a historically formed harmonious combination of nature and culture, which includes an established system of settlement and landscapes. Any "man-made landscape" changed by man is valuable as it includes mythology, spiritual culture, and the evolution of the historical development of the place (Livinskaya 2012). Therefore, it is necessary to take all available means to preserve wooden architecture monuments in their authentic environment.

If it is not possible to ensure the necessary transport communications and conditions for the proper preservation of a monument, it is required to transport it to the territory where measures for the maintenance, repair and restoration of a historical object are available. However, a monument must be placed in an identical landscape.

In her article "The landscape approach to conservation of the cultural heritage of the National Park "Kenozersky" and presenting it to the UNESCO World Heritage List", M.E. Kuleshova notes that cultural landscapes as a special category of cultural heritage have been considered by the World Heritage Committee since 1992. According to the typology developed by UNESCO, all cultural landscapes are divided into landscapes designed and created intentionally by man, organically evolved, and associative ones. The territory of the Zaonezhye Peninsula belongs to the organically evolved type. As a result of long-term agricultural, commercial, and sacred activities of people living there, the original natural landscape adapted to the purposes of their activity (Kuleshova 2015).

Yu.S. Ushakov, a researcher of architectural and planning traditions, villages in the Russian North, revealed changes in compositional techniques, spatial organization, and perception of development depending on the features of the landscape where a settlement was formed. In Zaonezhye, the settlement system traditional for the Russian North —villages, churchyards, small settlements, slobodas, and sketes — has been preserved. Groups of settlements were set around churchyards, and "nests" of villages were grouped next to them. Villages, as a rule, were united with some natural element (river bend, lake, island, peninsula), and the landscape composition of the entire Zaonezhye territory is related to Onega Lake, which is the center of the composition (Ushakov 1982).

In the 1970s, Yu.S. Ushakov noted the importance of a holistic and systematic analysis of folk architecture. He focused on the need to study monuments with account for their spatial relationship with each other and the natural environment. As a result of his surveys in the regions of the Russian North-West, he pitched the following idea: "…if parts of the whole (individual structures) were created by folk masters with great skills, why could not the entire unit (village, settlement, group of villages) created by the collective labor of the folk as the creator have the same high qualities?" In his research, he emphasized that when choosing a place, all its climatic, geological, transport features were taken into account, and the natural environment suggested planning and compositional techniques. Special attention was paid to the beauty of the selected area. The aesthetic value of nature was appreciated by our ancestors and was an integral part of life (Ushakov 1974).

Nowadays, in the globalization era, the organic unity, harmony of nature and traditional folk architecture are being lost rapidly.

Measures to protect the existing functional and cultural landscape in the territories of open-air museums were taken as far back as in the 19th century in the Scandinavian countries, where ethnographic open-air museums were established. One of the first museums was created by ethnographer Arthur Hazelius in 1891 on Djurgården Island on the outskirts of Stockholm. His goal as an ethnographer was to preserve elements of folk culture (eFinland 2020).

Following the Swedish museum, similar museums were created all over Europe for the same purpose. As a rule, they were located on the outskirts or near megacities, and turned into entertainment centers.

In the 1950s, the Kizhi reserve museum became the first architectural reserve in our country. According to the regulations on the state architectural reserve in Kizhi approved in June 1959, monuments of folk wooden architecture doomed to destruction started to be transported to the territory of 15 hectares. The Kizhi architectural ensemble became the center of architectural and ethnographic composition (Kizhi Museum 2020).

The measures taken to ensure the protection of monuments in an accessible place had positive and negative aspects. Breaking the bond between the brought monument and its former natural environment was one of the significant negative aspects of their placement in a new area.

The fusion of folk architecture with landscape makes it poetic and accurate in terms of adequate perception.

As a rule, the preservation of monuments takes place by their exposure in ethnic parks or open-air museums. They receive the necessary care there, but they are located outside of their historically formed functional landscapes (Chainikova 2018).

In the 1990s, experts turned to the issue of the comprehensive use of cultural heritage. Gradually, traditional ideas about heritage as individual material objects were interpreted differently. Currently, the preservation and development of cultural heritage are considered in relation to historical landscapes, technologies, and specific forms of management (Mironova 2020).

Today, in the southern part of the Zaonezhye Peninsula in Pryazhskiy District, we can visit the surviving historical villages Kinerma and Manga coming alive in the summer. Kinerma was put in order by the efforts of local residents, and in 2016, it became a member of the association of the most beautiful villages in Russia. It has buildings with an age of more than 100 years (Sivolap 2013).

Volunteers maintain miraculously preserved chapels in empty, abandoned, and inaccessible villages of the peninsula. But their efforts cannot solve the global problem.

For the analysis of the state and types of functional landscapes within the territory of the future Zaonezhsky TRC for the purposes of this study, and with the use of available sources, 20 objects, chapels (monuments of wooden religious architecture) were selected (as the most typical ones and preserved in a large number), which are located in Velikogubskoye rural settlement included in the southern part of the Zaonezhye Peninsula. The objects were considered in the context of their relationship with the type of functional landscape. The following types of landscape situations were identified as the most typical for this area:

– L-1 — on a lake shore. Example: chapel of St. George the Dragon Slayer, 18th century, Ust-Yandoma village (Figure 2).

Figure 2. Ust-Yandoma village, the chapel of St. George the Dragon Slayer, 18th century.

Figure 3. Samson's chapel in Kondoberezhskaya village, mid-19th century.

Figure 5. Roadside chapel of the Descent of the Holy Spirit, Tipinitsy village, end of the 19th century.

- L-2 — on a lake cape. Example: Samson's chapel, mid-19th century, Kondoberezhskaya village (Figure 3).
- L-3 — in a closed forest clearing. Example: chapel of Kazan icon of the God's Mother, 1780, Krasnaya (Gryaznaya) Selga area (Figure 4).
- L-4 — in a roadside clearing. Example: roadside chapel of the Descent of the Holy Spirit, Tipinitsy village, end of the 19th century (Figure 5).

Based on the visual assessment of functional landscapes, we can argue that they have different degrees of change:

- Completely changed original functional landscape. The object has lost its historical perception in the functional landscape. The landscape needs to be recreated (45% of the surveyed amount).
- Significantly changed original functional landscape. The object partially preserves the historical perception in the functional landscape. The landscape requires significant restoration works (40% of the surveyed amount).
- Slightly changed original functional landscape. The object mainly preserves the historical perception in the functional landscape. The landscape

requires no significant intervention to bring it to its historical appearance (15% of the surveyed amount).

The study establishes the following principles for placing historical chapels, which are suggested as recommendations for the formation of a wooden architecture museum in the southern part of the Zaonezhye Peninsula as part of the Zaonezhsky TRC area.

- Priority landscape situations should be defined in the selected territory, which should be presented in the form as close as possible to the historical composition and formed in the territory of the wooden architecture museum. They include L-1, L-2, L-3, and L-4 types.
- When transferring chapels, a new location in the territory of the future Zaonezhsky TRC should be chosen in accordance with the characteristics of historical functional landscapes for chapels transferred, so that they could retain their authenticity.
- If it is impossible to find a functional landscape in the territory of the future Zaonezhsky TRC, close to the historical one, for chapels transported from remote areas, it is necessary to recreate such type of landscape.
- If chapels are preserved in their original place, the historical functional landscape should be restored, and transport communications should be provided for the introduction of historical objects into cultural circulation.

REFERENCES

Chainikova, O.O. 2018. The evolution of preservation of monuments of traditional wooden architecture. *News of the KSUAE* 3 (45): 72–79. eFinland 2020. *Seurasaari — an island-museum under the open sky.* Accessed April 20, 2020. https://e-finland.ru/rest/all-year/seurasaari-ostrov-muzeyi.html.

Figure 4. Krasnaya (Gryaznaya) Selga area. Chapel of Kazan icon of the God's Mother, 1780.

Kenozero National Park 2016. *Kenozero readings.* Arkhangelsk Region.

Kizhi Museum 2020. *From the history of the Kizhi museum-reserve formation.* Accessed April 20, 2020. https://kizhi.karelia.ru/library/kizhi-40/403.html.http://kizhi.karelia.ru/http://kizhi.karelia.ru/library/kizhi-40http://kizhi.karelia.ru/library/kizhi-40/403.html.

Korablev, V.K. et al. (eds) 2001. *The history of Karelia from ancient times to the present day.* Petrozavodsk: Periodika.

Kuleshova, M.E. 2015. The landscape approach to conservation of the cultural heritage of the National Park "Kenozersky" and presenting it to the UNESCO World Heritage List. *The Heritage Institute Journal.* Accessed April 20,2020. https://cyberleninka.ru/article/n/landshaftnyy-podhod-k-sohraneniyu-kulturnogo-naslediya-natsionalnogo-parka-kenozerskiy-i-predstavlenie-ego-v-spisok-vsemirnogo.

Livinskaya, O.A. 2012. Concept of a cultural landscape of our country geography. *Pskovsky Regionologichesky Zhurnal* 14: 120–128.

Maxpark 2020. *Journey through the wooden Zaonezhye.* Accessed April 20, 2020. maxpark.com›Blog›content/6797230.

Mironova, N.I. 2020. *Socio-cultural space of Russian territories in the background of municipal reforms (Arkhangelsk Region, Chelyabinsk Region, Republic of Karelia). 4.3. Development of rural cooperation and cultural tourism in Karelia.* Accessed April 20, 2020. http://rykovodstvo.ru/exspl/22160/index.html.https://rykovodstvo.ru/https://rykovodstvo.ru/exspl/22160/index.html?page=10.

Musin, A.E. 1997. *Formation of Orthodoxy in Karelia (12th–16th centuries).* Accessed April 20, 2020. https://kizhi.karelia.ru/library/ryabinin-1995/153.htmlhttp://kizhi.karelia.ru/http://kizhi.karelia.ru/libraryhttp://kizhi.karelia.ru/library/ryabinin-1995http://kizhi.karelia.ru/library/ryabinin-1995/153.html.

Republican Center for State Protection of Cultural Heritage Sites 2020. *Cultural heritage sites in Karelia.* Accessed April 20, 2020. monuments.karelia.ru.

Sivolap, T.E. 2013. Museum specifics in regional development in the context of cultural heritage preservation. *Trudy SPbGIK* 193: 77–81.

SPOK 2020. *Concept of creating the Zaonezhsky tourism and recreational cluster.* Accessed April 20, 2020. http://spok-karelia.ru/uploads/f7b.pdf?v=1.

Ushakov, Yu.S. 1974. *Wooden architecture of the Russian North.* Leningrad: Znaniye.

Ushakov, Yu.S. 1982. *Ensemble in the architecture of the Russian North.* Leningrad: Stroyizdat, Leningrad Department.

Reconstruction and Restoration of Architectural Heritage – Sementsov et al (eds)
© 2020 Taylor & Francis Group, London, ISBN 978-0-367-65357-6

Specifics of rhythmic-and-metric as well as the architectural-and-spatial organization of Kamennoostrovsky Prospekt

S.P. Zavarikhin, M.A. Granstrem & M.V. Zolotareva
Saint Petersburg State University of Architecture and Civil Engineering, Saint Petersburg, Russia

ABSTRACT. Preserving the historical, urban-planning and architectural heritage of the central part of Saint Petersburg, developing these territories, and smoothly including them into the life of a modern city require a special approach to solving problems related to the reconstruction and renovation of the historic urban environment. The modern period characterized by the active reconstruction of the historic center puts forward new demands concerning the preservation of cultural heritage sites. If history is seen as the experience underlying modern practice, historical, architectural and urban-planning processes should be revealed by systematizing the existing and recovering missing information. This makes for analysis of spatial and temporal transformations that the city has undergone in its development. Identifying the inner laws of the genesis of these processes is important. The article gives an analysis of the modern state of the city landscapes within the boundaries of the historic part of the Petrogradsky District.

1 INTRODUCTION

Preserving the historical, urban-planning and architectural heritage of the central part of Saint Petersburg and including these territories into the life of a modern city require a special approach to solving problems related to the renovation of the historic urban environment. The modern period characterized by the active reconstruction of the historic center puts forward new demands concerning the preservation of cultural heritage sites.

Saint Petersburg is unique because it was built according to a single general layout within the minimum period of time. The historical architectural and spatial environment of Saint Petersburg is characterized by unique cohesiveness despite a variety of its components.

Historic and cultural continuity is a key element of the successful development of urban-planning processes since it is the embodiment of awareness of the lasting value of architectural heritage, careful attitude to historical and cultural landmarks.

2 MAIN PART

The Neva water area with a characteristic pattern of shorelines bordering large housing developments plays the leading role in shaping the central districts of the city. Avenues and streets that contribute to the perception of deep perspectives of environmental areas form a significant part of a single open space.

The Peter and Paul Fortress complex and, in particular, the Peter and Paul Cathedral have the greatest impact on the perspective expansion in the area of Petrovsky Island.

Petrovsky Island is the foundation of Saint Petersburg. It is there, near the fortress separated by the Kronverksky Canal where the first settlement of the city declared the capital of the Russian State originated from. Here, the first residential houses and the first streets appeared. Collegia, a wooden Trinity Church, a merchant harbor with a jetty, and a customs office were located at the square turned to the Neva River. The Cabin of Peter the Great, with buildings of the elite nearby, was located near the square.

As the city grew along the Neva River banks, Peterburgsky Island became the place where the Nevsky, Peterburgsky, Koporsky, Yamburgsky and Belozersky regiments were stationed. This shaped the structure of mutually perpendicular streets crossed by Bolshoy Prospekt. Later, Maly Prospekt and Bolshaya Pushkarskaya Street were laid in parallel to it. They were built up with wooden one- and two-story manors with gardens. The way of the future Kammenoostrovsky Prospekt crossed the island from Bolshaya Neva to Malaya Nevka. It connected the nucleus of the city with Kamenny Island and, then, with northern suburbs. There was a space near the Kronverk esplanade that remained unoccupied almost until the mid-19th century. A semicircular space adjacent to the esplanade in this place became the foundation for a radial street network, with the streets bumping into a regular network of regiment settlements located in the middle of the island (Zavarikhin 2012b).

The almost "natural" structures of the island streets were afterward captured in the regulation plan of 1861.

The volumetric and spatial environment of the Petrogradsky District and, in particular, the environment of the area under consideration is a result of two overlapping layouts (Sementsov 2003a) that were generated in different periods and, therefore, intended for different functional and design tasks.

Thus, the junction of two (future) avenues (Kamennoostrovsky and Bolshoy Prospekts) was laid over the structure of regiment settlements constructed in the 18th century (Kurbatov & Gorjunov 2013). At the end of the 19th century, when the city primarily developed to the north, and with the cancellation of the prohibition on the construction of private wooden houses on the Petrograd Side, these streets acquired the festive, entertaining and representative function expressed in their volumetric and spatial structure (Zolotareva 2015b).

Since there was no transport communication between Peterburgsky Island and other parts of the city, it hindered its development with regard to construction activities. Only adjacent islands (Kamenny, Yelagin, Krestovsky Islands) were actively built up with cottages and manors of the elite.

The construction of the Troitsky Bridge in 1903 moved the city boundary in the north, and Peterburgsky Island underwent a construction boom. Instead of small wooden houses remaining in the places of regiment settlements and local manors, multi-story stone residential houses were built, thus, turning the main streets into respectable highways. Kamennoostrovky Prospekt with a unique ensemble became the main representative street. The buildings of Kamennoostrovsky Prospekt that were erected primarily in the late 19th – early 20th centuries produces a holistic artistic impression (Granstrem & Zolotareva 2014).

Kamennoostrovsky Prospekt starts with the Troitsky Bridge where Troitskaya Square was established. Despite the fact that its space is made with structures erected at different times, it gives a holistic compositional and artistic impression. The square started with the construction of Kschessinska Mansion by architect Alexander von Hohen in 1904–1906. Despite a flexible layout, the architect fixed a corner with a rotunda, thus, making it a starting point of two streets — Kamennoostrovsky Prospekt and Bolshaya Dvoryanskaya Street (Zavarikhin 2012a).

Further development of the square started during the Soviet time. Its eastern part was fixed by the frontal areas of two buildings: the House of Political Convicts (architect G.A. Simonov) and the building of LenNIIproyekt erected in 1956 (architects O. Guryev, Ya. Lukin, A. Shcherbenok, N. Maksimov).

When speaking of Kamennoostrovsky Prospekt as an ensemble, it should be noted that it is the squares that set the rhythmics of its architectural and spatial arrangement. Getting over the green curtain of the park at the forward slope of the parapet we see another square. Its left side is formed by a semicircular building erected according to the

design of O. Guryev and V. Fromzel. The building is heavily decorated with a rhythmic line of pilasters at the windows of the third and fourth stories. A monument to Maxim Gorky is almost in the center of this semicircle. In order to include the cour d'honneur of the house of Fyodor Lidval at the odd side of Kamennoostrovsky Prospekt, the monument was placed in its axis (Shvidkovsky 2007).

The unique nature of the historic part of the Petrograd Side is that the environment of its main spaces was formed in the second half of the 19th — beginning of the 20th centuries. Arguably, the district became a testing site for new techniques of volumetric and spatial arrangements.

The architectural expressive means used in the system of dominants were also new. During the previous periods of the reign of the Classicism in its various forms, high-rise buildings fixing the main design points and serving as guide marks were the dominants (Zavarikhin 2015).

In the 18th century, the existing Prince St. Vladimir's Cathedral erected in 1765–1789 according to the designs by Rinaldi and Starov was the main dominant in the western part of the island. At the crossroads of Pushkarskaya and Vvedenskaya Streets, regimental Vvedenskaya Church owned by the military department was built in 1793–1810 by architect I.M. Lehm. Another regimental church — St. Matthew's Church — belonged to the Koporsky and Sankt-Peterburgsky regiments and was wooden until the beginning of the 19th century. The domes and bell chambers of the churches established visual connections to the central ensembles of the city (Zolotareva 2014).

In the second half of the 19th century, the split from the classical style with its universal and harmonized nature and the transition to the freedom of forms led to new design techniques, including with regard to visual guide marks.

Leo Tolstoy Square (until 1925 — Arkhiyereyskaya (Bishop) Square) belongs to two avenues (Kamennoostrovsky Prospekt and Bolshoy Prospekt) and is a junction that joins these two main axes of the island (Kefala 2015). Bolshoy Prospekt ends with a rental house designed by K. Rosenstein together with A. Belogrud. The house is a peculiar keystone of the square. A certain rhythm is set by green curtains of the public gardens on Kamennoostrovsky Prospekt. The green areas grow from the beginning of the prospekt and reach to Leo Tolstoy Square.

An analysis of the environment of Kamennoostrovsky Prospekt (from its beginning to Leo Tolstoy Square) points to the creation of a peculiar ensemble of a separate street in this part of the Petrograd Side. It is characterized by the alteration of individual buildings or groups of buildings with large-scale forms and details as well as green elements of public gardens.

Along the prospekt, from Leo Tolstoy Square to the Karpovka River embankment we see another square — Shevchenko Square. It is composed of Maly Prospekt, Levashovsky Prospekt and Ordinarnaya Street. The sharp avant-garde layout of the square emerged in the 1930s. Buildings erected during this period are also part of this layout (Kirikov & Stieglitz 2008). They include a residential House of Svirstroy workers (architect — I.G. Yavein) and the Promkooperatsiya Community Center, which should have become the design dominant of this area of Kamennoostrovsky Prospekt.

The architectural landscape of Kronversky Prospekt is quite unique. It brings a combination of dense houses on the one side (odd side) and dense curtains of trees in the Alexander Park on the other side. Besides, Kronverksky Prospekt is closely connected to the main dominant of citywide significance — Peter and Paul Cathedral (Kurbatov 2008). It is the geometric center of the circle of Kronverksky Prospekt. The cathedral's spire should have been seen from any point of Kronverksky Prospekt. In fact, this happens only during wintertime when the trees that form a green facade of the prospekt from the side of the Alexander Park become not so dense. During summertime, the spire of the cathedral can be seen from a limited number of areas (Granstrem et al. 2018). The visual perception of the Saint Petersburg Mosque, which is also a dominant of citywide significance, also depends on the time of the year.

The spatial-planning relationship between the Peter and Paul Cathedral and Kronverksky Prospekt is directly related to the history of the prospekt. It semicircles the crownwork of the Peter and Paul Fortress since it was developed with a regulated setback from the Peter and Paul Fortress and its fortifications (Sementsov 2003b).

Within Kronverksky Prospekt, there are visual dominants that have an impact both on the prospekt and the adjacent streets (Zolotareva 2015a). They include the observatory tower in the center of the ITMO University building in the built-up part of the prospekt. They also include the building of the former Community Hall of Emperor Nicholas II built in 1900–1902 and located in the Alexander Park. A long building that has been partially rebuilt currently consists of three units: the Baltic House Theater, the Planetarium, and the Music Hall.

There are local dominants in the built-up part of the prospekt as well (Shvidkovsky 2005).

One of the important components forming the environment of the historic city is its background systems of buildings connecting architectural ensembles, complexes and local dominants. It primarily consists of residential rental houses. The environment-forming development plays a special role for Saint Petersburg: it is a historical fabric of the city that creates its unique wholeness and, at the same time, gives an unmatched identity to different districts of the city (Sementsov 2012).

This environment (as opposed to city-forming ensembles) and a number of well-known accent buildings are the least studied. At the same time, it is the most vulnerable since the modern principles of monument protection do not include an integrated approach to the preservation of the environment-forming development.

This is one of the reasons for numerous urban-planning mistakes made in historic areas, continuous changes and even destruction of some fragments that, for several centuries, have created a holistic and complete environment of the historic city (Granstrem 2019).

Despite the "central" nature of the territory under consideration, it has areas with various degrees of environment completeness. The term "incomplete environment" hardly has its usual meaning in this case. To this date, this term, which has a negative connotation, denoted so-called lacunae intended for modern reclaiming (Granstrem 2013). Currently, it is certain that one of the parameters in the historical structure of the city space is its historical incompleteness that has design, aesthetical and functional meaning. The presence of numerous firewalls in Saint Petersburg that sometimes play a role of vivid accents exercising colossal emotional effect is an example of such incompleteness (Granstrem 2016). Thus, the notion of "historical incompleteness of the environment" can, in some cases, be regarded as one of the components of the structure of the historic city.

3 CONCLUSION

Spatial-planning specifics of the development of the territory under consideration is as follows:

- small quarters within the territory of the former regimental land;
- original unification of the land plots intended for residential houses;
- while primarily residential nature of the territory is preserved, new functions (industrial, warehousing, household, etc.) were introduced in the process of the inclusion of these land plots into the boundaries of the city;
- depending on the location of the territories, there are areas of dense rental houses (in publicly significant places); within the territories under consideration, these areas include areas adjacent to Kamennoostrovsky Prospekt and Bolshoy Prospekt;
- in publicly significant places, the scale of residential facilities becomes larger, and residential complexes are built;
- loose rental houses with areas of violated or incomplete environment are, as a rule, adjacent to the above territories;
- the territory of loose residential buildings in the historical part of the island includes facilities that are discordant with the historical environment of the adjacent territories.

REFERENCES

Granstrem, M.A. 2013. Revisiting analysis of environment-forming buildings on Vasilyevsky Island in Saint Petersburg. In *Modern Problems of Architecture and Civil Engineering: Proceedings of the 5th International Conference*, June 25–28, 2013. Saint Petersburg: Saint Petersburg State University of Architecture and Civil Engineering.

Granstrem, M.A. 2016. Is it possible tu scientific methods of reconstruction fragments of historical cities. *Arkhitekturny Almanakh* 1: 71–80.

Granstrem, M.A. 2019. Problems of preserving landscape and visual interconnections of the historic core of Saint Petersburg. *News of Higher Educational Institutions. Construction* 2: 98–109.

Granstrem, M.A. & Zolotareva, M.V. 2014. Research in the structure of historical housing development of Saint-Petersburg. *Zhilishchnoe Stroitel'stvo* 11: 23–25.

Granstrem, M.A. et al. 2018. High-rise construction in historical cities through the example of Saint Petersburg. *E3S Web of Conferences* 33: 01028.

Kefala, O.V. 2015. Development of the planning structure of the Petrograd Side in the first third of the XVIII century. *Bulletin of Civil Engineers* 6 (53): 21–29.

Kirikov, B.M. & Stieglitz, M.S. 2008. *Leningrad avant-garde architecture. A guide*. Saint Petersburg: Kolo.

Kurbatov, J. & Gorjunov, V. 2013. The fate of the creative legacy in the modern architecture in Russia. *World Applied Sciences Journal 23 (Problems of Architecture and Construction)*: 203–206.

Kurbatov, Yu.I 2008. *Petrograd. Leningrad. Saint Petersburg: lessons of architecture and urban-planning*. Saint Petersburg: Iskusstvo SPb.

Sementsov, S.V. 2003a. Stages of urban development of Saint Petersburg and typology of master plans of the 18th–20th centuries. In *Reconstruction of Saint Petersburg. International Scientific and Practical Conference*:

93–98. Saint Petersburg: State University of Architecture and Civil Engineering.

Sementsov, S.V. 2003b. Urban planning legislation of Saint Petersburg in the 18th – early 20th centuries. *Vestnik "Zodchiy. 21 vek"* 3 (11): 40–49.

Sementsov, S.V. 2012. Saint-Petersburg historic agglomeration as an urban planning world-wide object. *Internet-Vestnik VolgGASU. Series: Multi-Topic* 1 (20).

Shvidkovsky, D. 2005. The founding of Petersburg and the history of Russian architecture. State Academy of the Fine Arts of Russia 66: 79–97.

Shvidkovsky, D. 2007. *Russian architecture and the West*. New Haven, London: Yale University Press.

Zavarikhin, S.P. 2012a. About the silhouette and not just that. *Kapitel* 1: 34–37.

Zavarikhin, S.P. 2012b. *Saint Petersburg. Architectural stories*. Saint Petersburg: State University of Architecture and Civil Engineering.

Zavarikhin, S.P. 2015. Modern construction in the historical center of Saint Petersburg. In *Reports of the Scientific and Practical Conference "Modern Problems of History and Theory of Architecture"*: 115–122. Saint Petersburg: Saint Petersburg State University of Architecture and Civil Engineering.

Zolotareva, M.V. 2014. Identification of historical patterns in the development of the high-rise dominants system in the central part of Saint Petersburg (based on works of artists and graphic artists of the 18th – early 19th centuries). In *Proceedings of the International Scientific and Practical Conference "Modern science: theoretical and practical views"*: 105–109. Ufa: Aeterna.

Zolotareva, M.V. 2015a. Principles of spatial development of high-rise zoning of the center of Saint Petersburg. *Zhilishchnoe Stroitel'stvo* 11: 27–31.

Zolotareva, M.V. 2015b. Space planning features of the architectural landscape in the historic areas of Petrograd district in Saint-Petersburg. *Bulletin of Civil Engineers* 1 (48): 27–35.

Revitalization concept on historical area Near Zapskovye in Pskov

I.S. Zayats
Saint Petersburg State University of Architecture and Civil Engineering, Saint Petersburg, Russia

E.S. Bakumenko & A.S. Perepech
"Gruppa "Spektr" Ltd, Pskov, Russia

ABSTRACT: The article discusses basic principles of conservation and revitalization of the Near Zapsko-vye historical territory fragment in Pskov. Based on comprehensive scientific research, conceptual directions for working with existing fragments of multi-time development have been identified.

1 INTRODUCTION

The problems of preserving the historical cities environment are now becoming more acute against the background of negative trends in the loss of historical memory. Information is distorted or disappears under the influence of momentary decisions and modern society desires, incorrect and shallow reading of authentic elements, subjective assessments and unprofessional approaches. Insufficient research leads to the destruction unique information and historical material destruction. The city of Pskov, in which centuries-old history has formed the phenomenon of multi-layered urban planning structure, is no exception.

A complex "organism" requires special attention and a differentiated approach. Each element of the historical context is a unique marker, a sign symbol of a certain era. It helps to maintain the "spirit of the place", which promotes in finding harmonious solutions that tactfully include the historical environment in the modern city life, filling it with new functions. Therefore, an important task is historical environment genuine elements conservation in a context of their integration into modern existence conditions.

Pskov was first mentioned in the Lavrent'evskaya letopis' in 903 (Laurentian Chronicle 1377) and it is one of the oldest cities in Russia. In the 14th - 16th centuries the city was the capital of the independent Pskov Republic. Until the beginning of the 18th century, Pskov was one of the largest cities in Russia and Europe, the most important defense and trade center on the western borders of the country. After the founding of Saint Petersburg, Pskov lost its significance. In the 19th century, the provincial Pskov attracted many researchers of antiquity as a monument city.

The city had been growing; buildings instead of one-story became stone in two - three floors (Labu-tina 1985). Numerous, more than 50 churches, chapels, monasteries, medieval residential buildings, ancient fortress walls created a unique and attractive image. The 20th century has made its own corrections. Civil architecture became more diverse due to Industry developed.

During the Soviet period, a large number of multi-apartment buildings were erected in the central part of the city, schools, kindergartens, administrative buildings, university buildings were actively built. However, despite this, the territory of the historical center or the "Okolny gorod" has changed slightly. Some buildings of the Soviet era are organically integrated into the historical context and adorn the main streets. The Great Patriotic War did not spare the city either. The restoration of the city was carried out in accordance with planning and urban decisions that determined the city structure formation further direction. Today Pskov is an important tourist center of the north-west of Russia. Ancient Pskov monuments are included in the UNESCO World Heritage List (Committee for Cultural Heritage Protection in Pskov Region 2020b).

The urban planning environment within the boundaries of the "Historical Settlement of Regional Importance Pskov" (Committee for Cultural Heritage Protection in Pskov Region 2020a), which practically coincides with the territory of "Okolny gorod" within the outer ring of fortress walls, is unique and complex. Research and design were carried out as part of the training course for the retraining institute of SPSUACE under the supervision of teachers E.G. Bobrova and I.S. Zayats and concerned on a site in the Near Zapskovye, within the st. Leona Pozems-kogo, Shkolnoy st., Truda st. and a small name-less alley in the north-east (Figure 1). Diverse building includes elements of the 17th - 20th century, late constructions of the 60s and 70s of the 20th century, including those of low value or dissonant.

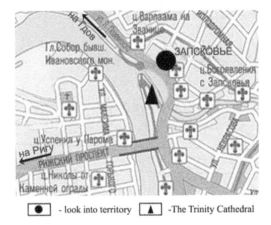

Figure 1. Situational scheme.

Urban planning analysis was carried out on the basis of archival materials, among which the 1778 map (Etomesto 2019a), the 1857 historical map of Pskov by I.F. Godovikov (Committee for Cultural Heritage Protection in Pskov Region 2020a), the 1928 map of the provincial city of Pskov with the settlements settled (Etomesto 2019b), taking into account the regulations and protection zones.

Historically stable directions of city streets and driveways were identified, including st. L. Pozemskogo, st. O. Koshevogo and Nabat st., which formed during the Middle Ages and coincided with modern directions of main city roads. Of course, their dimensions changed during the reconstruction process due to the increased traffic flow.

Among the elements of development were identified:

– especially valuable elements of the urban environment, including the main dominants - the Trinity Cathedral of 1699, the Church of Elijah the Prophet from the Wet Meadow (Ilyinskaya) of 1677, the Meyer House of 1901; environmental accents – the water tower of the Twine factory of 1954; panoramic views - on the Trinity Cathedral from the right bank of the Pskova river, to the Twine factory from the left bank of the Pskova river;
– stable and historically valuable elements of the urban structure;
– low value dissonant objects (Figure 6).

Changes that occurred during the building restoration after the military destruction were identified according to archival photographs by B.S. Skobeltsin 1966 - 1972 (Figure 3).

The investigated territory of Near Zapskovye clearly demonstrates the multilayered and different times of the existing structure of the city of Pskov. The current degraded state excludes it from today's active life, despite the socio-economic, cultural and tourist potential.

2 REVITALISATON CONCEPT

The fabric of the city is constantly changing in connection with modern requirements, the rhythm of life changing, the development of technology, the needs of society, etc. To adapt the historical elements of the urban environment to extend their existence in modern conditions, a detailed examination of the object from various points of view is necessary. Preservation of historicity and authenticity, economic feasibility, socio-cultural context are becoming important components. Not every monument of cultural heritage is able to adapt to a new function. In the current situation, it is necessary to take into account the negative impact of dissonant objects, which impedes the holistic perception of the historical environment and territories of cultural heritage objects. Working with lacuna allows to create

Figure 2. Draft offer the restoration of the gate. The Postnikov's yard, The retraining institute of SPSUACE.

Figure 3. Pskov View on the Near Zapskovye and the Kremlin from the northeast, 1966. 1 – Postnikov's Yard, 17th century, 2 - Andreev Manor, 19th century, 3 - the Meyer House, 20th century. Photo by B.S. Skobeltsin (Pskov State Unified Historical, Architectural and Artistic Museum Preserve 1966).

Figure 4. Draft offer the restoration of the southwest facade. Andreev Manor, The retraining institute of SPSUACE.

Figure 5. Draft offer the restoration of the Meyer House, The retraining institute of SPSUACE.

fragments of the original planning and architectural solutions aimed at emphasizing genuine monuments, harmonizing the visual perception of the historical context (Figure 2).

Zapskovye - a part of the city beyond the Pskov river, was first mentioned under 1323 in the annals as a posad in which the German army stood. Since 1465, this territory, fenced first with wooden, and by 1563 with a stone fortress wall, became part of the (Venice Charter for the Conservation and Restoration of Monuments and Sites 1964).

For a socio-cultural adaptation of the site Near Zapskovye in the historical environment, it is planned to create a tourist cluster "Zvannitsky", whose name

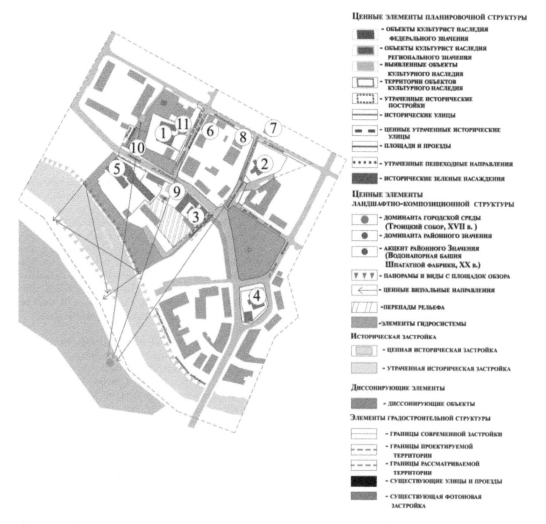

Figure 6. Reference plan. Objects of cultural heritage of federal significance: 1 - the Postnikov's yard, 17th century, 2 – Ilyinskaya church, 15th century, 3 - Trubinsky Chambers, 17th century, 4 - Church of Kozma and Damian from Primosti, 15th - 16th centuries, 5 - The building of the twine factory, beginning 20th century. Cultural heritage objects of regional significance: 6 - Andreev Manor, 19th century, 7- Residential house, 20th century, 8 - The sisters of mercy convent community house. Identified cultural heritage object: 9 - The Meyer House, 20th century. Other objects: 10 – Zhukova's House (in the cultural layer), 17th century, 11 – Spassky Nadlobin's monastery Church (lost).

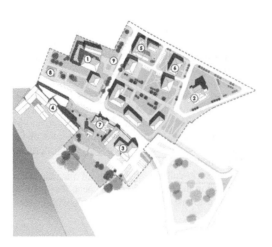

Figure 7. Revitalization concept of the Near Zapskovye historical environment. Objects of cultural heritage of federal significance: 1 - the Postnikov's yard,17th century, 2 - Ilyinskaya church, 15th century, 3 - Trubinsky Chambers, 17th century, 4 - The building of the twine factory, beginning 20th century. Cultural heritage objects of regional significance: 5 - Andreev Manor, 19th century, 6 - The sisters of mercy convent community house. Identified cultural heritage object: 7- the Meyer House, 20th century. Other objects: 8 - Zhukova's House (in the cultural layer) of the 17th centuries, 9 - Spassky Nadlobin's monastery Church (lost).

Figure 9. Walking route "In the wake of history". 1 - Trubinsky Chambers, 17th century; 2 - the Meyer House, 20th century. 3 – Zhukova's House (museumification), 17th century; 4 – Postnikov's Yard, 17th century; 5 - Andreev Manor, 19th century; 6 - Ilinskaya Church, 15th century; 7 – Church of Cosma and Damian from Primostya, 15th - 16th centuries, A - public transport stop, P - parking spaces for tourist vehicles.

corresponds to the historical st. L. Pozemskogo. A pedestrian tourist route connecting objects of various historical stages of the city's life: the Turbinsky chambers, the Meyer house (Figure 5), the archaeological museum of the Zhukova's House (Figure 8), the Postnikov's yard, Andreev Manor (Figure 4), the Ilyinskaya Church and the Kosma and Damian Church from Primostya become an important communication component of the tourist zone of the Near Zapskovye (Figure 9). The expansion of pedestrian zones is possible due to the reconstruction in historical dimensions of the Ilyinskaya dead-end square on the territory of the Ilyinskaya sisters of mercy convent community and the square in front of the Postnikov's yard.

Also become pedestrian existing driveways in the direction of Shkolnoy st. (Figure 7) The existing recreational areas are adjacent to the cluster: the embankment of the Pskov river, the territory of the hotel and restaurant complex of the former "Twine Factory", "Square of Democracy" at the crossroads of modern st. L. Pozemskogo and Truda st.

Figure 8. The Postnikov's yard and Zhukova's House museumification.

The regeneration of the existing square will convey the "spirit of the place" by marking with paving in the contours of the residential quarter lost in the 20th century, including pre-existing buildings and walkways. The creation of a pilgrimage center, with its own kindergarten, hotel, information points and parking in the southeastern part of the historical development area, adjacent directly to the Ilyinskaya Church from the east and located near the Kozma and Damian Church from Primostya, will increase the emphasis on temple complexes and the center of attraction for pilgrims.

The square around the existing clinic will become a transitional zone between residential quarters and the zone of museum complexes. New pedestrian communications between the objects of cultural heritage and objects of tourist attraction will allow regenerating the size of the quarters with a co-scale historical environment. The new functional zoning helps to logically emphasize and divide the territory into the zone of museum complexes and hotel facilities. Museums are being created on the basis of cultural heritage objects in accordance with the time of their construction: industrial production at the beginning of the 20th century will be housed in the Meyer twine factory, an exposition of the life of the boyar estate of the 17th - 18th centuries - in the Postnikov's yard and Trubinsky Chambers. The boundaries of the Postnikov's museum complex will expand significantly with the inclusion of the archaeological museum of the Zhukova's House (Figure 8).

The centers for studying the history of the city are supposed to organize museum and museuminteractive zones that create a network of tourist attractions in the quarter. So, the area in front of the Postnikov's yard can be used to hold city holidays and festivals related to historical reconstructions.

When designing in a historical environment, the need often arises for the additional layouts introduction, structures of a technical nature, facades revitalization and space-planning composition of existing buildings of late periods that are discordant with historical content. For these purposes, it is necessary to develop special regulations that should establish and adjust the use of materials and types of structures, color solutions, the use of certain types of advertising equipment, communication elements and engineering support for the urban environment.

A harmonious combination of old and new elements is achieved by scale and proportionality in relation to the historical environment.

The memory of significant lost monuments of history and culture is recorded in small architectural forms - symbols, on the site of the former Spassky Nadlobin's monastery, for example, taking into account the current regulations of the protection zone.

Provided the creation of a buffer zone around the quarter, which contributes to the protection, preservation and management, as well as maintaining the authenticity and integrity of cultural heritage objects.

3 CONCLUSION

Thus, the stated concept provides not only for the regeneration of the historical environment and objects restoration, it is also aimed at the development and addition of tourist attractions in Pskov. The creation of a tourist cluster in the Near Zapskovye can become an impetus for the revitalization of the historical environment of the city.

Making decisions on the basis of deep research and analysis of the characteristics of the urban environment should be aimed at involving cultural heritage sites in modern life. At the same time, the basic restoration principles should be strictly observed (Venice Charter

for the Conservation and Restoration of Monuments and Sites 1964). The choice of the method of working with the monument or territory, restoration or conservation, renovation or revitalization depends on the degree of preservation, authenticity of the object, the functional adaptation possibility in the conditions of economic and social development. It is necessary to use unique working methods for each object of cultural heritage, because they are unrepeatable. By identifying stable artifacts and variable structures, a spectrum of complex solutions is determined, in a harmonious combination of which it is possible to preserve the fragile "historical memory", invaluable to the ancient city.

REFERENCES

Committee for Cultural Heritage Protection in Pskov Region 2020a. Decree No. 674 dd. December 26, 2013 "On the approval of borders of the buffer zones of heritage sites, land use regimes, and urban-planning regulations within the borders of the buffer zones of the federal heritage site "Kremlin Ensemble". Accessed November 02, 2019. http://gkn.pskov.ru/distributivy-bes platnyh-programmdlya-prosmotra-vlozhennyh-faylov/zony-ohrany

Committee for Cultural Heritage Protection in Pskov Region 2020b. UNESCO cultural heritage sites in Pskov. Accessed May 15, 2019. http://gkn.pskov.ru/deya telnost/vsemirnoe-nasledieyunesko-v-pskove/

Etomesto 2019a. 1778 map of Pskov. Accessed August 26, 2019. http://www.etomesto.ru/mappskov_1778.

Etomesto 2019b. 1928 map of Pskov. Accessed May 15, 2019. http://www.etomesto.ru/karta824.

Labutina, I.K. 1985. Historical topography of Pskov of the 14th – 15th centuries. Moscow: Nauka.

Laurentian Chronicle 1377. Tale of bygone years.

Pskov State Unified Historical, Architectural and Artistic Museum Preserve. Archives. 1966 photo of Pskov. Fund 6026, List 66.

Pskov State Unified Historical, Architectural and Artistic Museum Preserve. Archives. 1856 map of Pskov. Fund 174.

Spegalsky, Yu.P. 1946. Pskov. Historical and artistic sketch. Leningrad, Moscow: Iskusstvo.

Venice Charter for the Conservation and Restoration of Monuments and Sites 1964.

Problems of urban and regional planning

Reconstruction and Restoration of Architectural Heritage – Sementsov et al (eds)
© 2020 Taylor & Francis Group, London, ISBN 978-0-367-65357-6

Employment of collaborative information technologies in urban planning: Case study of Alkut city, Iraq

A.M.H. Abokharima
Belgorod State Technical University, Belgorod, Russia

ABSTRACT: Cities worldwide face the challenges of climate change, air, water, and soil and inadequate housing conditions. One of the primary goals of urban planning is to safeguard and promote health through interdisciplinary work in the field of healthy and sustainable urban environments. This study is divided into three main phases that help to analyze the current urban environment of the city of Alkut in Iraq. The first phase is the analysis of land-use changes and the urban development model using remote sensing methods and GIS. The second phase involves steps to assess and adjust urban areas' microclimate, track Alkut growth, and resolve the issues of forecasting microclimate obstacles that the city will face in the coming decades. The third stage includes creating a statistical model and generating thematic maps for zoning the territory of the city, adopting the analytical hierarchy process methods.

1 INTRODUCTION

In Iraq, the urban population had increased dramatically from 68.87% (19 million out of 28.5 million) in 2007 to 70.28% (26.8 million out of 38.27) in 2017. Furthermore, this growth rate means that the impact of urbanization in Iraq has a faster pace, which can bring some chaos to the management of urban growth. This type of growth has a different impact on each development plan, which must be designed with the responsibilities arising from current and future development plans of the city. Concentrating on the local level and taking into consideration the regional and national plans. Besides, there are significant changes in climate variables in urban areas. Anthropogenic human activity, as a rule, entails a change in the ecological state. However, it is essential to maintain the balance that must be achieved through sustainable development and optimal resource management in urban planning (Abokharima 2020). In theory, sustainability should be achieved by establishing a balance between the economy, social development, and the environment. Sustainable urban development of urban areas can be achieved through the use of geoinformatics methods that will allow statistical modeling maps and track changes in climate variables in urban areas. This method will allow for developing ways to manage resources through municipal regulations. The General idea of this work is to apply a new methodology. The author proposes a new methodology for assessing the environmental aspects of Alkut city (Abokharima 2020). The methodology is divided into three

main phases that will help to describe the current state of the urban area, understand the existing environmental problems, and to highlight areas that should be developed or rebuilt (Figure 1).

The first phase is the analysis of changes in land use over the past decades. Simply put, it is a necessary step to understand how the urban area has reached the state in which it is at the moment. Because remote sensing data and the geographical information system are used as a source, the oldest available data was not older than 1975.

Even though there are older data of photogrammetry techniques that can be useful for this kind of study yet it is unavailable for security reasons besides the technical issues with this kind of data. In the second phase, climate change activities are carried out to ensure sustainability or control the development of the urban area. This model will provide a broad view of the study area of urban ecology. Also, this will allow making a forecast of the environmental problems that the city will face in three subsequentially periods: 2017 – 2022, 2022 – 2027, 2027 – 2032. After the acquirement of the results in the second phase, it is possible to proceed to the third phase. In the third phase, which is the analytical hierarchy process AHP, it is necessary to model and study various factors that directly affect the urban environment. These factors must then be given weight following their importance. The result should be a thematic map showing the locations of the areas to be developed or reconstructed. This step should set the direction for the city's growth in the future and play as a source of vital information for the decisionmakers.

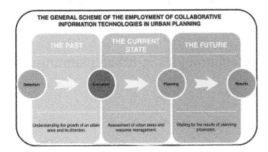

Figure 1. General scheme of the method.

2 PHASE 1

Land use is influenced by economic, cultural, political, and historical factors, as well as land ownership factors at various scales. Land use refers to human activities and various uses that occur on the land. Urbanization is inevitable when the load on land is high, agricultural incomes are low, and population growth is excessive (Perkova et al. 2015). In a sense, urbanization is desirable for human development. However, uncontrolled urbanization has caused many of the problems that our cities face today, leading to poor-quality habitats, such as acute problems with drinking water, noise and air pollution, waste disposal, traffic congestion, and other related issues. To improve the environmental condition of territories, technological progress in the relevant areas must solve the problems caused by rapid urbanization. Only then will the development of the city reach the necessary level of sustainability.

Recent advances in spatial technologies have had a significant impact on planning activities. Planning is of paramount importance for every country in the world. The purpose of using GIS is to provide data analysis to visualize complex models and relationships that characterize planning and policy issues (Abokharima 2020). Visualization of spatial models also supports the analysis of changes, which is essential when monitoring social indicators. In General, remote sensing takes into account the following four aspects:

– change detection;
– determining the nature of changes;
– measuring the degree of change;
– assessment of the spatial pattern of changes.

This study advises the employment of the neural network method, as shown in Figure 2. Using three satellite images of different periods to study changes in the studied territories.

Planning for changes in the urban landscape using applications and methods of remote sensing is effectively used to influence the ecological state of the area, which is the case in the study of urban land use. The remote sensing application is used to obtain data on changes in land use. Remote sensing

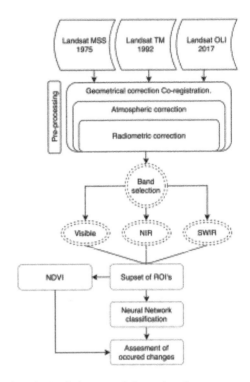

Figure 2. Applied process of change detection.

methods are beneficial for collecting and analyzing data in the process of quantifying urbanization and its growth (Abokharima 2020). Currently, this method is used to solve problems related to the development of thematic maps for the environment in the framework of periodic studies (Gazzaz et al. 2015). Remote sensing plays an essential role in the study of urban areas and makes it possible to track changes in the environment as a whole. Obtaining high-quality urban area data from remote sensing images is only possible with very high-resolution images. Typically, change detection applications require two sets of images taken at different times for comparison. The result can help in making decisions to solve some of the problems associated with city growth and urban planning against uncontrolled development (Abokharima 2020). On the other hand, the use of geographical information systems (GIS) in land use plays a vital role in detecting changes in cities and regions. Remote sensing and GIS methods have powerful tools that can be useful in processing and analyzing data sets. An essential characteristic of geographic information system applications is the simultaneous use of various functions. An excellent example of GIS software is the ArcGIS software. The ArcGIS software has a powerful tool for processing spatial data when detecting changes in an urban area. The quality of results depends on the resolution of the data source, taking into consideration the human factor.

2.1 Results and discussions of 1st phase

The stages of territorial development that have taken place over the past 42 years show that the processes of urban growth, resource management, and urban development have been affected by the struggles in Iraq of this period. For Alkut, this gives a clear idea of how years of war, famine, and political and social chaos can affect the overall living conditions of any urban area. The highest percentage of urban greening in the world is 25.9 percent in Vancouver, Canada, according to the world economic forum.

There are five types of land use of the earth's surface as a whole: built-up territory, high vegetation (trees in parks, squares, boulevards, woodlands), wetlands, low vegetation, shrubs, and small plants, and, finally, soil. In 2017, high vegetation of Alkut had 2.5% of land use, in 1992 0.65%, and 1975 0.7%. Over the years, the number of green spaces in the city has declined between 1975 and 1992 due to deforestation. The aftermath of the war between Iraq and Iran (1980-1988) was disastrous, besides the displacement of non-Iraqi citizens from the area in the late seventies and early eighties. However, there is also a less noticeable increase in the size of green areas or high stable NDVI areas compared to 1992 and 2017.

The 1975 classification, as shown in Figures 3 and 4, includes 30.6% of the urban area as built-up areas, 0.7% of woodlands, 6.2% of water bodies, 2.5% of shrubs, and 43.7% of bare soil. In the 1992 classification, the percentages differ significantly. The share of the built-up area was 21%, forests 0.65%, reservoirs 4.9%, shrubs 7.3%, and bare soil 40.9%. The numbers were not that encouraging in the sense of green areas; however, it is better than it was before. In 2017, 39.7% of the total area was built up, 2.7% of trees, 4.3% of reservoirs, 11.2% of shrubs, and 16.7% of bare soil (Figure 5).

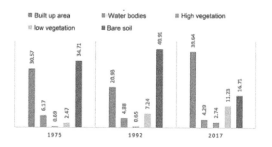

Figure 4. Results of the first phase by years.

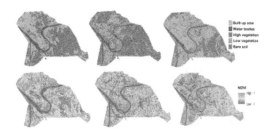

Figure 5. Classification results, A-1975, B-1992, C-2017; NDVI results, D-1975, E-1992, F-2017.

3 PHASE 2

CURB: Climate Action for Urban Sustainability was developed by the World Bank in partnership with AECOM Consulting, Bloomberg Philanthropies, and the C40 Cities Climate Leadership Group. As part of the development of this instrument, an advisory group of more than 30 World Bank staff and technical experts from private consultants and civil society experts was set up to provide input throughout the development process at critical moments. Since urban areas account for the bulk of energy consumption and greenhouse gas (GHG) emissions, cities play a crucial role in the fight against climate change. Cities will not only minimize their exposure to global GHG emissions by reducing their environmental impact but will also benefit from substantial local benefits such as enhanced air quality, improved public safety, and economic development. CURB is an interactive tool specifically designed to help cities take action on climate change, allowing them to draw up various action plans and assess their cost, feasibility, and impact on the environment. CURB is an interactive tool explicitly designed to help cities take action on climate change, allowing them to draw up various action plans and evaluate their cost, feasibility, and environmental impact. Figure 3 explains the steps to complete this type of simulation. It is quite vital for this kind of study to be implemented by cities to understand the current situation and what are the environmental barriers that will face the urban area in the future. In this study, this model plays the

Figure 3. Alkut city location.

leading role in giving a fair image of the current situation which will help decision-makers to make the most suitable decision that at least will keep the environmental difficulties within the minimum range.

3.1 Results of phase 2

The results of phase 2 for Alkut city has been affected by several factors like environment, location, size, and population. The most notable emission in Alkut city is the private buildings' energy emission with 779000 tCO2e/year. The other sectors have only 2 tCO2e/year for municipal buildings and facilities, 577000 tCO2e/year for transportation, 212000 for solid waste, and 28000 tCO2e/year for wastewater. The total of 2017 is around 1.5 million tCO2e/year expected to reach 2 million in 2032. Overall emission by energy type is for the production of electricity for the city with 708000 tCO2e/year, fuel combustion for vehicles with 645000 tCO2e/year, followed by liquified gas, diesel oil, and compressed natural gas. The hot weather of Iraq is the reason why the distribution of energy use by the end-user is the highest for cooling energy followed by lighting. For transportation in the city, 577000 tCO2e/year is only for on-road transportation. Natural gas nowadays considered a suitable replacement for gasoline since it is cheaper and eco-friendlier than gasoline. However, gasoline is still majorly used in cars in Iraq and evidentially in Alkut city. Passenger mode share is private cars, followed by taxis followed by buses. Private cars take a - higher percent since it takes more than 70% of transportation. For solid waste emission, solid waste takes more than 80% of the sector's emissions. Moreover, for solid waste emission by material, paper and cardboard are the highest, followed by food waste and then yard waste. Alkut is classified in the middle of the emission per capita chart compared to other cities in the world, as in Figure 2.

In Figure 6, it is possible to see that for 2017 target with the base year of energy use is

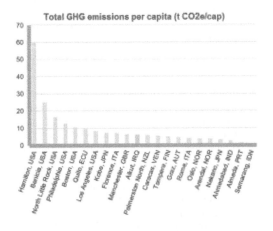

Figure 6. Total GHG emission comparison.

5.34 million MWh/year. In 2022 a target energy use with 5% is not going to change that much since the energy use still more than 5 million MWh/year. However, it is better than another 1 million in energy use, since the expectation is around 6.1 million as it is noticed in the graph of Figure 6. For 2027, with a 10% target, it is possible to decrease the need for energy with almost 2 million MWh/year. While in 2032, it is possible to decrease the need for energy by 3.5 million MWh/year with a target of 15%. The plan of the city of 5% in 2022, 10% in 2027, and 15% in 2032 are not fixed. The values are changeable, which the administration of Wasit province accepted them as achievable in the city. Nevertheless, it is entirely off the range of making the city sustainable in the long run.

4 PHASE 3

Land-use suitability is the ability of a given type of land to support a specific use by the land user. The process of analyzing land suitability involves the evaluation and grouping of specific land parcels regarding their suitability for a particular use. The principles of sustainable development analyze land suitability more complicated due to different requirements/criteria. It includes the consideration of not only the inherent ability of a land unit to maintain a specific land use for an extended period without degradation but also the socio-economic and environmental costs. In many cases, the relative weights of the different criteria involved in deciding whether a land mapping unit is suitable for a particular type of land use are challenging to determine. Therefore, it is necessary to adopt a methodology that allows the decision-makers to estimate the weight coefficients. One of these methods is the process of the analytical hierarchy of AHP. The geographic information system (GIS) is a powerful tool for entering, storing and searching, manipulating and analyzing, and displaying spatial and attribute data.

Meanwhile, land-use suitability analysis requires processing both spatial and attribute data in many data layers. Therefore, it is advisable to use GIS to use its capabilities in processing spatial data. Thus, the integration of GIS and AHP in land suitability analysis is expected to yield promising results. The primary purpose of this work is to identify areas that need to be changed, developed, or rebuilt within Alkut city. In the decision-making process, land uses are selected based on the opinions of stakeholders, which is the administration of the province of Alkut.

4.1 Results of the 3rd phase

Four criteria and 13 sub-criteria were considered for the city of Alkut. If the sub-criteria are ineffective within the city borders, then the sub-criteria are neglected. The values of the classification are according to the preferences of each administration of the city.

AHP values are calculated through a survey that is converted into a pairwise comparison that is converted into weights, Table 1. Two sub-criteria were neglected due to the case that they are not applicable in the city of Alkut. These sub-criteria are pollution sources and agricultural areas.

In order to execute the weighting process, the sub-criteria are weighted individually. AHP can be applied after the reclassification of each sub-criteria into a single system of classification. In this way, it is possible to get proper results. Other than that, the AHP process is not going to work. The sub-criteria nature is different, which makes it a delicate process to reclassify them into a single unique system of classification that can be applied for all the subcriteria, Figure 8. The following equation was applied in the Arc map environment to get the results.

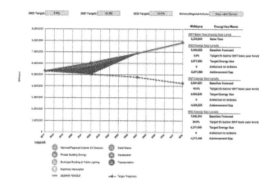

Figure 7. Target emission and expected emission level according to the plan of the city.

$$\Lambda = \sum_{i=1}^{n} C_{vi\,j} \qquad (1)$$

where C is the sub criteria, and v the sub criteria coefficient from Table 1.

After getting the results of the AHP using GIS software (ArcMap) (Figure 10), the results are going to be pixel features that can be converted into a point feature, Figure 9a. Each pixel is representing 900 m^2 of the city area of Alkut. Then the features of the points were processed to be calculated to get the center of the evidence, Figure 9b. The center of the evidence of occurrences, (the points), is an open area with little housing, as seen in Figure 7. Calculating kriging of the occurrences can give us an indication of the direction that the city should steer its growth toward it. It indicates the positive direction of growth. Due to the case that understanding and measuring the direction of urban growth is a more challenging process. According to the results (Figure 9c), the northwest of the city is the direction where the city should

Figure 8. The center of the occurrences (center of the points in Figure 9a).

Table 1. Weights values of AHP.

Sub criteria	Weights
Existing land use	0.0549
Pollution source	0.1099
Significant infrastructure	0.0879
Road traffic	0.0659
Water area	0.1319
Slope	0.0110
Elevation	0.0220
Heat island effect	0.1319
Ecological patches	0.1099
NDVI	0.0769
Agricultural areas	0.0330
Tourism resources	0.0659
Heritage sites	0.0989

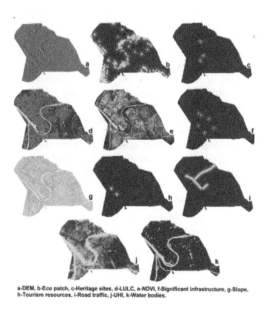

a-DEM, b-Eco patch, c-Heritage sites, d-LULC, e-NDVI, f-Significant infrastructure, g-Slope, h-Tourism resources, i-Road traffic, j-UHI, k-Water bodies.

Figure 9. Reclassified sub criteria.

167

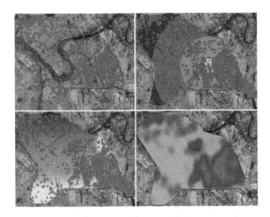

Figure 10. A-occurrence points, B-occurrence points with its center, C-occurrence points with kriging, D-areas that need change (in magenta), and areas that does not need changes (in cyan).

grow toward it. Furthermore, according to the results of the concentration of the occurrences, Figure 9d, the southeast of Alkut is where the most intensify area, where the changes should occur.

5 CONCLUSION

Geospatial tools are increasingly used for spatial planning, as well as transport and economic planning. They also help cities understand with much higher accuracy the interconnections of urban systems that affect their sustainable agenda. Examples of geospatial tools and capabilities available for cities include remote sensing analysis of the state of the urban ecosystem and ecosystem services, mapping of urban characteristics and models, mapping of infrastructure and vital public assets, predictive modeling, assessment of probabilistic risk with multiple hazards, mapping of land use based on highresolution images, and multi-criteria suitability analysis. It is possible to use geo-information systems, remote sensing, and decision support systems as tools to support sustainable urban development and management. They are not only working as tools for achieving sustainable urban development but also as decision support systems for planning processes. For cities that are out of the league of sustainable cities according to the UNEP, this system of work is affordable, applicable, and highly reliable. Of course, not all the cities in the world can afford

high-cost services of sustainability tools. This kind of work can be beneficial not only for the current period but for the future too. Due to the case that is using this process of work, it is possible to control the growth of the city adequately. The cities that are going to use this work should do the evaluation periodically. The sub-criteria are not fixed, which meant they change periodically and sometimes even daily. For example, UHI is not fixed; they are different in the morning than in the evening and so on. The author advises the involvement of another sub-criteria, which is air quality. The involvement of these sub-criteria can enhance the results, and the model can be more reliable.

Moreover, the author advises using data with more spatial and radiometric accuracy. The spatial resolution of the data in this work has not been more than 30m*30m pixels. This accuracy means that some sub-criteria cannot be highly accurate due to the case that when the reclassification system is used in phase 3, they will be less accurate than in reality. Of course, the high cost of photogrammetry, high spatial resolution RS data makes them hard to be acquired, and the processing of this type of data requires much training also. Also, it is necessary to evaluate various alternatives and scenarios and set realistic goals for the plans of Alkut city, and cities in general.

REFERENCES

Abokharima, A.M.H. 2020. The employment of collaborative information technologies in urban planning. *Architecture and Construction of Russia* 1: 8–15.

Gazzaz, N.M. et al. 2015. Artificial neural network modeling of the water quality index using land use areas as predictors. *Water Environment Research* 87 (2): 99–112.

Perkova, M.V. et al. 2015. Features of design of ecovillages in depressed areas in the city. *Research Journal of Applied Sciences* 10 (10): 608–619.

World Bank 2016. *The CURB tool: climate action for urban sustainability.* https://www.worldbank.org/en/topic/urbandevelopment/brief/the-curb-tool-climate-action-for-urban-sustainability#:~:text=CURB%20is%20an%20interactive%20tool,cost%2C%20feasibility%2C%20and%20impact.&text=Cities%20can%20use%20CURB%20at,technical%20support%20available%20upon%20request.

World Economic Forum 2017. *How green is your city? These maps compare green spaces around the world.* https://www.weforum.org/agenda/2017/01/these-interactive-maps-show-you-green-space-in-cities-around-the-world/

Reconstruction and Restoration of Architectural Heritage – Sementsov et al (eds)
© 2020 Taylor & Francis Group, London, ISBN 978-0-367-65357-6

New environmental consciousness

L. Baltovskij
Saint Petersburg State University of Architecture and Civil Engineering, Saint Petersburg, Russia

V. Belous & V. Volkov
Saint Petersburg State University, Saint Petersburg, Russia

ABSTRACT: The modern environmental social movement is not correlated with either a special theoretical concept or a new value system. The subject of this article is the search and formulation of basic concepts that reflect modern environmental consciousness. Until now, the study of the parameters and consequences of the ecological crisis has been carried out in the framework of the traditional worldview characteristic of the New Time. As a theoretical basis of the new environmental consciousness used to be the conceptual apparatus of classical ideologies (liberalism, conservatism, socialism) in their variable mix. The latest report to the Club of Rome (2018) radically changes the situation of conceptual understanding of the environmental value system, building a new worldview on the basis of the concepts of "full world", "new enlightenment" and the dilemmas arising from them. Answers to the challenges of the future arise based on new methodological principles and syntheses. Political ecology as a discipline of modern political science, points to such systemic values as an ecological state, ecological political sovereignty, justice, and responsibility.

1 INTRODUCTION

Modern ideologies have a conflicting genesis. Unlike classical ideologies, the stages of their development may differ somewhat in their sequence. Classical ideologies began their evolution as new systems of views and values in the writings of individual thinkers, who were able, beyond the variegation of historical events and changes, to see the key contradictions of the modern era and certain ways out. Such a traditional system of values had a decisive importance for public life in case of its ability to influence the political and spiritual life of society. When the influence of the system of ideas on social practice becomes historically significant, we are dealing with the views of a certain layer of society. Subsequently, the struggle for hegemony in the civil society space could result in the recognition of a particular ideology by the ruling elite. Thus, ideology reached the climax of its development, turning into a state one.

In the modern political discourse, with all the diversity of opinions in the scientific community, ideology comes, first of all, as certain system of ideas, a system of value judgments of a worldview nature. This is such a picture of the world that can become a sustainable motivation for social practices aimed at transforming existing social relations, or at their legitimation. One can agree with the Edward Shils' definition of ideology in the interpretation proposed by U. Matz: "… ideologies are only such value systems that, acting as a political worldview along with the power of faith, have especially great orientational potential and therefore are able to curb

the social anomie processes connected with the crisis" (Matz, 1992; 130). Social practice shows to what extent ideology proves to be historically successful in the development of specific states and peoples.

In the post-World War II period, the phenomenon of new ideologies arises, where the formation of ecology (precisely as ideology, i.e. eco-ideology) acquires special significance. It is born from an existential contradiction, namely, the possibility of self-destruction of mankind. A distinctive feature of the eco-ideology formation is that the first two phases of its evolution, typical for classical ideologies, in fact, merge in one-time interval. Environmental alarmist movements arose in the mid-20th century not so much on the basis of theoretical conceptual studies of philosophers and scientists, but on the basis of the spread of statements about the real possibility of destroying life on the Earth in case of a nuclear conflict between states by the media. Some seemingly local issues have become existential. For example, such problems as environmental pollution, climate change, habitats, and the reduction of biodiversity in terms of their consequences are comparable to nuclear warfare. The only difference is the speed of the disaster approach.

In order for the very possibility of these problems involving into the public discourse, the accumulation of certain scientific facts critical mass was required. Since 1866, when E. Haeckel gave it a name, ecology, being at first one of many biological disciplines, it subsequently began to rapidly acquire an interdisciplinary nature.

In the 1960s, the awareness of the ability of civilization to self-destruction led to the creation of social movement, called environmentalism. The concept of ecology began to be filled with a new expanded content. The German philosopher G. Jonas has already considered ecology as a science, which subject was the relationship between man and nature, but one in which man was the cause of changes within this relationship (Jonas, 1984; 26).

Around the same time, a global rethinking of environmental problems began as the foundation of a new environmental consciousness in the struggle for hegemony in the civil society space. This process took place both in the European countries and in the Western Hemisphere. One of the most important centers of the theoretical understanding of environmental threats is the Club of Rome, the organization created in 1968. Club members focused on the processes unfolding in the global world along with their consequences, including environmental problems.

2 MATERIALS AND METHODS

The theoretical base of eco-ideology in the initial period was formed on the basis of classical ideologies as its certain variation. According to generalized research by J. Clark, several ideological trends of this kind have emerged (Clark, 2012; 505–516), including: conservative environmentalism in libertarian (Anderson and Leal, 2001) and the traditionalist version of environmentalism (Bliese, 2002); liberal environmentalism (De-Shalit, 2000); natural capitalism (Hawken, 1999); eco-socialism (Dobson, Eckersley, 2006). Within the framework of eco-socialism, the "ecology laws" elaborated by the ecologist and politician B. Commoner in his book "The Closing Circle" (Commoner, 1971) have received a great recognition. The results of theoretical efforts to master environmental issues in the sphere of politics were reflected in solid compendiums (Bryant, 2015, Perreault, Bridge, McCarthy, 2015, Peet, Robbins, Watts, 2011). However, despite a thorough understanding of environmental problems and their political consequences, the development of the ecological worldview took place within the framework of classical ideological concepts and did not lead to the formation of an eco-ideology itself.

The main problem of old ideologies variations is their anthropocentrism. The idea of man's power over nature does not allow us to comprehend its active nature, and a man in spiritual unity with nature. The Anthropocene concept captures the enormous human geological influence on the existence and development of the Earth, but ignores the consequences of mankind activity. Modern society is like a child playing with matches. Awareness of the need to overcome anthropocentrism has led to the emergence of new theories that denied the special parameters and values of man. Based on this approach, Arne Naess formulated the concept of "deep ecology" as opposed to "surface ecology" (Naess, 1973).

The methodological foundation of our study is a critical approach based on a political interpretation of the system values of an environmental worldview. The authors' concept involves the consideration of the current principles of the new environmental consciousness in the general context of the contemporary cultural crisis, which is discussed in a number of previous collective publications (Baltovskij et al., 2015; Baltovskij et al., 2016).

3 RESULTS AND DISCUSSIONS

3.1 Environmental issues in the reports to the Club of Rome

The first document on the environmental issues of the international organization, which got the name the Club of Rome, was the 1972 report "The Limits to Growth" (Meadows, 1991), prepared by a group of Donella and Dennis Meadows. Based on mathematical modeling and analysis of five parameters (the use of unrenewable natural resources, environmental pollution, investment, population growth and food security), experts at the Massachusetts Institute of Technology predicted that at the beginning of the 21st century, mankind will reach the limits to growth in development. Recommendations were given to halt population growth through an active demographic policy, as well as to control production growth.

In 1992, the same authors published the report "Beyond Growth" (Meadows, 1994). Ten years later, the work "The Limits to Growth: the 30-Year Update" (Meadows, 2008) was published. These studies have shown that the development of civilization occurs in the context of the emerging environmental crisis. Under the auspices of the UN, a commission led by the Norwegian Minister of the Environment Gro Harlem Brundtland (Brundtland Commission) prepared a report "Our Common Future" (Our Common Future, 1989). As a result, the concept of "sustainable development" was introduced into scientific circulation. The search for reserves of economic growth in the framework of the concept of sustainable development in 1997 led to the publication of a new report, "Factor Four: Doubling Wealth, Halving Resource Use"(Weizsäcker, 2000), where a scenario for a four-fold increase in labor productivity that does not go beyond growth was proposed.

The development of theoretical foundations that could form the basis of eco-ideology was accompanied by self-criticism. The activity of the Club of Rome itself was criticized for pursuing the specific political goals of the transnational elite. Doubt was caused by the conclusions of reports and scientific research in the field of ecology. Some critics were very distrusted by the very fact of anthropogenic impact on climate change.

The opponents of ecological crisis researchers turned out to be the most diverse groups of the public, from individual scientists to representatives of business and politics. In the space of environmental problems discussion, different concepts encountered reflecting various worldview orientations, as well as the views of specialists in natural sciences and social disciplines. The problems of the ecological crisis, not being clearly theoretically justified, turned out to be quite deeply rooted in the public consciousness. Dividing society into opposing trends, it turned into a political problem itself.

3.2 The main theoretical provisions of the report to the Club of Rome (2018)

An analysis of current trends shows that eco-ideology needs its own theoretical foundation, formed in the context of a critical attitude to the worldview values of the entire previous development of human civilization.

Representatives of the Club of Rome attempted to find such a ground in the framework of a new report published in honor of the fiftieth anniversary of the Club of Rome in 2018 (Weizsaecker, Wijkman, 2018). The report was prepared by co-chairmen Ernst Ulrich von Weizsacker and Anders Wiikman, co-authored with thirty-four other members of the Club of Rome. The report claims to formulate the theoretical foundations of a new ecological worldview and a new eco-ideology. It consists of three large parts. The first ("C'mon! Don't Tell Me the Current Trends Are Sustainable!") is devoted to criticizing the state of the modern world. The second part ("C'mon! Don't Stick to Outdated Philosophies!") is devoted to worldview shifts in values and attitudes.

The title of the third part ("Come On! Join Us on an Exciting Journey Towards a Sustainable World!") speaks for itself. This section of the report, devoted to applied issues on the widest range of problems from global government to agriculture, we will leave out of brackets. The most significant for us is an attempt to create a conceptual theoretical base of the emerging environmental consciousness.

What is the actual ecological picture of the world and environmental worldview? First of all, the authors of the report offer a new picture of the world: "full world" as the opposite of the "empty world". The concept of "full world" was originally proposed by the American economist Herman Daly (Daly, 2005; 100–107). According to his position, current trends in the economy cannot be preserved in the future. The reality of the modern world is not what we imagine it to be. Unless radical changes are made, the world will face a total loss of wealth and the possibility of environmental disaster.

Mankind has been forming for a long time in the space of the "empty world" with uncharted lands, seas, inexhaustible resources. In such a world, extensive development was prevailing.

The worldview and value systems of the "empty world" were expressed by the ideals of anthropocentrism, progress, and the European Enlightenment of the 17th – 18th centuries. In the 20th century, mankind found itself in a "full world", encountering borders in its activity and existence. There are no more unknown lands and continents, the scarcity of resources becomes obvious, the economy overcomes all national borders and turns to be global. The economy must be transformed so that the world can become sustainable in the long run.

First, it is necessary to limit the use of all resources to the level at which the resulting waste can be absorbed by the ecosystem. Secondly, the exploitation of renewable resources is possible only to a level that does not exceed the ability of the ecosystem to re-create resources. Third, the depletion of unrenewable resources should not exceed the pace of development of renewable substitutes.

he report to the Club of Rome "The Limits to Growth" (1972) already showed the systemic nature of the crisis in the modern world. The growth of the Earth's population from one billion to almost eight billion causes a problem with food security, which entails the depletion of land, water and energy consumption. These changes entail climate change and increased pollution of the Earth, oceans, and air by industrial waste. "Natural and man-made capital are more often complements than substitutes and that natural capital should be maintained on its own, because it has become the limiting factor. That goal is called strong sustainability. For example, the annual fish catch is now limited by the natural capital of fish populations in the sea and no longer by the man-made capital of fishing boats. Weak sustainability would suggest that the lack of fish can be dealt with by building more fishing boats. Strong sustainability recognizes that more fishing boats are useless if there are too few fish in the ocean and insists that catches must be limited to ensure maintenance of adequate fish populations for tomorrow's fishers"(Daly, 2005;102).

From the point of view of E. Weizsacker and A. Wijkman, the very concept and practice of capitalism were formed in the "empty world", and the ongoing financial crises in the 21st century are of a different nature than those described in the classic textbooks of political economy. "Capitalism as we know it, with its focus on short-term profit maximization, is moving us in the wrong direction – towards an increasingly destabilized climate and degraded ecosystems. In spite of all the knowledge we have today, we seem unable to change course, literally driving planet Earth to destruction"(Weizsaecker, Wijkman, 2018; VII-VIII).

The inconsistency of the current situation can be illustrated by the example of such an important macroeconomic indicator as the gross domestic product, which determines the policy of all states in the world, but in fact cannot adequately reflect sustainable development in the "full world", since GDP is aimed at a positive measurement of costs, even if they only compensate for the effects of destruction and disaster. "For example, an oil spill increases GDP because of the associated cost of clean-up and remediation, while it obviously detracts from overall well-being. Examples of other activities that increase GDP include natural disasters, most illnesses, crimes, accidents and divorce"(Weizsaecker, Wijkman, 2018; 55).

The concept of a "full world" requires a change in a person's views on his own place in the world. In the ecological discourse, the concepts of environment and nature often distort the real state of things. The idea of a "full world" allows us to see modern problems in a new way.

It should be noted that ideas about the world can have several dimensions. There is an objective world as a result of knowledge of the natural sciences. It is opposed by the subjective world as a complex of ideas, values, experiences of an individual person. The life world is thematized by the original meanings and defined by the circle of human problems. This is the world of everyday life, and going beyond it. The concept of a "full world" should clarify the relationship between different ideas, enriched with new content. At the same time, it already sets a certain metaphysics of eco-ideology.

E. Weizsacker and A. Wijkman propose to consider the attitude of religious denominations to the environmental problem using the example of the second encyclical Laudato si '(Praise be to you) of Pope Francis, (June 18, 2015). A message from this essential encyclical claims that humanity is on the path of self-destruction if rules are not adopted that seriously limit the utilitarian tendencies of the current economy. Anthropocentrism is erroneous even in relation to all living beings possessing the same perfection as a person created in a divine image and likeness. The anthropocentrism of the Enlightenment stems from the phenomenon of unhappy consciousness, which puts man in the place of God. Francis calls for a decisive cultural revolution.

The distorted perception of the modern world is based on many sources and their incorrect interpretations. It was the installations of the "empty world" that led to the dominance in the previous era worldview of the principles of the "invisible hand of the market"preference over state legislation; free trade based on comparative advantages, mutually beneficial for all market participants; competition, which is, supposedly, the evolution and progress driving force.

In the modern world, these principles are turned to be taken out of their context. For quite understandable reasons, the classics of political economy did not have the opportunity to apply the theory of comparative advantages to the transnational capital, the IMF with capitals of national proportions, and also to globalization processes. In the latter case,

transnational capital has an absolute advantage over nation-states. C. Darwin's natural selection and competition neither in nature, nor in the economy, exhausts the mechanisms of evolution and development and is supplemented by the need to preserve diversity, solidarity and protection of small and medium-sized producers (Weizsaecker, Wijkman, 2018; 76–82).

The core of the philosophy of Enlightenment and its epistemology are logical induction and empiricism. This approach, however, has its limits; it does not work in relation to biology, psychology, social reality. The criticism of reductionism is multifaceted. The authors suggest refer to other traditions and authors, such as Gregory Bateson (Bateson, 2016), and Fritjof Kapra (Kapra, 2004).

Changes in the worldview should be so fundamental and systemic that the authors of the report discuss the necessity of a new Enlightenment. It should be based on the experience of various civilizations. The ancient traditions of eastern culture are based on the idea of balance. First of all, a balance between reason and feelings should be achieved, that would result in a holistic worldview. The Chinese symbols Yin and Yang are an example of a balance of opposites. Western and Islamic traditions tend to distinguish and choose between good and evil, although the dialectic philosophy of George Hegel, or the integral psychology of Ken Wilber show that the tradition of balance is inherent in European philosophy likewise (Weizsaecker, Wijkman, 2018; 95).

In the new Enlightenment, the principle of synergy should be the basis of the balance: between man and nature; between short- and long-term perspective; between speed and stability; between private and public; between women and men; between justice and reward for achievements; between state and religion. The balance between humanity and nature is designed to become the core of the new Enlightenment. If in the "empty world" the relationship between man and nature was natural, then in the "full world" the balance between them is a huge problem. Using the remaining oases of nature as resources to meet the needs of a growing population leads the planet to death, not balance. The New Enlightenment implies a rejection of anthropocentrism, but retains the ideal of humanism (Weizsaecker, Wijkman, 2018; 95–97). The concept of "New Enlightenment" is the central ideological point of the report "Come On! Capitalism, Short-termism, Population and the Destruction of the Planet", both in meaning and in location.

Despite the profound impression that the report makes, it is still dominated by the old enlightening intention of relying on the mind, depoliticizing the essential contradictions in the modern world. The conceptual position of the authors of this work on the issues discussed is presented below.

The basis of eco-ideology should be founded on a system of values developed within the framework of political ecology. An original attempt to justify the political ecology is made by B. Latour in his work "Politics of Nature". According to his position, there is no single picture of nature in front of us. There are quite active actors who claim their own vision, and therefore "political ecology does not promise peace. It's only beginning to understand in which wars it should take part and whom it should consider an enemy" (Latour, 2018; 244).

From our point of view, political ecology is a field of knowledge that explores the environmental crisis as the basis on which the modern world is divided into opposing political unities (Borisov, Volkov, 2014; 56). What is the essence of the environmental crisis? The environmental crisis is understood as such human activity, with the continuation of which in a limited foreseeable time, the conditions for the reproduction of this activity would be destroyed.

The concept of "eco-political justice" is ambiguous. Only general discussions around the topic of "justice" will allow policy makers to identify themselves adequately with respect to the human-nature system. This concept should enter the system of human values and get its certainty. In fair relations, subjects are free and responsible. Eco-political justice involves the certainty of law, state and economy.

The consequence of the institutionalization of eco-political justice is that politically divided countries must inevitably become friends. Political pressure and violence are becoming unproductive in the world. The effect of environmental glocalization unites the local places of possible environmental disasters and the entire space of global consequences in a single system.

4 CONCLUSIONS

The formation of eco-ideology is undergoing a new (third) stage, associated with the understanding of the value system relevant to existing threats. In the formation of modern environmental consciousness, the alarmist stage can be distinguished, when attention to the self-preservation of life on the Earth was the main motive of social movements. The second stage can be described as the search for sustainable development. It was associated with the production of projects of a social and technological nature, aimed at containing the expansion of human civilization into nature, preserving the environment as an unconditional value. The third stage is associated with the development of political and environmental concepts, with its inherent potential to create a value system that can become a conceptual basis for the development of eco-ideology.

Today, perhaps for the first time, there is a trend when political interests orient people toward solidarity, rather than political dissociation in resolving crises. Political responsibility becomes relevant. This value is global in nature, although its carriers are turned to be individual political institutions. The state must take on another characteristic, becoming, along with legal and social, also an environmental one. A feature of the ecological state is the responsibility policy within the framework of the principle according to which, bearing responsibility for the planet, the state protects itself, and not vice versa. The development of a value system in the space of political ecology should lead to such concepts as a life world, eco-political sovereignty, eco-political justice, political responsibility. Along with the themes of "full world", "new enlightenment" (discussed in a report to the Club of Rome), they can constitute an ensemble of values that will form the basis of a new environmental awareness.

REFERENCES

Anderson, T. and Leal, D. 2001 Free market environmentalism. New York: Palgrave Macmillan.

Baltovskij L., Belous V. and Kurochkin A. 2015 Society and authorities: New mechanisms of communication in conditions of the network world. Journal of Applied Sciences. 15(3): 538–544.

Baltovskij L., Abalian A., Belous V. and Eremeev S. 2016 Applied Aspects of Politics in Russia. Social Sciences. 11(5): 631–638.

Bateson G. 2016 Mind and Nature: A Necessary Unity, Nyköping: Philosophical arkiv.

Bliese, J. 2002 The greening of conservative America. Boulder, CO.: Westview Press.

Borisov N. A., Volkov V. A. 2014 In search of a new paradigm: a sketch of the political ecology, IPTS SZIU RANKHiGS.

Bryant R. L. 2015 The International Handbook of Political Ecology. Cheltenham; Northampton, MA: Edward Elgar.

Clark J. P. 2012 Political Ecology. Encyclopedia of Applied Ethics, 3, San Diego: Academic Press.

Commoner B. 1971 The Closing Circle: Nature, Man, and Technology. New York: Knopf.

Daly H. E. 2005 Economics in a full world. Scientific American. 09:100–107.

De-Shalit, A. 2000 The environment: between theory and practice. Oxford: Oxford University Press.

Dobson A., Eckersley R. 2006 Political theory and the ecological challenge. Cambridge: Cambridge University Press.

Hawken, P., et al. (1999) Natural capitalism: creating the next industrial revolution. Boston: Little, Brown and Co.

Jonas H. 1984 Das Prinzip Verantwortung. Frankfurt am Main.

Kapra F. 2004 Skrytye svyazi[the hidden connections, Sofiya.

Latour B. 2018 Politics of Nature. Ad Marginem Press; Moskva.

Medouz D. i dr. 1991The Limits to Growth, MGU.

Medouz D. KH., Medouz D. L., Randers J. 1994 Beyond growth. To prevent a global catastrophe, to ensure sustainable development, Panageya.

Medouz D. KH., Rehnders J., Medouz D. L. 2008 The limits to growth: the 30-Year Update, Akademkniga.

Naess A. 1973 The Shallow and the Deep, Long-Range Ecology Movement.

Peet R. Robbins P. Watts M. 2011 Global Political Ecology. London; New York: Routledge.

Perreault T. Bridge G. McCarthy J. 2015 The Routledge Handbook of Political Ecology. London; New York: Routledge.

Reconstruction and Restoration of Architectural Heritage – Sementsov et al (eds)
© 2020 Taylor & Francis Group, London, ISBN 978-0-367-65357-6

Urban planning model of ecologically balanced development of Siberia South

P.V. Skryabin

Saint Petersburg State University of Architecture and Civil Engineering, Saint Petersburg, Russia

ABSTRACT: The ecological capacity of the natural landscape will determine the scale and direction of urban development of any territory in the XXI century. The purpose of the research: to develop an urban planning model of ecologically balanced settlement in the conditions of unique natural landscapes. For this purpose, we used methods of spatial grid overlay, which in this case implies the overlay of the grid of the ecological framework on the framework of settlement, while applying a systematic approach that is, considering the territory as a system at three large-scale levels (Federal, regional and municipal). Results: the area of natural territories disturbed and not disturbed by economic activity is calculated, the basis for the formation of an ecological framework and a settlement framework is proposed, which together with urban planning and environmental regulations makes up the urban planning model for the environmentally balanced development of the southern part of Siberia.

1 INTRODUCTION

The author conducted a study of landscape features, revealed the ratio of disturbed and undisturbed natural landscapes within the boundaries of the studied territory-the South of Siberia.

The problem of the research is caused by the city-level contradiction characteristic of this territory, which consists in the need to implement two mutually exclusive processes: the process of urban development of natural territories for social and economic development, for the development of recreational, agricultural or industrial activities, and the process of preserving the ecological balance of these territories for environmental sustainability.

Specialists and professors from universities in Europe and the United States of America are working to preserve the ecological qualities of native landscapes in spatial planning: Christina V. Haaren (V. Haaren et al. 2008),, Craig W. Johnson (Johnson and Buffer 2008), Ellen Hawes (Hawes and Smith 2005), McHarg (McHarg 1971), Niko Balkenhol (Balkenhol 2016), Simone Allin (Allin and Walsh 2010) and a number of others.

2 MATERIALS AND METHODS

The subject of the research is the possibility of an environmentally balanced urban development in the South of Siberia. Research objectives:

– determine the ratio of the area of untouchable natural landscapes to the area of used territories.
– identify the potential and limitations of urban development of the territory

– determine the direction of urban development of the territory, that is, the functional use and spatial distribution of regulations for urban development.

Urban planning regulations, along with functional zoning, are proposed to be established based on the qualities of the natural landscape, that is, taking into account the natural biochemical processes that ensure the stable functioning of the ecosystem. In an effort to understand the mechanisms of ecosystem functioning, the author argues for the need to build a new urban planning model focused not on the material and technical solution of social, economic, engineering and transport issues, but on the balanced development of the territory without disturbing the natural landscape (McHarg 1971).

Research methods include the spatial grid overlay method: the grid of the ecological framework is superimposed on the settlement framework, which consists of settlement nodes (cities and towns) and transport links between them. The combination of axes and nodes divides the territory into cells, each of which should be considered as a space for functional filling, the space of land for various purposes that make up the urban planning model. To build the urban planning model, a hierarchical system approach is used, which considers a hierarchically subordinated grid system, in the context of three taxometric scales: Federal (several bordering administrative entities-regions), regional (within a separate entity – region, Republic or province) and municipal (within the borders of a municipal district).

3   RESULTS AND DISCUSSION

The ecological framework includes natural land-scapes of varying degrees of value and productivity, which are formed under the influence of four elements (water, temperature, atmosphere and soil) caused by the interaction of two forces: solar radiation and gravitational force. Under the influence of gravity, water flows from the tops of watersheds to lowlands and basins, transferring soil particles, plant seeds and biologically chemical elements, forming the soil. The soil accumulates chemical elements that are nutritious for plants, organic compounds are mineralized in the soil (fallen leaves, withered grass, fallen tree trunks and plant stems), these chemical processes require a certain temperature regime and an abundance of oxygen from the atmosphere. The atmosphere not only contributes to the flow of chemical reactions, the grinding of solid inorganic rocks (rocks and stones), but also, by distributing the flow of cold and warm air, provides a temperature regime for biological life forms. Temperature, changing cyclically over time, provides acceleration and deceleration of all chemical processes occurring in each natural landscape.

In any natural landscape, three spatial components should be distinguished: a lowland, a slope, and a peak. From the top, under the influence of wind and water runoff (rain, snowmelt), the particles of eroded rocks, chemical compounds and plant seeds are transferred down the slopes to the basins. Thus, there is a distribution of material, that is, leaching and weathering from the peaks (eluvial process), flowing down the slopes (deluvial process), accumulation in the lowlands of intermountain basins (accumulative process). In accordance with this Professor Bolshakov A. G. (Bolshakov 2003) proposed the Distribution of land plots by ecological value:

- mountain tops and upper elevations of watershed mountain ranges as ecologically poor eluvial donor sites represented by dark coniferous taiga and Alpine meadows;
- average environmental values of the slopes of hills and mountains covered with dark coniferous taiga forest and deluvial areas;
- the most valuable in ecological terms accumulative areas of lowlands, intermountain basins and bottoms of river valleys, talvegi streams with the highest biological productivity.

This scheme of the spatial structure of natural landscapes can be traced on all taxometric scales, of which there are three in the natural landscape. The simplest taxometric element is a facies that occupies several tens of meters in height and several hundred meters in length (for example, a forest of dark coniferous taiga in the chemala valley). The second large-scale level is a landscape area that covers an area from several hundred meters in elevation to several tens of kilometers in length (Uymon valley,

Kuray valley, Ukok plateau). The largest level includes the entire landscape, covering an area of several kilometers in height in mountain conditions or several hundred meters in plain conditions and extending up to several hundred kilometers (for example, Mountain Shoria).

This division is chosen for the convenience of environmentally-oriented planning of territory development at any administrative-territorial level according to the natural boundaries of natural landscapes and their elements, which are designated by the kinks of the relief. In other words, the tops of watershed ridges that separate one river basin from another, and the edges of intermountain river valleys are the boundaries of landscape elements. At the same time, a network of transport and communication links is superimposed on the landscape map, the axes of which are largely laid according to the geological shape of the terrain, simultaneously dissecting the earth's surface into fragments of different scales. Following this logic, the author considers the territory in the southern part of Siberia bounded on the North by the TRANS-Siberian railway transport and communication corridor, on the East by the Kemerovo-Novokuznetsk transport and communication link, on the South and South-West by the state border with Kazakhstan, Mongolia and China (Figure 1).

Within these borders, there are three broad natural landscapes specific to Russia. The forest-steppe landscape along the TRANS-Siberian highway (Barabinsk and Ubinsk steppes) determines the production and agricultural specialization of the morphotype corresponding to this territory and allocated in the perimeter of the axes of transport and communication corridors Novosibirsk-Barnaul-Biysk – Novokuznetsk – Kemerovo – Novosibirsk. The analysis of the ratio of areas of the natural

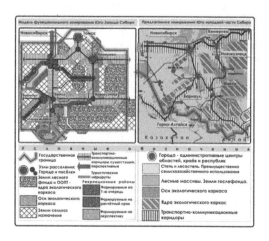

Figure 1. Urban planning model on the scale of the Western Siberian Federal district (Novosibirsk – Barnaul-Gorno-Altaisk-Biysk – Novokuznetsk – Kemerovo).

landscape changed by economic activity (63% or 44 thousand km^2) to undisturbed areas (37% or 26 thousand km^2) allows us to speak about the relative ecological value of this landscape, which is inferior to the neighboring landscapes of the Altai territory.

The fertile steppes of the Altai territory, along the Chui tract and in the foothills of the Sayan mountains, where the steppe landscape is divided by belt forests and large woodlands covering the hillsides in the area of Srostkov and Belokurikha with balneological springs, determines the agricultural and recreational specialization of the territorial landscape morphotype in the perimeter of the axes Barnaul-Biysk-Belokurikha-Rubtsovsk-Barnaul. The ratio of areas of the natural landscape changed by economic activity (64% or 90 thousand km^2) to undisturbed areas (36% or 50 thousand km^2) allows you to attribute this area to an ecologically valuable landscape.

The highest ecological value is represented by the mountain landscape of the Siberian Sayans, which is cut by an extensive network of deep and narrow river basins. This territory includes the South of the Kemerovo region (Gornaya Shoria) and the Altai Republic, defining their recreational economic specialization (the entire Republic is a historical and landscape reserve, 22% of the area of which is occupied by specially protected natural territories). The total area of undisturbed natural landscapes according to the author's calculations is 85% (78.7 thousand km^2), the area of natural territories transformed by economic activity of people is 15% (13.9 thousand km^2).

To identify the potential of urban development of the Altai mountains, it is proposed to start with the calculation of the ecological capacity of its natural landscape. The methodology for such calculations was proposed by academician V. V. Vladimirov (Vladimirov 1996). Somewhat rethinking the formula proposed by him, the author offers his equation for calculating the ecological capacity of the natural landscape, which is measured in the total area of natural territories not disturbed by economic activity, which is necessary to preserve the ecological balance, compensating for the excessive environmental load on the landscape.

$$E_{terr} = \left(\frac{S \times H_{pop} \times T_{wat} \times 2.5}{\sum_{i=1}^{n} O_i} \right)$$

Where E_{terr} is the total area of undisturbed natural territories necessary for maintaining the ecological balance and comfortable living of the population;

S – the area of the settlement area under consideration, in km^2;

H_{pop} is the number of the existing population;

T_{wat} – annual demand for fresh water (lit^3 per inhabitant) or oxygen (m^3 per person);

O_i-value of water volume in rivers – volume of water flow in rivers (thousand m^3) or average value of oxygen reproduction by forest vegetation (thousand m^3);

2.5-transition coefficient for oxygen withdrawn from the atmosphere or water from surface sources.

Applying this expression to, for example, the territory of recreational morphotype with a mountain landscape within the administrative boundaries of the Republic of Altai, the maximum required area of the strict natural areas, component 76,455 thousand km^2 with the available non-renewable natural re-Surah and landscape conditions (area of Gorny Altai 92,6 thousand km^2, the forest area of 60,0 thousand km^2, the total volume of clean water in the river of 20 km^3, the number of inhabitants according 2020 is 220 thousand people). Thus, this landscape has a certain potential for settlement not by extensive urban development of the territory (not by removing new free areas from nature), but by resource-saving urban transformation of economically developed territories. However, there are certain legal and environmental restrictions.

Environmental restrictions are caused by a dense network of small rivers and lakes that have coastal strips and water protection zones (200 meters from the banks of the Katun river), whose borders include all areas that are scarce for development with a relatively calm terrain. This is related to the problem of congestion of buildings, institutions for servicing tourists and transit traffic along coastal strips that accumulate biologically valuable substances for the ecological reproduction of the landscape. Therefore, it is necessary to move as far as possible inland from the banks of rivers building towards the slopes. However, the slopes covered with dark coniferous taiga belong to the lands of the forest Fund and are not subject to development. Legal restrictions are related to the land use regime established for forest lands, for water protection zones (article 65 of the Water code of the Russian Federation 2006), for agricultural land, as well as for specially protected natural territories (PAS). Specially protected natural territories occupy the largest area in the Altai mountains. There is a special legal regime within the boundaries of protected areas that prohibits the construction of any engineering networks and highways, as well as the construction of any (civil, industrial, economic) objects (article 94, Land code of the Russian Federation 2001). In this regard, it is proposed to consider protected areas on a regional scale as the cores of the ecological framework. The cores should include a dedicated Scheme of territorial planning of the Republic of Altai (2016) existing protected areas: Nature Park "Katun" (in the Northern half of the Republic along the right Bank of the Katun), Sumultinsky reserve (in the Central part of the Republic), natural Park Uch-Enmek (Eastern half), Saylyugemskiy and Katun preserves (near southern boundary). In addition, it is proposed to designate the cores of the ecological framework proposed by the territorial planning Scheme: natural parks and reserves. These territories, which are particularly valuable in terms of ecology, occupy a deep position in relation to natural (rivers) and transport axes (roads), include the high – altitude zones of the middle mountains and highlands (500-3000 km above sea level) and are separated by lanes of transport and communication corridors. Taking into account the need for unhindered flow of natural biochemical processes in the landscape (the transfer of valuable chemical elements by water down the slopes to the river valleys, the movement of air masses, the accumulation of soil), the author suggests linking all the cores of the ecological framework with axes:

– linking the natural Park Uch-Enmek natural Park "Katun" through the valley of the river Katun;
– linking the natural Park of the Katun with Sumultinsky reserve through the valley of the river Sumulta;
– Bunch Smolinska reserve natural Park Uch-Enmek and Shavlinski reserve through the valley of the river Chuya.

In the space between the axes of the ecological framework, there are axes of transport and communication corridors: Chemalsky tract, Chuysky tract, Cherga – turata – Ust-Kan connection (highway 84K-96), Cherga – Yabogan connection (highway 84K-121), Ust-Kan – Ongudai – Aktash – Kosh-Agach connection, Ust-Kan – Ust-Koksa – Uymon valley connection – access to the Chuysky tract. In addition to the axes of transport and communication corridors, it is necessary to distinguish natural axes perpendicular to them – small rivers. Further, on a more approximate scale, the author considers a landscape area of about 2000 km^2 within the boundaries of the mountain landscape, located along the Katun River, extending from the village of Ust-SEMA in the North to the village of Chemal in the South (Figure 2). The Eastern border of this territory is the upper line of the Seminsky mountain range, the Western border is the Iologo Range. Squeezed between the slopes of the Ridges, the narrow and long valley of the Katun river is a composite axis of settlement, along which the transport and communication link Chemalsky tract passes. Along the highway, at the points where small mountain rivers and streams that flow down from mountain slopes intersect with transport links, there are localities with the same name: the villages of Ust-SEMA, Cheposh, Askat, Anos, Uznezya, Elekmonar, Ayula and Chemal.

Using the equation for calculating the ecological capacity of the territory in relation to this landscape area, we get the maximum possible area of natural territories that are not subject to urban development to preserve the ecological balance – 1.8 thousand km^2. It should be noted that there are limited areas available for urban development in difficult mountainous terrain. Potential opportunities for urban development

Figure 2. Urban planning model on the scale of an administrative and economic entity of the Russian Federation (the Republic of Altai).

are available only in the bottoms of narrow intermountain river valleys located in a perpendicular direction from the main axis of the transport and communication framework of settlement-the Chemalsky tract, in the direction along small rivers and streams.

However, it should also indicate certain restrictions: along the coast of small mountain rivers there are strips of valuable, water-protected and protective forests, where it is legally prohibited to conduct forestry, place capital construction projects, carry out tree felling, use of chemicals and mining (art. 111-114 of the Forest code of the Russian Federation 2006). Horse-walking and Cycling routes are allowed here. At the same time, given the regime a 100-meter protection zone, even the capital construction lightweight construction is possible with the engineering equipment, prevent ingress of untreated sewage into the reservoir (item 15 and 16 of art. 65 of Water code of the Russian Federation 2001). Based on this, the offset of development from the river Bank is a necessary condition for urban development in such an area, spots and stripes for development will be located between the borders of water protection zones and the axes of the ecological framework.

The ecological framework will be formed on the basis of existing specially protected natural territories. Katun nature Park on the right side of the intermountain river valley is proposed to connect strips of forest plantations with a natural Park on the opposite side. The deep direction of the urban development vector away from the main natural axis of Katun will allow you to unload the coast from self-standing structures for tourists, redirect the flows of self-organized auto travelers away from the water protection zone, using free areas of small river valleys and streams. In this space you should allocate typical landscape facies along a mountain stream, Elekmonar, a tributary of the Katun river (Figure 3).

The untouchable area for urban development natural areas will be according to the calculations by the proposed equation of 16.2 km^2 Sumultinsky reserve). At the same time, the share of natural territories not changed by economic activity is 70 % (12.5 km^2) and 30 % (37.5 km^2) of economically developed territories.

The potential for urban development has sections of wills Creek upstream from the built-up area of the village. The Western part of the development of the existing village falls within the boundaries of the water protection zone of the Katun mountain river and is located dangerously close to the water cut, which requires protective measures. Attribution of development from the coast line to 200 meters is a necessary condition not only for the course of natural ecological processes (purification and accumulation of biologically valuable elements), but also for safety from unforeseen natural phenomena (erosion and crumbling of the coast or flooding). The most environmentally safe sites are located parallel to the banks of the Elecmonar stream, one of which is located three kilometers upstream of the Elecmonar and the second one is 600 meters South of the village. They are proposed for urban development by the territorial planning Scheme of the Chemalsky district (developed under the guidance of Professor S. B. Pomorov).

The author of this article sees the possibility of deep development of the territory in the format of a linear-nodal system with mainly recreational specialization in areas free of forest vegetation on the slope. At the same time, taking into account the need to pass a transport link (road, horse-walking route) outside the borders of the water protection zone, it is proposed to allocate several more sections for the construction of these nodes on the axis of this connection. It is possible to place a three-stage system of recreational service nodes. The first node, which includes a group of objects of year-round operation (tourist chalets, campsites, holiday cottages) around the capital core (tourist hotel), is proposed to be located on a free plot, on the right Bank of the Elecmonar within the boundaries of the village, where meadow vegetation with stony soil on a relatively calm terrain has a relatively high resistance to anthropogenic influences. The second node consists of non-capital structures (yurts, villages, portable modules) with a stationary core (camping), it is proposed to place it in the high-altitude landscape zone of the coniferous taiga. The third node, which is a group of non-capital structures for temporary stay of tourists and travelers, is proposed to be located in the upper reaches of the stream in the high-altitude landscape zone of Alpine meadows on environmentally poor eluvial areas.

A number of zones parallel to the natural compositional axis – the Elecmonar stream-with different urban planning regulations are proposed:

– coastal protection strip,
– water protection zone,
– zone of limited urban development
– zone of untouchable natural territories that form the ecological framework for urban development.

The ecological framework covers not only the upper reaches of the watershed ridges and their spurs, connecting specially protected natural areas on opposite sides of the intermountain basin, but also includes water protection strips along the banks of rivers. The space between the axes of the ecological framework is a limited area of land with relatively calm terrain at the foot of the slopes and at the lower elevations of the slopes of the mountains, at the same time remote from the river Bank.

4 CONCLUSIONS

Summing up the proposals for the construction and environmental model of spatial development in the South of Siberia, it is possible to identify a specific approach to planning the spatial development of this part of the country at three administrative and territorial levels (municipal, regional and Federal).

At the level of a separate municipality, it is proposed to first allocate territories that are not disturbed by economic activity and combine them into an unbroken network of ecological framework, the area of which should be sufficient for the natural renewal of vital resources (water, oxygen, soil). Territories in the space of cells of this ecological framework are subject to urban development, with the exception of coastal strips and water protection zones. As the simplest element, it is proposed to consider the landscape facies, which is

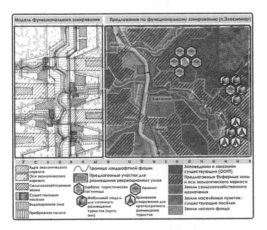

Figure 3. Urban-ecological model on the scale of the municipal district (Chemalsky district of the Altai Republic).

part of the landscape area – a larger structural element corresponding to the regional level of settlement. Spatial development planning at this level is carried out in the same way by combining all specially protected natural territories with ecological axes into a single network, while at the same time allocating lanes of transport and communication corridors along which urban development of the territory will take place. The territory across borders of administrative-economic entities (Federal level) include different natural landscapes, are also joining in the network of ecological frame, the nodes which occupy the deep areas between the grid transportation and communication corridors (TRANS-Siberian railway, Chuya, communication Kemerovo-Novokuznetsk, and Barnaul – Biysk – Novokuznetsk – Abakan). At the intersection of the axes of transport and communication corridors with major Siberian rivers, there are urbanized nodes-cities.

REFERENCES

Christina V. Haaren, Carolin Galler, Stefan Ott. 2008. Landscape planning. The basis of sustainable landscape development. Leipzig: Federal Agency for Nature Conservation.

Craig W. Johnson and Susan Buffer. 2008. Riparian Buffer Design Guidelines. For Water Quality and Wildlife Habitat Functional on Agricultural Landscape in the International West. Washing-ton: United States Department of Agriculture Forest Service.

Ellen Hawes, Markelle Smith. 2005. Repair Buffer Zones: Functions and Recommended Widths. Yale School of Forestry and Envi-ronmental Studies.

McHarg. 1971. Desing with nature. New York: Natural History Press Doubleday & Company, Inc.

Niko Balkenhol, Samuel A. Cushman, Andrew T. Storfer, Li-sette P. Waits. 2016. Landscape Genetics: Concepts, Methods, Appli-cations. Wiley & Sons Ltd.

Simone Allin, Cormac Walsh. 2010. Strategic Spatial Planning in Eu-ropean City-Regions: Parallel Processes or Divergent Trajecto-ries? Maynooth: National University of Ireland.

Bolshakov A. G., 2003. Land-shafta urban Planning organization as a factor of sustainable development of the territory.

Vladimirov V. V. 1996. Settlement and ecology. Moscow: stroizdat.

The water code of the Russian Federation of 03.06.2006 № 74-FZ (URL: http://www.consultant.ru/document/cons_doc_LAW_60683 (accessed 25.04.2020).

Land code of the Russian Federation of 25.10.2001 № 136-FZ URL: http://www.consultant.ru/document/cons_doc_LAW_33773 (accessed 25.04.2020).

Forest code of the Russian Federation from 04.12.2006 N 200-FZ URL: http://www.consultant.ru/document/cons_doc_LAW_6429/(accessed 25.04.2020).

Pomorov S. B. 2008. Recreation and tourism in the mountains and foothills of the Altai Architectural and urban planning organization of recreation facilities. Barnaul: publishing house of Altai state technical University.

Scriabin P. V. 2019. On the methodology of urban planning in the South of Siberia, Bulletin of TSASU, 21(6): 59–69.

Scheme of territorial planning of the Republic of Altai. Decree of the government of the Republic of Altai No. 566 of 22.11.2018, URL: https://fgistp.economy.gov.ru/?show_document=true&doc_type=npa&uin=8400000002 0102201811282.

Territorial planning scheme of the Chemalsky district of the Altai Republic. Administration of the Chemalsky district of the Altai Republic URL: http://www.chemal-altai.ru/index.php/gradostroitelstvomo/skhemy-territor ialnogo-planirovaniya-chemalskogo-rajona.

Problems of engineering reconstruction, performance of repair and

reconstruction works on monuments

Reconstruction and Restoration of Architectural Heritage – Sementsov et al (eds)
© 2020 Taylor & Francis Group, London, ISBN 978-0-367-65357-6

Linear multi-factor regression models in management system of Russian construction industry

I.N. Geraskina & M.S. Egorova
Saint Petersburg State University of Architecture and Civil Engineering, Saint Petersburg, Russia

ABSTRACT: The paper describes a linear multi-factor regression model based on outcomes of interest for the last 27 years with trend forecasting regarding the construction industry of the Russian Federation. The authors prove that it is impossible to use the model when forecasting results of innovation and investment solutions since it does not consider the following important properties: stochasticity, non-linear relationships, system potential, high dynamism of the structure, and a tendency for information misrepresentation. They provide a rationale for the relevance of the regression and differential approach when modeling the investment and construction sector and strategic decision support.

1 INTRODUCTION

The coming social and economic crisis will be characterized by significant structural dynamics, growth of entropy and chaos in the system, fluctuation waves breaking old global economic relations and forming new ones. Under these conditions, the top leadership of the country will need development programs and managerial decisions substantiated by objective forecast data to keep the economy afloat and ensure the survival of its subsystems. It is obvious that the solution to such strategic problems pertaining to the national economy requires comprehending and developing methodological bases for economic and mathematical modeling. To ensure sustainable development of the Russian economy, it is necessary to develop an investment and construction sector (ICS) characterized by a significant share in the GDP structure, resource and energy efficiency, technical and technological innovations. At present, the investment and construction sector represents a self-regulating social and economic system in a slow recession, with a prominent monopoly, complicated by project financing of residential development, which, in aggregate, restricts investment flows and necessary modernization of manufacturing technologies.

1.1 Purpose of the study

The purpose of the study is to prove that it is impossible to use multiple linear regression (MLR) models when forecasting the ICS trend since such models do not consider the following basic properties: cyclic nature and inertia, high level of openness and stochasticity, non-linear relationships between parameters, and underlying processes forming system potential, as well as changes in response to the dynamics of contributing factors, etc.

1.2 Methods

In the course of the study, the following methods were used: approximation of statistical data, regression and phase analyses, system synthesis, linear modeling, and visualization of results in a graphical form.

2 MAIN PART

A comprehensive analysis of the ICS and key factors of its dynamics made it possible to identify phase variables, i.e. indicators characterizing its properties and trend. First of all, these include such statistical indicators as "Commissioning of buildings, structures, individual production facilities, houses as well as social and cultural facilities" measured in mln m^2, "Specific weight of construction in GDP", and "Share of profitable companies in the total number of construction organizations" measured in % (Federal State Statistics Service 2016, Geraskina & Zatonskiy 2017, Geraskina et al. 2017). Moreover, a set of control parameters having the maximum effect on ICS phase variables was identified (Table 1) (Geraskina et al. 2017).

In a linear multi-factor model (LMM), the outcome indicator "Commissioning of buildings, structures, individual production facilities, houses as well as social and cultural facilities" measured in mln m^2 is used. The time interval for modeling and forecasting the study subject is divided into large periods (Tables 2–4) corresponding to the bifurcation points in the social and economic development of the USSR and Russian Federation: 1) 1990–1998; 2) 1999–2007; 3) 2008–2017. When modeling and forecasting trends of complex economic systems, multi-factor models are used where the value of a social and economic indicator or group is

Table 1. ICS outcome of interest.

Name	Indicator
Phase variables	
Y1	Commissioning of buildings, structures, individual production facilities, houses as well as social and cultural facilities, mln m^2
Y2	Specific weight of construction in GDP
Y3	Share of profitable companies in the total number of construction organizations, %
Control parameters	
X1	Fixed investments, mln RUB (before 1998, bln RUB)
X2	Volume of mortgage loans, bln RUB
X3	Costs per ruble of work, kopecks
X4	Population of the RSFSR/Russian Federation, people
X5	Average monthly salary of employees in construction organizations, thous. RUB
X6	Production index in the construction materials industry
X7	Price composite index regarding construction materials and works
X8	Per capita income, RUB per month (before 1998, thous. RUB)
X9	Availability of proprietary resources in construction organizations
X10	Share of expenditures for property acquisition in population expenditures
X11	Refinancing rate, %
X12	Annual inflation in the Russian Federation, %
X13	Volume of residential construction, thous. m^2
X14	Index of labor productivity
X15	Level of profitability in construction, %
X16	Average price for 1 m^2 of total area in the primary housing market, RUB
X17	Availability of fixed assets, bln RUB
X18	Number of operating construction organizations, units

determined by other factors that simultaneously and to a different extent affect the outcome (Mitsek 2011).

Let us apply this approach (Akayev et al. 2012, 2016, Malkov 2011) to build an LMM and verify its applied capabilities in the economy. All factors and criteria were standardized according to equation (1) to be reduced to the single dimensionality:

$$y = \frac{y(t) - \min_t y(t)}{\max_t y(t) - \min_t y(t)} \tag{1}$$

where min $y(t)/t$ is the minimum value of criterion Y depending on the year, max $y(t)/t$ is the maximum value of criterion Y depending on the year.

Standardized factors and criteria for the periods specified are given in Tables 1–3. Then, model factors

were analyzed in terms of their mutual correlation to eliminate their apparent interrelations. Pair correlation of data time series was determined by equation (2):

$$\rho = \frac{\sum (x - x')(y - y')}{n\sigma_x\sigma_y} \tag{2}$$

where x, y are the values of exposure and outcome of interest, respectively; x', y' are the mean values of the corresponding indicators, σ_X, σ_Y are the mean square deviations (standard deviations of x and y variables) that can be checked by equation (3):

$$\sigma_x = \sqrt{\frac{\sum (x - x')^2}{n}}, \sigma_y = \sqrt{\frac{(y - y')^2}{n}} \tag{3}$$

where n is the number of observations.

The LMM results made it possible to determine factors with a high mutual correlation:

1. As for the period of 1990–1997, $X9$ and $X18$ factors correlated with coefficient 0.959823. In this situation, it is logical to eliminate $X18$ since it demonstrated a low correlation with the criterion.
2. As for the period of 1998–2007, the following relationships were identified: a) a high mutual correlation between $X1$ and $X5$, $X8$, $X13$, $X16$ and $X17$. Therefore, it makes sense to eliminate $X1$; b) regarding $X5$ and $X13$ factors, the former can be eliminated due to the fact that $X13$ had the maximum weight in 1990–1997 and 2008–2016; c) regarding $X5$ and $X16$, it is expedient to eliminate the latter to maintain $X17$, which is more important; d) regarding $X8$ and $X4$, to minimize the number of factors eliminated, it is logical to eliminate $X8$; e) regarding $X12$ and $X11$, $X12$ is eliminated since in 2008–2016 it was highly significant.
3. As for the period of 2008–2016, choosing from $X8$ and $X9$, $X8$ was eliminated by analogy with the period of 1998–2008. It should be noted that other indicators characterized by a low mutual correlation of the factor and criterion were also eliminated: a) 1990–1997 — $X7$, $X12$ and $X14$ model factors; b) 1998–2007 — $X18$ model factor; c) 2008–2016 — $X10$ and $X11$ model factors.

We built an LMM with regard to the "Commissioning of buildings, structures, individual production facilities, houses as well as social and cultural facilities" (4):

$$y_{est}(t_i) = a_0 + \sum_j a_j x_j(t_i) \tag{4}$$

where a_0 is the independent coefficient, a_i represents the coefficients of influence of the ith factors of $x_i(t)$ at the moment (year) t on the criterion value.

Table 2. Values of the phase variable and control parameters in 1990–1998.

Indicator	Year				
	1990	1991	1992	1993	1994
Y1	61.7	49.4	41.5	41.8	39.2
X1	249.0	211.0	2670.0	27,125.0	108,810.0
X3	85.0	84.0	83.0	80.0	81.0
X4	147.7	148.3	148.7	148.5	148.3
X5	0.4	0.7	8.1	78.0	283.3
X6	102.3	101.3	100.1	95.2	87.2
X7	115.0	112.3	110.5	131.2	120.5
X8	0.2	0.5	47.8	542.1	2476.1
X9	50.0	50.2	50.4	51.1	52.3
X10	1.2	1.2	1.5	1.3	1.5
X11	25.0	25.0	80.0	210.0	180.0
X12	160.7	160.4	2508.8	840.0	214.8
X13	61,694.5	49,422.6	41,518.3	41,808.2	39,224.3
X14	101.0	100.0	99.6	101.5	100.3
X15	20.1	20.2	20.4	18.6	20.2
X16	-	-	-	-	-
X17	187.1	189.5	190.3	200.0	210.0
X18	7086.0	6975.0	6905.0	9594.0	12,491.0

Indicator	Year			
	1995	1996	1997	1998
Y1	41.0	34.3	32.7	40.8
X1	266,974.0	375,958.0	408,797.0	407,086.0
X3	82.0	90.0	90.0	94.0
X4	148.1	148.3	147.9	147.7
X5	595.1	967.4	1221.0	1.3
X6	92.0	74.5	95.9	94.8
X7	140.1	105.0	112.0	106.8
X8	515.9	769.5	940.6	1010.2
X9	52.0	52.7	53.0	53.2
X10	1.7	1.7	1.9	2.5
X11	160.0	48.0	28.0	60.0
X12	131.6	21.8	11.0	84.5
X13	41,036.4	34,300.9	32,702.6	30,684.6
X14	100.5	100.6	101.0	101.1
X15	23.3	11.6	11.2	6.5
X16	1012.0	1846.0	2161.0	3075.0
X17	221.0	521.0	434.0	446.5
X18	12,776.0	13,462.0	13,697.0	13,716.0

It is reasonable to determine LMM coefficients, minimizing the square deviation of the statistical data from the estimated data using equation (5):

$$s = \sum_t (y(t) - y_{est}(t))^2 \rightarrow \min \qquad (5)$$

This will allow us to identify LMM regression and determine which factors in their linear combination have the lowest/highest effect on the system response (Tables 5–6). Based on the data analysis, we can conclude that in the first LMM option, X13

was a determinative factor, which is quite obvious since it is a residential stock that was constructed and commissioned. With a view to this, we can eliminate it from the LMM and repeat calculations.

The following results appear to be objective and quite interesting for the ICS top leadership (Chainikova et al. 2018, Geraskina & Kuligin 2019, Panibratov et al. 2019, Scott & Gough 2003):

1. Factor X2 (volume of mortgage loans provided) had a high stable significance starting from 1999. The emergence and development of the mortgage

Table 3. Values of the phase variable and control parameters in 1999–2007.

Indicator	Year				
	1999	2000	2001	2002	2003
Y1	42.1	44.7	47.7	49.6	53.7
X1	67,044.0	116,523.0	150,471.0	176,241.0	218,636.0
X2	1.6	8.2	15.3	50.0	100.0
X3	93.0	92.0	92.0	93.0	94.0
X4	147.2	146.6	146.3	145.2	145.0
X5	1.8	2.8	4.2	4.8	6.2
X6	110.2	113.1	105.5	103.7	107.0
X7	131.2	137.4	145.0	124.0	111.2
X8	1658.9	2281.1	3062.0	3947.2	5170.4
X9	52.4	47.5	49.4	45.0	45.2
X10	5.3	7.5	8.9	10.9	12.7
X11	55.0	25.0	25.0	21.0	16.0
X12	36.6	21.0	18.8	15.1	11.9
X13	32,016.7	30,295.8	31,703.2	33,832.2	36,449.3
X14	101.4	102.0	101.4	101.2	105.3
X15	7.0	7.2	6.0	6.2	5.7
X16	4000.0	8254.9	9724.8	9014.8	15,275.3
X17	413.7	301.5	337.2	363.1	403.3
X18	130,846	129,340	118,374	112,971	113,578

Indicator	Year			
	2004	2005	2006	2007
Y1	60	66.3	75.6	98.1
X1	286,501.0	361,111.0	473,002.0	671,622.0
X2	214.0	56.3	263.6	564.6
X3	94.0	88.0	87.0	91.0
X4	144.3	143.8	143.2	142.8
X5	7.3	9.0	11.3	14.3
X6	106.1	102.7	103.1	103.1
X7	112.0	112.0	114.9	121.2
X8	6410.3	8088.3	10,154.8	12,540.2
X9	45.4	44.5	42.1	40.4
X10	11.0	10.4	10.3	9.6
X11	13.0	12.0	11.0	10.0
X12	11.7	10.9	9.0	11.9
X13	41,040.1	43,559.5	50,552.1	61,221.3
X14	106.8	105.9	115.8	112.8
X15	4.2	3.9	5.1	5.8
X16	19,660.8	23,771.6	31,473.6	43,883.2
X17	431.7	604.9	711.3	992.9
X18	114,464	112,640	122,441	131,394

lending mechanism after the collapse of the USSR had a positive effect on the growth of residential construction in the ICS structure in the Russian Federation irrespective of other dynamics of the economic environment.

2. The weight of $X4$ factor (population of the country) increased linearly from -0.6 to +0.1, which marked the beginning of changes in the public opinion regarding the necessity and volume of privatized housing, as well as income. In the years since the RSFSR, the major part of the society sought to improve living conditions in households and obtain the ownership of larger areas, which stimulated the development of the housing market.

3. Factor $X6$ (production volume of construction materials) affected indicators of residential construction to a greater extent after the 2008 crisis. It is obvious since most part of construction materials' production can be, first of all, attributed to residential construction and civil infrastructure, and only after that — to other facilities.

Table 4. Values of the phase variable and control parameters in 2008–2016.

Indicator	Year				
	2008	2009	2010	2011	2012
Y1	102.5	95.1	91.5	99.0	110.4
X1	878,162	797,601	915,210	110,356	1,258,609
X2	655.8	152.5	400.0	716.9	1028.9
X3	91.0	92.0	90.0	92.0	90.0
X4	142.8	142.7	142.9	142.9	143.0
X5	18.6	18.1	27.2	30.8	34.2
X6	103.4	103.7	103.4	103.0	103.2
X7	115.8	114.9	121.2	118.6	105.7
X8	14,863.3	16,895.0	18,958.4	20,780.0	23,221.1
X9	39.5	37.1	41.0	41.9	44.5
X10	5.4	13.9	14.8	10.4	9.9
X11	13.0	8.7	7.7	8.0	8.3
X12	13.3	8.8	8.8	6.1	6.6
X13	64,058.4	59,891.6	58,430.7	62,264.6	65,741.5
X14	109.1	94.4	99.6	105.2	101.6
X15	5.6	5.0	4.5	4.3	5.0
X16	51,333	48,865.4	48,122.8	42,581.4	46,740.0
X17	1220.8	1221.3	1434.8	1499.9	1553.0
X18	144,036	175,817	197,507	209,185	205,075

Indicator	Year			
	2013	2014	2015	2016
Y1	117.8	138.6	139.4	130.2
X1	1,345,024	1,355,752	124,354	114,221
X2	1353.9	1762.5	1147.3	1776.0
X3	95.0	95.0	94.0	95.0
X4	143.3	143.7	146.3	146.5
X5	36.7	36.5	29.9	30.0
X6	103.4	103.8	104.0	104.0
X7	103.7	105.2	107.6	105.2
X8	25,928.2	27,767.0	30,224.5	27,670.0
X9	45.2	45.7	50.2	46.7
X10	9.8	6.9	14.1	13.7
X11	8.2	8.2	8.2	10.0
X12	6.4	11.4	12.9	5.4
X13	70,484.9	84,191.4	84,962.6	82,895.0
X14	99.8	98.5	95.4	95.0
X15	8.3	5.7	4.8	4.5
X16	49,573.3	50,921.9	52,010.8	53,563.2
X17	1581.9	1685.3	1710.6	1727.7
X18	217,961	226,838	210,437	209,354

4. On the contrary, after the 2008 crisis, price index $X7$ was a limiting factor during the housing market transformation. Indeed, before 2008, most citizens had the means to improve their living conditions and purchase additional property.

5. The influence of the availability of fixed assets $X9$ on the dynamics of residential construction decreased and deceleration in growth stopped, which is typical for the period before 1998. This is similar to the conclusion regarding factor $X6$ (the largest construction organizations completed the construction of industrial facilities according to USSR plans prior to 1998). With the optimization of tax legislation in the Russian Federation, the use of subcontract works was streamlined, which was manifested in neglecting the level of the availability of proprietary fixed assets in the general contractor company. A similar trend was observed in the case of factor $X17$, which had a high stable significance before the 2000s. From the moment when tender committees appeared and tendering came into operation, priorities in the construction industry

Table 5. LMM coefficients.

Time interval	A0	A1	A2	A3	A4
1990–1997	0.241	-0.069	-	0.158	-0.258
1998–2007	0.070	-	0.190	-0.034	-0.054
2008–2016	0.019	0.136	0.164	0.142	0.070
	A5	A6	A7	A8	A9
1990–1997	-0.112	-0.026	-	-0.033	-0.118
1998–2007	-	0.055	0.074	-	-0.002
2008–2016	0.003	0.088	-0.282	-	0.073
	A10	A11	A12	A13	A14
1990–1997	-0.183	0.022	-	0.569	-
1998–2007	0.074	-0.119	-	0.314	-0.079
2008–2016	-	-	0.175	0.300	0.037
	A15	A16	A17		
1990–1997	0.197	-0.061	0.096		
1998–2007	-0.044	-	0.466		
2008–2016	-0.157	-0.041	0.061		

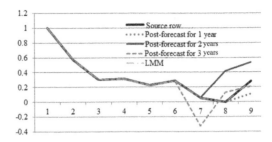

Figure 1. LMM and post-forecasts for 1990–1998.

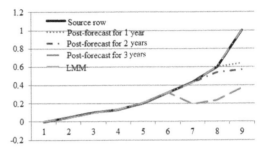

Figure 2. LMM and post-forecasts for 1998–2008.

Table 6. LMM coefficients without X13.

Time interval	A0	A1	A2	A3	A4
1990–1997	0.589	-0.131	-	-0.336	-0.567
1998–2007	0.131	-0.285	-	-0.046	-0.096
2008–2016	-0.003	0.144	0.255	0.231	0.106
	A5	A6	A7	A8	A9
1990–1997	-0.368	-0.044	-	-0.077	-0.203
1998–2007	-	0.056	0.076	-	-0.024
2008–2016	0.005	0.133	-0.359	-	0.093
	A10	A11	A12	A14	A15
1990–1997	-0.348	0.049	-	-	0.414
1998–2007	0.085	-0.155	0.068	-0.056	-0.073
2008–2016	-	-	0.261	0.056	-0.244
	A16	A17	A18		
1990–1997	-0.084	0.225	-		
1998–2007	-	0.615	-		
2008–2016	-0.053	0.101	0.009		

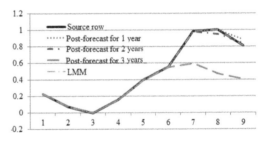

Figure 3. LMM and post-forecasts for 2008–2016.

changed and now vendor selection is based on the competitiveness of applicants reducing the cost of works and contract performance time, committing to improve construction quality, etc.

6. Factor *X12* (inflation) demonstrated the acceleration of changes in the main statistical indicators of residential construction. Moreover, under unstable economic conditions, investment of financial resources in private property is not only one of the most common but also one of the most convenient ways of keeping available cash assets. We are confident that in this instance actual and official inflation figures might deviate.

7. In the years since the RSFSR, factor *X15* (level of profitability) had a continuous restraining influence on residential construction. We believe that this is due to the implementation of shadow financial schemes and mechanisms, which was quite natural in the ICS (residential, industrial, road engineering, etc.) of that time.

Thus, the analysis of the LMM results makes it possible to explain events occurring in the ICS with some degree of probability. However, to assess the possibility of their use when forecasting the influence of organizational-and-managerial, technical, technological and other innovations on the ICS, it is required to evaluate the forecasting properties of the model by applying the widely used post-forecast method for 1, 2 and 3 years (Figures 1–3). The graphical representation shows that it is difficult to make a forecast for the first and second periods even for a year, and, as for the post-forecast for three years, the model of the third period significantly differs from the actual data.

3 CONCLUSION

The results of assessing the possibility to obtain the maximum achievable effect of the linear regression indicate that it makes no sense to use an LMM when forecasting the ICS dynamics. LMMs cannot take into account immanent properties of large economic systems: high level of openness and complexity; cyclic nature, fluctuation and hysteresis processes, constant entropy dynamics.

REFERENCES

Akayev, A.A. et al. (eds). 2012. *Modeling and forecasting global, regional and national development*. Moscow: LIBROKOM.

Akayev, A.A. et al. 2016. *Structural and cyclic processes of economic dynamics*. Saint Petersburg: Publishing House of Polytechnic University.

Chainikova, G.R. et al. 2018. Development of foreign language lexical competence on the basis of a learner's terminological thesaurus and dictionary. *European Journal of Contemporary Education* 7 (1): 51–59.

Federal State Statistics Service 2016. Construction in Russia. Accessed February 05, 2020. http://www.gks.ru/free_doc/doc_2016/stroit_2016.pdf.

Geraskina, I.N. & Kuligin, K.N. 2019. Methodological aspects of providing balanced innovative development of construction complex of the Russian Federation. *IOP Conference Series: Materials Science and Engineering* 698 (7): 077046.

Geraskina, I.N. & Zatonskiy, A.V. 2017. Modeling trends of investment and construction activities of the Russian Federation. *Vestnik MGSU* 12 (11 (110)): 1229–1239.

Geraskina, I.N. et al. 2017. Modeling of the investment and construction trend in Russia. *International Journal of Civil Engineering and Technology* 8 (10): 1432–1447.

Malkov, S.Yu. 2011. Hierarchical system of global dynamics modeling. In Akayev A.A. et al. (eds), *Projects and risks of the future. Concepts, models, tools, forecasts*: 208–231. Moscow: KRASAND.

Mitsek, Ye.B. 2011. Econometric modeling of fixed investments in the economy of Russia. Dissers.ru free e-library. Accessed February 15, 2020. http://dissers.ru/avtoreferati-dissertatsii-ekonomika/2/a45.pdf.

Panibratov, Y. et al. 2019. Developing sustainable competitive advantage of a firm through human resource management practices: a competence-based approach. *Global Business and Economics Review* 21 (1): 96–119.

Scott, W. & Gough S. 2003. *Sustainable development and learning: framing the issues*. London, New York: RoutledgeFalmer.

Supply air jet simulation with machine learning

I.D. Kibort

Ukhta State Technical University, Ukhta, Russia

ABSTRACT: The article considers an example of using machine learning to simulate supply air jet flow coming from a circular grille. Experimental setup and process are described. The article addresses the specifics of machine learning development, adjustment and application to obtain simulation outputs, the values of which have satisfactory convergence with experimental data. The main advantages and disadvantages as well as prospects of the proposed simulation method are analyzed.

1 INTRODUCTION

Currently, there are many methods to calculate microclimate parameters and numerous tools to predict changes in the state of indoor air. Computer simulation offering wide possibilities for a designer has gained widespread use. Such computing systems rely on software algorithms based on the mathematical apparatus. The more microclimate variable parameters this apparatus covers, the more accurately the algorithm operates. However, with an increase in the volume of data, the hardware load, and, therefore, the duration of computations increase as well. It is possible to decrease the hardware load by means of algorithm optimization, which, obviously, introduces some conditionality in the programming interpretation of a mathematical model. Under certain conditions, this factor can be manifested in the reduction of algorithm operation accuracy.

2 MATERIALS AND METHODS

Machine learning with a virtual neural network (NN) represents a method of software computing free from this disadvantage. It relies on the operation of a logic algorithm instead of an algebraic one. This significantly simplifies the process of direct computing but requires a special approach when developing the algorithm (Russel & Norvig 2007).

Let us review the operation of a distributed NN and specifics of its structure. In terms of operation, NNs can be compared with biological neural connections. However, they have fundamental differences. In a neural network, a neuron has some intermediate value, which is determined by mathematical operations (addition or multiplication) with the weights of synapses connected to this neuron (Stepanov 2017).

Figure 1 shows a local connection between neurons in a neural network designed to simulate the operation of a test bench developed. Neurons 1 (N1) and 2 (N2) pass their values to neuron 3 (N3) using synapses A (S1) and B (S2). The weight of a synapse is a coefficient that is multiplied by the corresponding value of neurons. The value of neuron 3 is determined by the following equation (Ignatjev 2001):

$$N_3 = N_1 S_1 + N_2 S_2 \tag{1}$$

In its turn, the N_3 value can be based on any number of connections. In general, the more neurons and connections are, the more accurately the NN operates (Coelho & Richert 2016).

Irrespective of NN architecture, an initial (input) layer of neurons (i1), which determines input values, and an output layer (i3) are obligatory. Depending on NN logic design, the output layer can be represented by one neuron (with its numerical value as a particular output) or a set of neurons with numerical values. Computing layers (i2) determine the accuracy of NN operation (Cherkasov & Ivanov 2018).

NN operation implies computing neuron values in each layer sequentially, depending on the weights of synapses, which results in obtaining neuron values in the output layer. It is regulated by changes in the weights of synapses, which is actually a learning process (Grebennikova 2017). The basic principle of NN learning includes the following:

1. an input layer of neurons is formed in accordance with learning array inputs;
2. after NN operation, output layer values are compared with learning array outputs;
3. if the convergence of values is not satisfactory, the weights of synapses are adjusted and Step 2 is repeated. If the convergence is satisfactory, the process is repeated starting from Step 1 for the next "inputs/outputs" ratio.

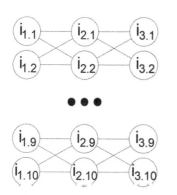

Figure 1. Throttling simulation NN architecture.

Table 1. Characteristics of the experimental measurements.

Measurement No.	Fan electric motor power %	Air flow m3/h	Purpose of measurement
1	5	50	Learning array
2	10	75	Reference array
3	15	115	Learning array
4	20	145	Reference array
…			
19	95	625	Learning array
20	100	650	Reference array

The process is repeated to ensure stable satisfactory convergence for all learning array elements. With an increase in the number of computing layers and iterations, higher convergence can be achieved. The goal is to determine a reasonable relationship between network complexity, duration of learning, and output accuracy (Ignatjev & Madrakhimov 2003).

Let us consider NN application when simulating speed values (in isovels) of supply air jets coming from circular outlets.

Figure 2 shows the scheme of a test bench. The test bench consists of a circular supply outlet with a diameter of 150 mm (1, Figure 2) that allows for a relatively symmetrical straight air jet, and a system of reference points to measure air-jet speed (2, Figure 2). This bench makes it possible to perform consecutive measurements of air-jet speed in the plane of the jet symmetry axis.

3 RESULTS AND DISCUSSION

The test procedure includes the following stages: starting a blower fan, regulating its delivery, and performing consecutive measurements of air-jet speeds in each measurement point.

During measurements, a testo 425 thermal anemometer was used. It has a thin sensitive probe, which minimizes the influence of the researcher and instrument on jet behavior. It also makes it possible to determine the average air-jet speed based on long-term measurements at the same point.

We performed 20 experimental measurements at different values of electric motor power with an increment of 5% (Table 1). Half of the arrays obtained are used as data for NN learning. The other half are used as reference values to determine the degree of NN operation adequacy (Ignatjev 2003).

Interpolation of the values between the measurement points was performed by the derivation of an approximating function and the subsequent computation of the speed value.

NN learning is based on changes in synapse coefficients, which are repeated to ensure satisfactory convergence between the network output and experimental data for all experiment options. As for the case analyzed, the more options of air consumption are considered by the NN, the more accurate the simulation model is. Figure 3 shows the distribution for speeds of air coming from a circular supply outlet, at air consumption L = 650 m3/h, in the XY plane. During the tests, 20 XY planes were plotted with corresponding points. These outputs are used for NN learning.

Figures 4, 5, 6 and 7 show XY planes with air speeds, obtained as a result of NN operation. They

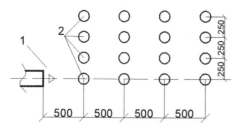

Figure 2. Location of air-jet speed measurement points.

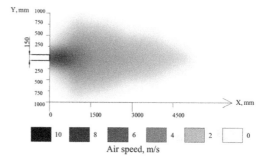

Figure 3. Distribution for speeds of air coming from a circular supply outlet, obtained experimentally, in the XY plane.

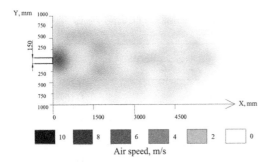

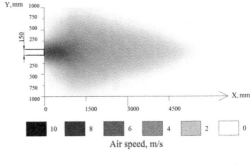

Figure 4. Distribution for speeds of air coming from a circular supply outlet, obtained with the NN (after 100 iterations), in the XY plane.

Figure 7. Distribution for speeds of air coming from a circular supply outlet, obtained with the NN (after 100,000 iterations), in the XY plane.

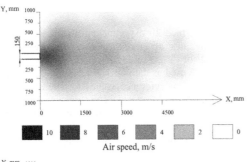

Figure 5. Distribution for speeds of air coming from a circular supply outlet, obtained with the NN (after 1000 iterations), in the XY plane.

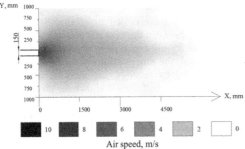

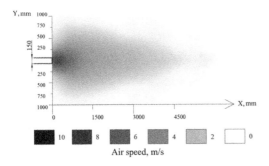

Figure 6. Distribution for speeds of air coming from a circular supply outlet, obtained with the NN (after 10,000 iterations), in the XY plane.

demonstrate how the number of learning cycles affects NN operation adequacy and simulation accuracy. Based on this visual representation, we can argue that it is expedient to analyze the convergence of experimental data and data obtained with the NN only after 10,000 learning cycles (Logunova 2018).

As a result, after the significant number of iterations (Figure 7), simulation accuracy of the system under consideration is at a satisfactory level. The residual between the experimental data and data obtained with the NN in individual measurement points does not exceed 8.65%. The obvious disadvantage of this simulation method is the need for a significant experimental base and long-term learning process (Callan 2018).

4 CONCLUSIONS

It is evident that NNs cannot be used when computing microclimate parameters with a large number of unknown variables. In the case described, only one supply outlet option was considered. The NN computing capabilities can be extended with the learning base development. However, it is important to find a balance between the input data required for learning and the level of NN learning. We assume that some typical situations can be described by grouped trained NNs. Then, depending on situation complexity, an algorithm for the transmission of data computed from one NN to another can be developed, which will ensure their operation within a single system. The advantage of this method is in the enhancement of simulation capabilities due to the reduction of hardware load and computing time.

REFERENCES

Callan, R. 2018. *The essence of neural networks*. Translated by Sivak A.G. Moscow: Williams.
Cherkasov, D.Yu. & Ivanov, V.V. 2018. Machine learning. *Science, Technology and Education* 5 (46): 85–87.

Coelho, L.P. & Richert, W. 2016. *Building machine learning systems with Python*. 2nd edition. Translated by Slinkin A.A. Moscow: DMK Press.

Grebennikova, I.V. 2017. *Methods of mathematical processing of experimental data*. 2nd edition. Moscow: Flinta Publishing House, Publishing House of Ural University.

Ignatjev, N.A. 2001. Selection of the minimum configuration of neuron networks. *Computational Technologies* 6 (1): 23–28.

Ignatjev, N.A. 2003. Extract of implicit knowledge from the various type of data by using the neural networks. *Computational Technologies* 8 (2): 69–73.

Ignatjev, N.A. & Madrakhimov S.F. 2003. Some methods for neural networks transparency increase. *Computational Technologies* 8 (6): 31–37.

Logunova, O.A. et al. 2018. *Processing of experimental data with computers. Textbook*. Moscow: INFRA-M.

Russel, S.J. & Norvig, P. 2007. *Artificial intelligence: a modern approach*. 2nd edition. Translated by Ptitsyn K.A. Moscow: Williams.

Stepanov, P.P. 2017. Artificial neural networks. *Young Scientist* 4: 185–187.

Reconstruction and Restoration of Architectural Heritage – Sementsov et al (eds)
© 2020 Taylor & Francis Group, London, ISBN 978-0-367-65357-6

Reconstruction of historical bridge using modeling and 3D printing

N.V. Kozak, D.A. Vabishchevich, A.V. Kvitko & M.P. Klekovkina
Saint Petersburg State University of Architecture and Civil Engineering, Saint Petersburg, Russia

ABSTRACT: This article presents the results and describes the process of reconstruction of the historical appearance of the Volodarsky bridge in the Leningrad (now - Saint Petersburg, Russia). The reconstruction was carried out in order to preserve the appearance of the lost historcal object and to popularize this urban environment object, which is both extraordinary from the engineering point of view (unique reinforced concrete arches spans) and urban planning point of view (connection of architectural ensembles of two city eras). An analysis was made of the solutions existing in the world and in Russia to preserve the appearances of lost architectural and engineering structures. The reconstruction of the historical appearance of the bridge was carried out by creating a polygonal model in the Trimble SketchUp, models of reinforced concrete prestressed arches of side spans were draw with high-precision. Plastic layout made using three-dimensional printing technology was also designed and manufactured. To prepare the bridge model layout, a search and study of architectural, engineering and operational documentation was conducted. Based on the results of the this project done, an analysis was made of the advantages and disadvantages of these methods and an assessment was made of their prospects when used for historical reconstruction of architectural and engineering objects.

1 INTRODUCTION

The cultural heritage of Leningrad is an important part of the history of Saint Petersburg. A number of objects, both engineering and architectural, were lost in the course of operation, development and transformation of the urban environment. One of these objects is the first Volodarsky bridge, opened in 1936, the southernmost of the city bridges of Saint Petersburg across the river. Neva (Figure 1).

The idea of reconstructing the old Volodarsky bridge appeared at the Department of highways, bridges and tunnels of SPBGASU in 2018. Volodarsky bridge, along with other Soviet buildings of this time, became a landmark object of engineering and architectural thought. The first Volodarsky bridge was part of the ensemble of the new district of the city, an important object of the environment. The side arched spans are made in the form of unique reinforced concrete flexible arches with rigid puffs. However, the result of innovation was a short service life and the need for reconstruction with subsequent changes in appearance. Thus, today on this place you can see the reconstructed bridge in 1985.

The relevance of reconstruction of the historical appearance of the bridge is due to a number of objective reasons:

1. The materials stored in the archive are not Accessible to a wide range of users. As a result, the unknown historical appearance for the younger generation;
2. Limited term of preservation of paper sources- drawings and photos. Thus, when analyzing the

safety of documentation, a significant part was identified as lost due to domestic and other circumstances;
3. The Importance of preserving the history of the city in the memory of residents of the district, popularization of local lore.

HISTORICAL INFORMATION

The crossing at this point was approved in the draft General plan for shifting the city center to the South, in the area of the modern Moscow area. The bridge played an important role in connecting new promising areas of the South-East with the future center.

The bridge was designed by such well-known architects and engineers of that time as Nikolsky A. S., Dmitriev K. M, Perederiy G. P., Kryzhanovsky V. I., Kachurin V. K. the Bridge was designed at the dawn of the iron-concrete era, so the bridge with a reinforced concrete flexible arch was chosen from a number of proposed variations, taking into account the need to overlap significant spans (more than 100 m) for construction (Figure 2).

Approaches to the bridge were designed with intersections in two levels, this solution was used in Leningrad for the first time and became the prototype of modern interchanges. Easing the visual perception of the arch structures was achieved by using the minimum possible cross section of 0, 62x1, 57 m. The scale of the project and the involvement of a large number of engineers and builders, among others, was one of the reasons for the creation Of an office for the construction of embankments at the Volodarsky bridge, which was later transformed into the modern JSC "Trust Len-moststroy" (Bogdanov

Figure 1. Photo of the old Volodarsky bridge, view from the left Bank from the lower side. The 1960s.

Figure 2. The process of installing arch, 1936.

2016). The construction of the mo-STA was the basis for many scientific and engineering studies at that time. The uniqueness and experimentation of the structure, unfortunately, seriously affected its characteristics during operation. In 1985, due to the detected overload in the armature, it was decided to close the bridge and put it for reconstruction (Novikov 1991), (Punin 1971), (Bunin 1986).

2 METHODS

Reconstruction of lost objects of the urban environment is carried out in various ways. The basis for all methods is the study and analysis of archived data. One of the first methods is graphical reconstruction – i.e., the image of an object based on descriptions and drawings. The most common method is layout-creating a reduced material model of an object. This method is easy to use, but in the modern world it is not the most accurate and informative and most often serves for public presentation and presentation. Computer modeling is a better method. The accuracy and detail of the model depends only on the time of creation and quality of execution. Various software programs in the field of architecture and construction allow you to create information models for specific tasks. 3D printing is a modern way to transform a model into a layout. Photogrammetry and 3D scanning allow you to capture the base or part of an object if it is not completely lost, as well as to take

into account the environment and terrain (for example, during archaeological excavations) (Berthelot et al. 2015). The method of 4D modeling is also gaining popularity, i.e. creating a model taking into account the stages of historical evolution (Guidi and Russo 2011).

A computer simulation method was chosen for the engineering reconstruction of the Volodarsky bridge. This is due to the desire to combine the preservation of appearance and information about structures, which is of immediate importance for such an innovative engineering facility. The following algorithm was used for reconstruction.

1. Study of drawings, descriptions and photographs on the database of the archive of SPb GUP Mosto-Trest, as well as other sources located in the public domain.
2. Development of a digital model based on archived data (drawings, diagrams, photos) in the Trimble SketchUp 2017 environment. The arches of the side spans, as the most interesting objects from the engineering point of view, were proposed to be restored with high accuracy for detailed transmission of engineering solutions. The remaining elements were restored in general sizes to convey the architectural and urban planning ideas.
3. To create a plastic layout in the 1:150 scale, the following sub-steps were highlighted for visual demonstration: simplifying the digital model for 3D printing, printing and assembling the layout.

3 RESULTS

3.1 *Computer model of the bridge*

Given the uniqueness of the reinforced concrete arch structures, the authors set the goal of creating a layout with the most accurate repetition of the structures (Figure 3). Based on the original drawings, the final model displays all the structural features and details of the bridge, including cross sections of reinforced concrete elements, junctions, and combined wood-asphalt pavement of the roadway (Figure 4).

Figure 3. The arch model in Trimble SketchUp.

Figure 4. Oblique section of the arched bridge span.

Figure 5. Reconstruction of the bridge in a modern environment. The view from the top.

Since the first Volodarsky bridge was also an important part of the architectural ensemble, visualization of the historical appearance in the modern context of the environment from various viewpoints was developed (Figure 5, 6).

3.2 *Bridge layout*

A plastic model of the bridge at a scale of 1: 150 was made for mass familiarization with the appearance of this engineering and urban development object (Figure 7,8,9). The model was made using three-dimensional printing technology, and the geometric model of the structure developed at the previous stage was adapted to prepare the initial models. The bridge elements were printed on 3DQ Mini printers of the interdepartmental laboratory of SPBGASU, as well as on the CREALITY 3D Ender-3 Pro 3d printer. White PLA-plastic Filaments were used as the printing material. In total, more than 200 parts were printed, which were later assembled into a single model.

4 DISCUSSION

The manufactured model of the bridge was subsequently successfully presented at the thematic exhibition of the center for the visually impaired of the Nevsky district of Saint Petersburg, which makes it possible to consider such reconstructions an important step in getting acquainted with the architectural and engineering history of a wide range of people, including those with various restrictions.

Layout of lost objects is a good way to reproduce architectural and urban planning ideas,

Figure 6. Reconstruction of the bridge in a modern environment. Waterfront view.

Figure 7. General view of the printed bridge model.

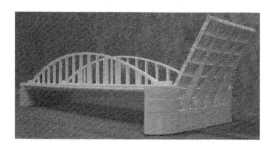

Figure 8. Printed model of the bridge.

Figure 9. Details of the printed bridge model.

which was confirmed by the results of the project. However, from the point of view of engineering heritage, such layouts – unassembled and made of the same material – are very uninformative. For the tasks of transmitting engineering information, a more appropriate solution in the context of the development of modern technologies is the development and subsequent provision of wide access to it of a computer model of the object.

A promising development of the ideas of classical education of geometric computer modeling is the use of full-fledged information models for these purposes. These solutions will allow you to transmit detailed information about the materials and stages of the design lifecycle.

5 CONCLUSION

Reconstruction of lost elements of the environment as a way to preserve cultural heritage is widely used in the modern world. When applied to objects located at the intersection of engineering and architecture, such as the Volga bridge, reconstruction in the form of a high-precision digital model is also a way to preserve information. Recreating the layout of the first Volodarsky bridge allows you to see a three-dimensional model of the lost object, contributes to the popularization of knowledge about the history of the district and the city as a whole, as well as the evolution of urban construction. The method of modeling and 3D printing allows us to improve the interaction between related areas in terms of heritage preservation, including bridge construction and architecture.

6 ACKNOWLEDGEMENT

For the provision of historical materials and access to the archive, the authors of the article Express their gratitude to Saint Petersburg state unitary enterprise "Mostotrest"; for assistance in printing the elements, we Express our gratitude to the staff inter-departmental laboratory of SPBGASU; for the assistance in the organization of Assembly Express our gratitude to the General Director of "CCM" LLC Yaroshutin D. A.; for help in assembling the layout, we Express our gratitude to the employees of "CCM" LLC, As well as to V. F. Kozak.F.

REFERENCES

Bogdanov G. I., 2016. Despite all obstacles: lenmostostroy Trust: 80 years in the service of Saint Petersburg, 1936–2016. Saint Petersburg: Branko Publishing house.
Novikov Yu. V., 1991. Bridges and embankments of Leningrad. Leningrad: Lenizdat.
Punin A. L., 1971. The story of the Leningrad bridges. Leningrad: Lenizdat.

Bunin M. S., 1986. Bridges Of Leningrad. Essays on the history and architecture of bridges in Saint Petersburg-Petrograd — Leningrad. Leningrad: Stroizdat.

Berthelot, M., Tony, N., Gigi, L., A. Bishop, A., De Luca, L., 2015. 3D Virtual Reconstruction and Visualization of Complex Architectures. Avila, Spain, The International Archives of the Photogrammetry, Remote Sensing and Spatial Information Sciences.

Guide, G., Russo, M., 2011. Diachronic 3D reconstruction for lost cultural heritage. Trento, Italy, International Archives of the Photogrammetry, Remote Sensing and Spatial Information Sciences.

Preservation of historical buildings during the development of underground space in an urban environment

R.A. Mangushev, A.I. Osokin, F.N. Kalach & S.A. Podgornova
Saint Petersburg State University of Architecture and Civil Engineering, Saint Petersburg, Russia

ABSTRACT: The authors study the applicability of various drill-injection systems when strengthening the foundations of historical buildings with grouts to perform compensation compaction of soil and get a monument out of the emergency state. In the course of the study, tests are performed with the use of grouts based on cement and a polymer binder.

1 INTRODUCTION

Intense construction activity related to space development in the built-up historical part of Saint Petersburg requires close attention to the assurance of safe geotechnical works on construction sites in close proximity to adjacent buildings. It is especially important when construction is performed near architectural monuments and historic buildings.

It should be noted that the geotechnical complexity of operations associated with the construction of foundations and underground structures (including the adaptation of architectural monuments to modern needs, with the underground space development) is mainly determined by soil conditions in the historical part of Saint Petersburg (Mangushev et al. 2018). The engineering and geological profile of the historical districts of the city is distinguished by a thick layer of water-saturated silty-clayed soils. Such fine-dispersed water-flooded soils include: fill soils, silty sands, and sand clays, layered and varved loams (very soft and liquid in terms of liquidity). It is not possible to erect essential structures having high loads on these soils without pile foundations and special geotechnical measures of soil mass stabilization to improve soil stability and load-bearing capacity (Mangushev et al. 2012). This applies particularly to urban historic areas with dense development.

2 ENGINEERING AND GEOLOGICAL CONDITIONS OF THE SITE

The geological profile of the site under consideration, located in the central part of the city (up to the drilling depth of 40 m) consists of postglacial (lacustrine and marine), lacustrine and glacial as well as glacial deposits of the Quaternary stratigraphic sequence, covered from the surface with asphalt and fill soils. In particular, the engineering and geological structure of

the site is characterized by the following strata (Figure 1).

The technogenic deposits (Tg IV) are represented by asphalt and fill soils: medium coarse sands, sand clays (more rarely), with construction debris in the form of crushed bricks and wood chips; 1.8–3.0 thick.

The lacustrine and marine deposits (lm IV) are represented by peats, peaty soils, liquid loams, and sand clays with very soft interlayers, with admixed organic materials, silty sands of medium density, loose sands of medium density, loose coarse and gravel sands.

Upper Quaternary lacustrine and glacial deposits (lg III) are represented by liquid loams with interlayers of very soft varved loams, plastic sand clays.

The Upper Quaternary glacial deposits (g III) are represented by plastic sand clays, tight gravel sands, firm loams with interlayers of stiff loams, plastic sand clays with interlayers of stiff loams.

The Vendian deposits of the Kotlin horizon are represented by very stiff clays (including dislocated ones).

In terms of hydro-geological conditions, the site under consideration is characterized by the development of one aquifer limited to the bottom of the technogenic deposits, the layer of lacustrine and marine sand and sand clay soils as well as silty-sand interlayers in lacustrine and glacial loams. It is a water-table aquifer with the free water surface, which is recharged through the infiltration of precipitations.

The groundwater level was recorded at a depth of 1.5 m from the day surface. Table 1 shows the standard and estimated values of soil characteristics.

3 TECHNICAL CONDITION OF THE FOUNDATIONS IN THE BUILDING STRENGTHENED

A historic building of cultural heritage is located in close proximity to the restoration site, where adaptation to modern needs (with the underground space development) is performed. It is a structure of 2–5

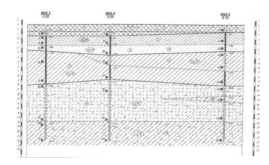

Figure 1. Characteristic engineering and geological profile of the restoration facility site.

Table 1. Physical properties of soils.

EGE	G. In.	We	Il	e	ρ g/cm³	E MPa
1	tg IV	–	–	–	–	–
2		1.0	4.0	2.4	1.2	3.0
3	m, I, IV	0.3	–	0.7	1.9	14
4		0.3	1.2	0.7	1.9	11
5	Lg III	0.4	1.3	1.0	1.8	7.0
6		0.2	0.8	0.4	2.1	12
7	G III	0.2	0.5	0.6	2.0	15
8		0.2	0.2	0.5	2.1	19

floors with two inner yards. Its structural design is represented by a wall system with load-bearing longitudinal and transverse walls. A technical inspection revealed cracks with a width of 20 mm in the vaulted ceiling of the first floor, beam corrosion, deflection of wooden beams with a sag in a span (together with the filler) (Figure 2). The ceiling structures are partially operational or in a critical condition.

The foundations are qualified as partially operational. The materials of the structural foundation

survey show that the foundation in the upper part (semi-basement) is composed of red medium-burned clay brick with dimensions of 260 × 120 × 65 mm. The walls of the building are plastered with sand-lime mortar (in a layer 40 mm thick) both from the outside and inside. It is noted that brickwork moisture indicators are more than 2.5 times higher than the permissible values. The capillary ascent of water in the brickwork is also observed. The building has shallow strip foundations constructed on a natural bed. The foundations are supported by wooden mud sills. Below the brickwork, the foundation is composed of broken limestone with longitudinal dimensions of stones from 300 × 80 mm to 510 × 160 mm on sand-lime mortar, and from a depth mark of -1.5 m to a depth mark of -1.9 m (from the day surface), it is composed of granite stones with longitudinal dimensions of 200–380 mm on sand-lime mortar. The foundation is distinguished by wooden mud sills with a diameter of 220 mm located below the granite masonry, which are placed along the layer of granite stones 140 mm high.

The foundations of the transverse walls are composed of broken limestone with dimensions of stones from 370 × 120 mm to 400 × 120 mm on sand-lime mortar. The lower part of the foundation, from a depth mark of -1.6 m to a depth mark of -2.0, is composed of granite stones with dimensions of stones from 250 × 160 mm to 380 × 290 mm on sand-lime mortar. Under the transverse wall foundation, a wooden board with dimensions of 260 × 80 mm is placed with wood chips and plant debris underneath.

The joints between the stones of the rubble stone foundation vary in thickness from 20 to 50 mm. The foundation has a uniform masonry with even courses; stones were selected by size. The foundations of the longitudinal wall and transverse walls have a bonding. The foundations of the longitudinal wall are trapezoidal-stepped, and the foundations of the transverse walls are rectangular.

During the survey of the foundations, the groundwater level in a test pit was recorded at a depth of 2 m from the day surface. As of the moment of the survey, the mud sills were located below the groundwater level (Figure 3).

Based on the survey performed, the following defects were revealed:

- no masonry mortar in individual joints of the foundation masonry from 30 to 60 mm, in the lower part of the foundation — at a depth of the outer course, throughout the foundation depth to its bottom — at a depth of the outer course (2–4 masonry courses);
- leaching of the foundation masonry mortar at a depth of 15 mm throughout the foundation height;
- no horizontal waterproofing between the wall and the foundation;
- no vertical waterproofing of the brickwork (in areas of contact between the walls and soils);
- no waterproofing of the floor on the first story;

Figure 2. Room with damaged wooden ceilings (a) and a tell-tale installed on a crack (b) to perform geotechnical monitoring.

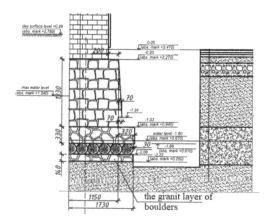

day surface level +0.29
(abs. mark +2.760)

0.00
(abs. mark +2.470)
-0.20
(abs. mark +2.270)

max water level
(abs. mark +1.540)

70

-1.34

-1.63
(abs. mark +0.940)

water level -1.80
(abs. mark +0.670)

30
-1.66
(abs. mark +0.610)

-2.08
-2.22
(abs. mark +0.250)

1150
1730

the granit layer of
boulders

Figure 3. Foundation layout based on the structural survey results.

– direct interaction between the bottom of the brickwork and soil; brickwork moisture indicators are 1.8–2.6 times higher than the permissible values. The capillary ascent of water in the brickwork is observed;
– the groundwater inflow through the granite masonry courses at a depth mark of -1.80 m is observed;
– discoloration of the mud sill surface and their decay to a depth of 1/4 cross-section (up to 25 mm). As compared to standards, the mud sill wood strength is decreased by approximately 50%.

According to the results of field and laboratory soil tests, silty sand clay serves as a base for the foundations.

In terms of regulatory requirements, the building is in an emergency state and no additional deformations are allowed.

According to the results of numerical simulation using PLAXIS 3D software, the maximum estimated values of additional displacements in the architectural monument (Figure 4) adjacent to the site of geotechnical works do not exceed the maximum permissible values specified in regulatory documents (provided the foundations are strengthened prior to any construction works).

4 MAIN METHODS OF PREVENTIVE FOUNDATION STRENGTHENING

To perform preventive strengthening of the foundations in existing buildings affected by underground construction, various geotechnical approaches are used: compensation grouting using tube a manchettes (TAMs), incremental grouting, jet grouting, construction of drill-injection piles with the transfer of load (in full or partially) from the building strengthened, construction of geotechnical barriers (Mangushev & Nikiforova 2017).

Grouting (including jet grouting) technologies are tools ensuring soil stabilization and imparting water-tight properties to soil masses. The latter is especially important for the safe excavation of construction pits near operating buildings. Let us list several mobile drilling rigs and hydraulic drilling systems that can be used indoors:

– Russian Sterkh SBG-PM3 drilling rigs have a mass of 572 kg and dimensions of 1700 × 750 × 2040 mm, which makes it possible to use them indoors (Figure 5). A drilling string is used to drill boreholes. Its running down and rotation are ensured by a hydraulic head with a penetration spindle having a torque of 2500 Nm. This drilling rig can be used to drill injection wells and construct drill-injection piles, or ensure jet grouting.
– Russian SBU-100 drilling rigs are used outdoors and can move around a facility by means of a caterpillar chassis (Figure 6). The drill head power is 4.0 kW. The mass of the rig is about 5.0 tons. The drilling rig dimensions are 5350 × 4000 × 2286 mm. These units are distinguished by a wide range of process capabilities: drilling of boreholes for TAMs to perform grouting, strengthening of foundations, and construction of drill-injection piles.

Total displacements uₓ
Maximum value = 0,01862 m (Element 67721 at Node 90072)
Minimum value = -7.342*10⁻³ m (Element 1347 at Node 131890)

Figure 4. Estimated area of the impact of additional displacements at the surface level, and additional displacements of existing buildings, caused by pit development.

Figure 5. View of a Sterkh drilling rig during the drilling of injection wells on one of the restoration sites.

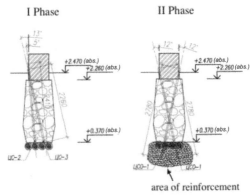

area of reinforcement

Figure 8. Scheme of foundation strengthening.

Figure 6. SBU-100G drilling rig drilling boreholes for TAMs at a cultural heritage site in the center of Saint Petersburg for the subsequent strengthening of the foundations.

– Italian Beretta 43T drilling rigs (Figure 7) have a mass of 2.5 tons and a motor power of a 55 kW. The rig torque is 6200 Nm at a rotation speed of the drilling string of 30–200 rpm. The unit dimensions are 2510 × 1500 × 2368 mm. A hydraulic power unit is located separately from the rig and attached to it with hydraulic hoses, which improves rig mobility and makes it possible to use it in various confined conditions.

To perform grouting using various technologies, other drilling rigs (able to operate in different soil conditions and drill through rubble stone foundations of existing buildings) can be used as well. The following indicators are crucial when choosing drilling

Figure 7. View of a Beretta 43T drilling rig on one of the engineering restoration sites in Saint Petersburg.

equipment: mass, dimensions, the possibility to operate indoors, motor power, torque, and the possibility to change drilling tools when drilling in different environments.

Among the grouting technologies, a technology with the use of TAMs and double packers (obturator rings) proved to be efficient. To implement it in different soil conditions, TAMs and double packers of special designs were developed, as well as special casing grouts with adjustable properties, based on local Cambrian clays, to secure TAMs in a well (Yermolayev et al. 2008).

Various solutions were analyzed. Considering that works performed should have a reduced impact, the following sequence (Figure 8) was selected:

1. foundation body consolidation with cement;
2. grouting of the contact area between the foundation and soil;
3. construction of wells for strengthening using TAMs;
4. grout feeding (compensation grouting) carried out during pit digging in case of additional deformations.

The purpose of performing consolidation grouting, filling voids in the foundation body and under its bottom, stabilizing the foundation soil, creating a soil-cement mass under the foundations to avoid deformations in structures in the form of deflections and cracks, and improving physical and mechanical properties of the foundation soil is to ensure safe and reliable building operation and accommodation of a part of foundation loads.

The main objective of the solutions suggested (considering the historical and architectural significance as well as the protection status of the facility) is to preserve the building without deformations when performing earthworks in the adjacent territory.

5 SPECIFICS OF USING VARIOUS COMPOSITE GROUTS FOR SOIL STABILIZATION IN THE AREA OF MUD SILLS

Based on the experience in work with different facilities in the center of Saint Petersburg, we can conclude that during soil grouting (Figure 9) with the use of Portland cement, abrupt injection pressure jumps may occur and it may be difficult to determine the moment when pressure exceeds six atmospheres, resulting in an uncontrolled rise of the building strengthened due to hydraulic fracturing (Osokin et al. 2020). When a cement-based grout is injected, soil particles start moving in relation to one another (due to the size of cement particles), thus reducing pore channels and preventing the grout from further distribution in the soil mass, which leads to an increase in feed pressure and switching from the permeation technology to the technology of hydraulic fracturing.

This problem can be solved by changing the injection grout composition from Portland cement to polygel without changes in the construction technology. Since polygel particles are significantly smaller than cement particles and polygel includes an integrated system of additives, it has a better ability to penetrate soil pores, compacting and binding the weak soil mass.

However, the process of choosing the main parameters of modern injection materials, which affect the efficiency of soil mass permeation, has some peculiarities. Let us consider a negative experience of using a grout not adapted to particular soil conditions. During a laboratory study, we determined the influence of permeating samples of fine-grained silty sands with a polymer solution on their deformation characteristics. The study focuses on the modified physical and mechanical characteristics of silty sand strengthened during permeation with the use of grout based on a polymer binder.

The laboratory experiment included:

1. maximum saturation of fine-grained silty sand with water (3 l);
2. introduction of injection grout to it (1.075 l);
3. curing of the strengthened soil sample for 24 h (Figure 10);
4. collection of several samples from the artificially strengthened soil mass with their subsequent testing in a stabilometer under triaxial loading according to the consolidated drained testing procedure.

Based on the results of the laboratory tests, we determined modulus of deformation for both water-saturated fine-grained silty sand and the soil mass artificially strengthened during permeation with the use of grout based on a polymer binder. Table 2 shows the test results for different effective pressures in the device chamber (Figure 11).

Based on the test results, it is possible to conclude that permeation with grout based on

a

b

Figure 10. Sample of strengthened fine-grained sand: a — curing of the soil sample for 24 h; b — a soil sample taken out of the container with the clear segmentation of silty sand and grout.

Figure 9. Injection wells in a test area.

Table 2. Results of laboratory experiments.

No.	Pressure in the chamber	Vertical pressure	Effective pressure in the chamber	Modulus of deformation
	MPa	MPa	MPa	MPa
	Undisturbed soil			
1	0.100	0.598	0.100	4.096
2	0.200	1.031	0.200	5.963
3	0.300	1.549	0.300	7.819
	Grouted soil			
4	0.100	0.601	0.100	4.396
5	0.200	1.036	0.200	6.126
6	0.300	1.551	0.300	8.024

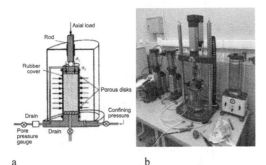

a   b

Figure 11. Testing under triaxial loading: a — a structural scheme of the stabilometer chamber; b — a stabilometer in the laboratory of the Geotechnics Department in the Saint Petersburg State University of Architecture and Civil Engineering.

a polymer binder in the amount of 36% of the volume of the tested water-saturated samples of fine-grained silty sand did not show any significant changes in deformation characteristics (modulus of deformation). Its increase was less than 5%. To increase the efficiency of the technology applied, it is required to adjust the injection grout composition by changing parameters affecting its ability to penetrate so as to increase deformation characteristics of water-saturated fine-grained silty sands. It is also required to study the soil, which is to be further strengthened, in laboratory conditions in detail to determine its grain-size distribution, moisture, and permeability coefficient.

6 CONCLUSION

Based on the studies and tests performed, we can conclude that when performing engineering works to adapt architectural monuments to modern needs (with underground space development in the built-up part of the city), it is necessary to predict relevant risks by means of geotechnical substantiation and take them into account in a design solution, providing a rationale for the optimal construction technology, as well as provide R&D support for restoration works and safety control with geotechnical monitoring.

The analysis of experience in construction in the historical center of Saint Petersburg shows that preventive strengthening of foundations in affected existing buildings prior to construction works ensures safe, reliable and successful implementation of a construction project. Under these conditions, geotechnical monitoring shall be started in advance, preferably two months before any geotechnical or construction works.

REFERENCES

Ahmed, A. 2015. Compressive strength and microstructure of soft clay soil stabilized with recycled basanite. *Applied Clay Science* 104: 27–35.

Ilyichev, V.A. & Mangushev, R.A. (eds.) 2016. *Geotechnical engineer's reference book. Bases, foundations, and underground structures*. Moscow: ASV.

Kalach, F.N. et al. 2019. Characteristics of ultrafine permeation grouting for foundation soil of Northern River Terminal in Moscow. In R.A. Mangushev et al. (eds), *Geotechnics Fundamentals and Applications in Construction: New Materials, Structures, Technologies and Calculations*: 109–113. Leiden: CRC Press.

Kharchenko, I.Ya. & Alekseyev, S.V. 2013. Combined soil grouting during underground space development under restrained urban conditions. *Metro and Tunnels* 5: 18–20.

Mangushev, R.A. & Nikiforova, N.S. 2017. *Technological settlements of buildings and structures near underground construction sites*. Moscow: ASV.

Mangushev, R.A. et al. 2012. *Methods of made-ground preparation and construction*. Moscow, Saint Petersburg: ASV.

Mangushev, R.A. et al. 2018. *Geotechnics of Saint Petersburg. Experience of construction in soft soils*. Moscow: ASV.

Osokin, A.I. et al. 2020. Value of additional vertical deformations of foundations depending on injection grouting conditions. *IOP Conference Series: Materials Science and Engineering* 775: 012144.

Shakirov, I.F. & Garifullin, D.R. 2015. The research of bearing capacity and deformation of sandy ground, reinforced by pressure cementation. *News of the Kazan State University of Architecture and Engineering* 4 (34): 200–205.

Yermolayev, V.A. et al. 2008. Grouting works under the foundations of residential buildings. In R.A. Mangushev (ed.), *Geotechnics: Scientific and Applied Aspects of Constructing Overground and Underground Structures in Complex Soils*: 151–156. Saint Petersburg: Saint Petersburg State University of Architecture and Civil Engineering.

Reconstruction and Restoration of Architectural Heritage – Sementsov et al (eds)
© 2020 Taylor & Francis Group, London, ISBN 978-0-367-65357-6

Influence predicting of vibro-immersion and vibration removal of sheet piles on additional deformations of new construction object

R.A. Mangushev, V.M. Polunin & N.S. Nikitina
Saint-Petersburg State Architecture and Construction University, Saint Petersburg, Russia

ABSTRACT: The paper presents the results of numerical modeling of vibration immersion and vibration extraction of sheet piles in the geotechnical finite-element software package Plaxis using the module "Dynamics". It is noted that, in the process of vibrating elements in dispersed water-saturated soils, a change in the structure of the soil occurs in the near-groove space, which must be taken into account in the process of predicting the deformation of buildings during construction work. Calculations are made to determine the propagation of vibrations in the soil, and the values of vibration accelerations in the Plaxis dynamic module are obtained. The values of additional precipitation of the building under consideration, determined during the static calculation, are compared with the values of vertical deformations according to the results of numerical calculation and the monitoring results of this structure at the construction site.

1 INTRODUCTION

New construction in dense urban conditions is impossible without the development of underground space both on the open areas of the house premises and under the buildings themselves (Mangushev et al. 2013). At the same time, when fencing pits, the most used are metal fences, such as the "Larssen sheet piling", pipe sheet pile, etc. which account for more than 50% of all types of possible fences. In many ways, this is determined by the factor of their possible reuse (Figure 1).

Vibro-immersion and vibration removal of sheet piles by high-frequency loaders in conditions of dense urban development are currently referred to as "gentle" technologies, but, as construction practice shows, in conditions of thick soils, there are no absolutely safe technologies and any effect on the foundation soil should be taken into account in the process prediction of additional deformations of buildings of the surrounding development.

Often this happens due to the limited criteria that are controlled during the dynamic impact at the construction site. Thus, TSN 50-302-204 regulates only the magnitude of the maximum permissible fluctuations in the structures of buildings that fall into the zone of influence. There are no specific values for the soil (Government of Saint Petersburg 2004).

2 METHODS

In the northern part of Saint Petersburg, during the vibration extraction of test sheet piles (profile - O45-072 pipe sheet pile, length - 18 m), significant deformations of the already constructed part of the new building occurred (8 out of 14 floors were completed).

Figure 2 shows the engineering and geological conditions of the construction site used in further analytical and numerical calculations

The upper part of the soil base is a large thickness of water-saturated sands. Engineering geological element IGE-10 is dusty sands that have thixotropic properties.

The 14-story building has a pile foundation made of bored piles with a diameter of 520 mm and a length of 7.2 m, while the tips of the piles are located much higher than the mark of the bottom of the sheet pile fence of the pit of 18 m length. The building's ground structures are monolithic reinforced concrete; the supporting structures are columns, walls and overlapping.

In the process of geotechnical monitoring, it was revealed that the building draft, in places adjacent to the dowel vibration extraction zones, reached 3-3.5 cm within several hours, and then stabilized at 5.1 cm (Figure 3).

In our opinion, there can be several reasons for the development of large deformations: the influence of dynamic effects on the soil of the base and foundations of the building, possible deformations of the sheet pile wall in the soil mass, the development of significant friction forces in the locks of sheet piling, siltation of the sheet pile during the time spent in the soil, soil sticking to the surface of the sheet piles, as a result, the appearance of the attached mass and additional soil volume.

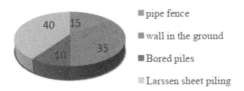

- ■ pipe fence
- ■ wall in the ground
- ■ Bored piles
- ▨ Larssen sheet piling

Figure 1. Distribution of types of foundation pit fences in the construction market.

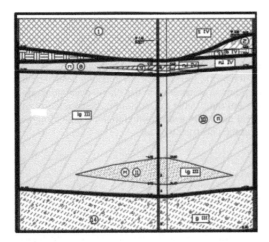

Figure 2. Engineering and geological section of the construction site.

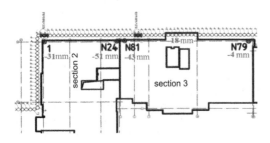

Figure 3. The value of deformations by sedimentary grades during monitoring in the process of sheet piling vibration extraction.

The above factors lead to the occurrence of significant accelerations of the surrounding soil mass. Since the base is composed of water-saturated sands, the effect of liquefying the soil takes place.

Common soil models, such as Mohr-Coulomb (MC), Hardening Soil (HS), Hardening Soil Small Strain (HSS), etc., can be used for dynamic calculations (Brinkgerve et al. 2006). The disadvantages of these soil models include the lack of proper consideration of liquefaction of sandy soils.

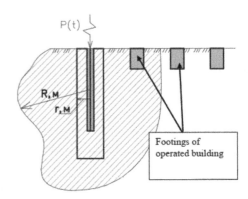

Figure 4. Calculation scheme taking into account the effect of changes in stress-strain state of the area near the sheet piling in the process of vibro-immersion or vibro-extraction.

The works by Savinov & Luskin (1960), Barkan (1959), Ershov (1970), Kovalevsky (1967) provide a description of the processes occurring in soils during vibration immersion of rigid elements in them (Figure 4).

The processes when the sheet piles is vibrated in the soil can be characterized by two zones (Figure 3): r, m is the zone in the near-pile space in which the soil passes into a state of heavy viscous fluid; R, m - the distance at which there is a decrease in the strength and deformation characteristics of the soil characteristics of the soil, which should be taken into account when calculating the additional precipitation foundations of neighboring structures.

An assessment of the distribution boundaries of these zones under specific ground conditions can be obtained using numerical simulation of pile vibrations (Figure 4).

To determine the zones of wave propagation during the vibration extraction of the tongue for the building under consideration, the principal zones of variation in the strain-stress state of the soil mass were estimated. For the dynamic load, the maximum driving force of the vibrator (900 kN) and a frequency of 33 Hz, which are characteristic of high-frequency vibration absorbers (Verstov & Belov 2007), were adopted.

In the course of the calculations, it was found that the zone where the soil becomes a heavy viscous fluid is at a distance of $3dp$, where dp is the diameter of the sheet pile. The area where the soil has reduced soil characteristics can be pre-assigned at a distance where the acceleration of the soil exceeds 0.15 m/s2, appendix N (Government of Saint Petersburg 2004). According to numerical calculation, it is $5dp$.

As the results of calculating the dynamic problem, it is convenient to obtain an accelerogram – the

Figure 5. Accelerogram obtained in the course of numerical simulation. Blue - the value of soil accelerations at a distance of 1 m from the source of oscillations. Orange - the value of acceleration of soil vibrations at a distance of 5m from the source of oscillation.

development of acceleration of oscillations in the soil over time for 0.2 s (Figure 5)

Based on this, it is proposed to consider the static problem that takes place at the time interval for sheet pile extraction and at the stage of restoration of soil properties (Figure 6).

To account for the actual stiffness of the building, all wall elements were modeled by Plate elements with actual stiffness. Piles were modeled by the elements of the "Embedded beam". The load from the eight floors already built was applied as evenly distributed over the slab of a grillage.

The near-sheet pile ground was assigned the characteristics of a heavy fluid, the Linear Elastic (LE) model. This zone is highlighted in pink in the design diagram. The zone where the strength and deformation characteristics were reduced by 30% is highlighted in yellow. In the near-pile space, soil compaction occurs during the vibration of the sheet piles, as a result of which a trough of subsidence occurs, which can be observed as a result of the calculation in Figure 7.

Figure 8 presents a comparison of the results of numerical modeling and monitoring. Note that the nature of the distribution of the building's settlement

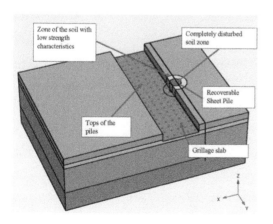

Figure 6. Calculation scheme for static calculation.

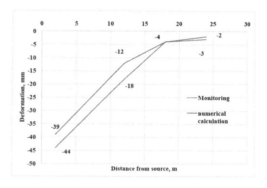

Figure 7. Deformation scheme of the calculation model.

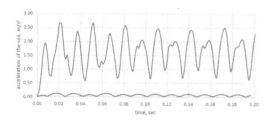

Figure 8. Graph comparing the results of numerical modeling and field observations.

and its value satisfactorily coincides with the results of field observations.

3 CONCLUSIONS

Based on a numerical study, the following conclusions can be made:

1. The behavior of soils under dynamic loads is a complex phenomenon, which can only be modeled using a number of assumptions.
2. A great influence on the surrounding soil mass during vibration immersion and vibration removal of the sheet piles is caused by a violation of the technology of work and other external factors, such as: changes in soil properties, the occurrence of significant friction forces in sheet pile locks, sheet pile curvature, etc.
3. In the geotechnical substantiation of new construction, it is necessary to take into account the influence of vibration immersion and vibration removal of sheet piles as serious technological impacts on additional deformations of buildings in the surrounding buildings, which requires

improvement of existing and development of new methods for their forecast (Ilyichev & Mangushev 2016).

REFERENCES

Barkan, D.D. 1959. *Vibromethod in construction*. Moscow: Gosstroyizdat.

Brinkgerve, R.B.J. et al. 2006. *Plaxis: finite element code for soil and rock analyses. Users manual*. Rotterdam: Balkema.

Ershov, V.A. 1970. *Dynamic properties of sandy soils and their consideration in assessing the stability of soil structures. PhD Thesis in Engineering*. Leningrad.

Government of Saint Petersburg 2004. *Regional Construction Regulations TSN 50-302-2004. Designing the foundations of buildings and structures in Saint Petersburg*. Saint Petersburg: Government of Saint Petersburg.

Ilyichev, V.A. & Mangushev, R.A. (eds) (2016). *Geotechnical engineer's reference book. Bases, foundations, and underground structures*. Moscow: ASV.

Kovalevsky, E.D. 1967. Results of observations of fluctuations in sandy soils and building settlements when driving a sheet pile. In Mechanics of soils, bases and foundations. Scientific Conference of Leningrad Civil Engineering Institute: 34–38. Leningrad.

Mangushev, R.A. et al. 2013. *Design and installation of underground structures in open pits*. Moscow, Saint Petersburg: ASV.

Savinov, O.A. & Luskin, A.Ya. 1960. *Vibration immersion method and its application in construction*. Leningrad: Gosstroyizdat.

Verstov, V.V. & Belov, G.A. 2007. Perfection of technological decisions on immersing and extraction of steel sheet piles by vibrating method. *Bulletin of Civil Engineers* 4 (13): 38–44.

Renovation of coastal industrial zones with possibility of using engineering geodesic dome structures made of wood and polymer materials

O. Pastukh & D. Zhivotov
Saint Petersburg State University of Architecture and Civil Engineering, Saint Petersburg, Russia

ABSTRACT: As part of the modern development of the urban environment in many countries of the world, comprehensive programs are being developed and implemented that cover the issues of urban revitalization and improvement. Today, much attention is paid to the issues of architectural appearance and functional purpose in the renovation, development and reconstruction of industrial coastal areas in the largest cities of the world. In this article, the authors will consider innovative concepts using the latest achievements of science and technology for buildings and structures on coastal territories. They will prove the rationality of using non-standard shell structures, including geodesic domes made of wood and modern polymer materials. The use of engineering structures of this type in the construction of buildings and structures for various purposes during the renovation of industrial coastal areas will not only create a unique silhouette of the coastline of major cities, but also take care of the historical value of cultural heritage objects. At the same time, their historical appearance will be preserved, which can be delicately integrated and adapted to the modern needs of society, without forgetting about ecology and environmental protection.

1 INTRODUCTION

Renovation of coastal areas in large cities is one of the urgent tasks that city authorities have to solve, taking into account not only the economic component, but also the aesthetic and environmental one. Due to obsolescence or re-profiling of production facilities located in coastal industrial zones, it is possible to rationally use large areas of land within the city limits. Renovation of coastal industrial territories positively solves transport problems and contributes to improving the environmental situation of megacities. The city authorities pay close attention to the preservation and repurposing of historical buildings that are under UNESCO protection, located on industrial territories (FL No 73, 2020).

When developing projects for reconstruction and development of these territories, functional zones are identified:

1. Residential zones that facilitate the connection of water and residential development, positively affecting the psychological and physical health of residents;
2. Additional space for the reorganization of transportation hubs;
3. Recreational areas to maintain a favorable environment in the city.

If there are objects of historical and cultural heritage in the territories under consideration, a delicate and professionally decided integration and adaptation of buildings to the modern needs of society is required, while preserving the historical or other value of the objects.

2 METHODS

The relevance of the study of mesh shells that serve as the supporting backbone of a spherical structure, which is, for example, a lightweight antenna structure has a multi-dimensional character, due to the automation of complex calculations and intelligent production systems (Fuller R.) (Figure 1).

Geodesic dome-an architectural structure in the form of a sphere formed by connecting rods in triangles on the cellular principle. The principle of building a dome-shaped frame was developed by the American architect Richard Fuller based on the geometric shape of the Earth in the 1950s. The surface of such a dome consists of steel ribs of different lengths, which when combined form the shape of a dome. The practical application of the building contour geometry proposed by Fuller is based on dividing space by vectors. The basic unit of this division is the tetrahedron. The above separation allows you to achieve optimal space filling and the most complete use of the structural strength of materials.

The main advantages of geodesic domes are: high load-bearing capacity; high speed of installation compared to traditional frame and frameless methods of building construction; weight of the dome elements reduces the cost of materials and zero-cycle work; reduced construction time (Fuller R.); ideal

а.　　　　б.

Figure 1. Radio telescope with a lightweight antenna structure, closed by a dome. USA, high-Tech (http://bourabai.kz/physics/0155.html).

Figure 2. A-assembly; b-joint testing by Instron 5998. (Photos by D. A. Zhivotov, 2019).

aerodynamic shape with high resistance to seismic, wind and hurricane impacts (Yue Guan, Lawrence N. Virgin, Daniel Helm, 2018); cellular configuration of many available Diamatic dome templates is particularly convenient for converting with mutual support of elements; coating structures using renewable energy sources from natural sources (Porta-Gándara M.A. & V. Gómez-Muñoz. 2005; Zhivotov D.& Tilinin Y. 2020).

The disadvantages of geodesic domes include the fact that the production of modern building materials is primarily aimed at the construction of buildings made of rectangular materials (plywood, glass, hard insulation mats).

The authors of the article, based on their own experimental studies of the nodal dome connection with the application of various loads, in order to simulate real processes in the construction and determine the actual indicators of the bearing capacity of nodes and elements, consider the use of these structures to be the most optimal solution for preserving cultural heritage objects in the renovation of coastal industrial territories.

These tests were carried out on specialized equipment in the mechanical laboratory of Saint Petersburg state University of Architecture and Civil Engineering at a normal temperature of 220°C (Figure 2 a, b). From this point of view, we can recommend using a combination of wood and polymer materials in any large-span spatial structures.

3 RESULTS

From a technical point of view, the design of a geodesic dome is the optimal solution for covering large spaces, which are the most rational for the renovation and adaptation of coastal territories. Because the main focus in solving the constructive issue is on protecting cultural and architectural heritage objects from the destructive effects of water.

Covering historical buildings with a large-span geodesic dome solves this problem, giving a unique architectural appearance to the place, thereby increasing its tourist and economic attractiveness.

Currently, more than three hundred thousand significant buildings and structures have been constructed in the world using the geodesic dome design. Among the most famous are the dome set on the South pole, the magnificent "Klimatron" (greenhouse) in the Botanical garden of St. Louis (Missouri, USA), "Golden dome" of the American exhibition in Sokolniki Park in Moscow in 1959, fishnet USA pavilion at the world exhibition in Montreal 1967, with a height of over 60 meters and a diameter of 75 meters.

Fuller's Biosphere, S. Berg, 1967, Montreal, Quebec, Canada

In 1967, Fuller realized his ideas for spherical houses by assembling a pavilion for the United States exposition at the international exhibition in Montreal. This geodesic structure was named the Fuller Biosphere and is now one of the symbols of the city of Montreal. The sphere consists of steel rods and has a diameter equal to 70 meters. During the exhibition, polymer fabric was used as the covering of the structure. Inside the structure, 4 steel platforms were installed, which were divided into seven levels (Figure 3).

At the end of the 20th century, Fuller's architectural ideas received a second wind. They are used to create eco-friendly and economical homes, build cinemas and planetariums. In addition, due to the lightness of the frame and the ability to mount it in the shortest possible time, the Fuller sphere is used as a mobile structure.

Stockholm Globen Arena, S. Berg, 1988, Stockholm, Sweden

The 85-meter-high Stockholm Globen Arena in Stockholm is the largest spherical structure in the

Figure 3. Fuller Biosphere, 1967, Montreal, Canada.

Figure 5. Aquarium Building, 1988, Genoa, Italy.

Figure 4. Stockholm Globen Arena, 1988, Stockholm, Sweden.

world, a striking example of a harmonious combination of a bright architectural idea and a non-standard engineering approach to solving the problem. The structure of the dome consists of 48 curved steel columns, the inner shell of the sphere is made of latticed aluminum, the outer one is made of thin metal plates covered with white lacquer, 140 mm thick, laid out exactly on the aluminum grid. Additional support for the fragile dome is provided by pipes-pillars also made of aluminum. The diameter of the arena is 110 meters and is intended for various uses. In 2010, the SkyView Elevator was opened on the outer South side of the arena, transporting visitors to the top of the arena (Figure 4).

Aquarium in Genoa, Renzo Piano, 1992, Italy

The Aquarium building in Genoa, located in the old port area of Genoa, is considered the second largest aquarium in Europe (3,100 square meters). Today, the aquarium consists of 70 tanks with a total volume of 6 million liters, which are home to a variety of fish, reptiles and invertebrates. Every year, the aquarium in Genoa is visited by about 1.2 million people (Figure 5).

4 DISCUSSION

The structures of the triangular components that cover the surface of the sphere become more robust with increasing size, as they redistribute stress over the entire surface. This theoretically allows you to build spheres of colossal size. In his project "Cloud nine", 1940, Fuller proved that as the size of the sphere increases, the volume of space it encloses grows faster than the volume of the structure itself enclosing this space.

Based on the above, the authors suggest using large-span shells of geodesic domes and other engineering spatial structures made of modern polymer materials and wood for the reconstruction and development of coastal industrial territories. The main purpose of such buildings is considered not only industrial enterprises and factories, warehouses, logistics centers, but also public buildings for various purposes, while taking care of high energy efficiency and environmental friendliness of buildings, in accordance with international quality standards and green construction (BREAM, LEED, GREEN ZOOM).

5 CONCLUSIONS

1. The use of spatial structures, including geodesic domes, made of modern materials in the formation of the coastline of large cities creates a unique architectural silhouette, becomes a bright and memorable object.
2. The use of geodesic domes made of wood and modern polymer materials in the implementation of projects for the reconstruction and renovation of industrial coastal areas is appropriate and effective from an engineering and architectural point of view. Such solutions will not only allow you to realize the most daring architectural fantasies, but also fit into the historical environment of the city and will demonstrate the possibility of combining modern structures, environmentally friendly technologies and the latest design in a full-fledged architectural object.

3. The possibility of the use of engineering structures made of wood and polymer materials opens wide horizons for the transformation and development of coastal industrial areas, taking into account economic and environmental aspects of infrastructure development worldwide.

REFERENCES

About objects of cultural heritage (historical and cultural monuments) of the peoples of the Russian Federation, Accessible: https://rg.ru/2002/06/29/pamjatniki-dok.html (02.03.2020).

Kalvebod Waves, Accessible: http://jdsa.eu/kal/(02.03.2020).

Shubin V. I. & Zhivotov D. A. Connection of load-bearing rods for geodesic domes and other spatial structures, Moscow, Russia, patent no. RU170483U1.

Fuller R., Geodesic domes, Accessible: https://ongreenway.org/2014/12/geosfera-fullera-geodezicheskij-kupo/ (02.03.2020).

Yingxin Wu & Masahiro Takatsuka. 2006, Spherical self-organizing map using efficient indexed geodesic data structure Neural Networks. 19: 900–910 DOI: 10.1016/j.neunet.2006.05.021.

Yue Guan, Lawrence N. Virgin, Daniel Helm, 2018, Structural behavior of shallow geodesic lattice domes. International Journal of Solids and Structures, 155: 225–239. DOI: 10.1016/j.ijsolstr.2018.07.022.

Rizzuto J.P.. 2018. Experimental investigation of reciprocally supported element (RSE) lattice honeycomb domes structural behavior. Engineering Structures. 166: 496–510. DOI: 10.1016/j.engstruct.2018.03.094.

Porta-Gándara M.A. & V. Gómez-Muñoz. 2005, Solar performance of an electrochromic geodesic dome roof Energy. 30: 2474–2486. DOI: 10.1016/j.energy.2004.12.001.

Zhivotov D.& Tilinin Y., 2020. Experimental studies of the strength of nodal joints of geodesic domes made of wood and fiberglass made on a 3D printer for the Arctic and Northern territories. The Publication Series of LAB University of Applied Sciences, 2: 57–65.

Reconstruction and Restoration of Architectural Heritage – Sementsov et al (eds)
© 2020 Taylor & Francis Group, London, ISBN 978-0-367-65357-6

Buckling of shell roof structures under different loads

D.S. Petrov, A.A. Semenov & A.Yu. Salnikov
Saint Petersburg State University of Architecture and Civil Engineering, Saint Petersburg, Russia

ABSTRACT: This paper demonstrates an approach to analyzing the strength and stability of cylindrical shell panels, which are used in construction as coatings and overlaps of buildings and structures. The ANSYS Mechanical APDL 19.2 software package is used for calculations. In the first two calculation models, the shell structures were fixed on two longitudinal straight sides with a fixed hinge, and on the other two sides with a curvature, they had a free edge. In the third considered variant of the shell design model, the structure was fixed pivotally-immobile along the entire contour. Concentrated force and evenly distributed load were considered as load variants. The calculations took into account the effect of the shell structure's own weight. For all the considered variants of structures, graphs of the dependence of normal displacements on the load were obtained, the values were calculated taking into account geometric nonlinearity. The values of critical loads of stability loss were obtained for the variants of structures with evenly distributed load, while the other structures did not lose stability. Stress intensity distribution fields are obtained.

1 INTRODUCTION

Each type of building structure has its own requirements for the building structures that they consist of. These requirements directly depend on the main functions performed by the object. Building structures may differ from each other in how they work under current loads, construction technology, and operation features. Also, when designing building structures, the economic feasibility of using a certain type of structure in each specific case is analyzed.

One of the most common requirements is the need to cover a large operational area of the structure without the use of intermediate supports. For example, such a need arises in the construction of public buildings: stadiums, concert halls; in the construction of industrial buildings: warehouses, hangars for machinery and equipment, factory buildings; in the construction of power facilities: machine halls of power plants that cover the cores of nuclear power plants.

This problem can be solved using such elements of building structures as prestressed reinforced concrete beams and beams, steel trusses, cable-stayed structures, as well as shell structures, which are the most complex from the point of view of designing and calculating the stress-strain state under the influence of loads.

Due to the shape and parameters of the materials used, shell structures can cover large areas with a relatively low material consumption, which affects the economic efficiency of the project.

Shell structures can be classified according to various characteristics: the method of building (rotation surface, surface of the slope); functional purpose (covers, canopies and canopies, closed shells, coatings and walls, walls, wall bases); the type of support (at individual points, along lines, on the surface, a combination of these types); Gaussian curvature (parabolic (zero Gaussian curvature), hyperbolic (negative Gaussian curvature), elliptical (positive Gaussian curvature)).

For example, double-curvature shells are used for overlapping structures with a large square grid of columns, overlapping warehouses, machine shops and other production facilities where it is necessary to ensure free movement of transport or place oversized equipment (Bekkiev et al. 2018, Sankar et al.2016, Trushin 2016). Such shells are used both in single-span and multi-span buildings.

Conducting comprehensive studies of the process of shell deformation using the most accurate mathematical models will allow us to assign a documented safety factor, which will contribute to their safe operation, as well as reduce the material capacity of the structure and reduce its cost-effectiveness.

The purpose of this work is to test a method for studying the stability and stress-strain state of a cylindrical shell structure under the influence of various loads in the ANSYS software package.

2 THEORY

2.1 *Type of structures under consideration*

We will consider cylindrical panels of thickness that are under the influence of an external concentrated force F or evenly distributed transverse load. This type of shell is also widely used in construction (Ashok et al. 2015, Chen et al. 2017, Iskhakov & Rybakov 2016, Sengupta et al.2015).

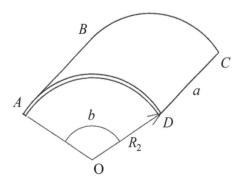

Figure 1. Schematic image of a cylindrical shell.

Let' take the middle surface of the shell as the coordinate surface. The axes are directed along the lines of the main curvature of the shell, the axis is normal to the median surface in the direction of concavity (Figure 1).

2.2 *Software*

Among the tools of SAE (computer-aided engineering), an important place is occupied by software packages based on the finite element method, which allow simulating the operation of the structure under study based on a detailed description of its geometry, the physics of the simulated processes, the properties of the materials used, performance characteristics, and other user-specified source and initial data (Ashok et al. 2015).

Modern software systems based on FEM allow conducting research for various application areas, and the most commonly used are NASTRAN (Ashok et al. 2015, Li et al. 2013), ABAQUS (Calladine 2018, Błachut 2016, Feng & Hu 2017), ANSIS (Trushin & Goryachkin 2016, zeibek, Topkaya & rotter 2017, Levchenya, Kolesnik & Smirnov 2019, Sonat et al. 2015).

To conduct this research, we will use the ANSYS Mechanical APDL 19.2 software package. It implements the function of calculating the movement of points of elements of the structure using the arc - length method, which allows us to study the movement of nodes of finite elements in the critical loading region, after the loss of stability of the structure.

3 METHODS

The shape of the structure will be set geometrically, by points, using the built-in tools of the software package. The formed geometric model of the shell structure is divided into a finite element grid. in this case, the Shell 181 element type was selected (the shell element with finite deformations; it has four nodes, each of which has six degrees of freedom; it allows for a full set of nonlinear effects, including large deformations (Ashok et al.2015). The grid of finite elements is selected as regular, and the elements have a rectangular shape.

Next, we will create relationships of elements that correspond to the selected method of fixing the outline of the structure. We will also set the self-weight load and applied external loads.

Further design calculation is performed using the following algorithm:

– calculation of forces in the structure from a single force (Static), taking into account small displacements (Small Displacement Static);
– calculation of the eigenvalue coefficient of structural stability (Eigenvalue Buck-ling);
– calculation of forces in the structure based on the force multiplied by the obtained coefficient (Static), taking into account large displacements (Large Dis-placement Static), analysis of the structure behavior;
– selection of the load at which the structure loses stability by gradually increasing the applied load and calculating the forces in the structure (Static), taking into account large displacements (Large Displacement Static) and analyzing the behavior of the structure according to the load – displacement schedule.

4 NUMERICAL RESULTS

4.1 *Cylinder shell Panels*

Table 1 shows three variants of shells for calculation, which differ between the applied load and the method of securing the contour. Shell material parameters: steel (linear, elastic, isotropic) with parameter $E = 2.1 \cdot 10$, MPa, $\mu = 0.3$, $\rho = 7800$ kg/м^3.

Table 1. Geometric parameters of the considered structures.

Var.	Linear size a, m	Radius of curvature R_2, m	shell thickness h, m	Corner b, (°)	Conditions fastenings	Load parameter
1	16.00	16.00	0.08	1 (28.65)	Hinged AB & DC	Concentrate forces
2					Hinged AB & DC	Distributed load
3					Hinged by contour	Distributed load

Setting the geometric shape of the structure, pay special attention to the center point (node no 57) and the quarter point (node no 173).

The finished geometric model of the shell structure is divided into a 32x32 finite element grid.

The shell structure for variants 1 and 2 is fixed pivotally along the straight faces AB and DC (Figure 2, a) (links are created against linear movement along three axes: X, Y, Z, while angular movements remain free). For option 3, the structure is hinged on all sides of the shell (Figure 2, b).

The gravity parameter is set as the initial loads to account for the structure's own weight: acceleration of free fall along the vertical y axis. For acceleration of free fall, the value is 9.8065 N/kg.

Under the influence of their own weight, the studied structural units receive the following maximum displacement and stress intensity (Table 2, Figure 3).

Consider the process of deforming the structure under the action of a concentrated force (option 1). Let's calculate the forces in the structure as

a)

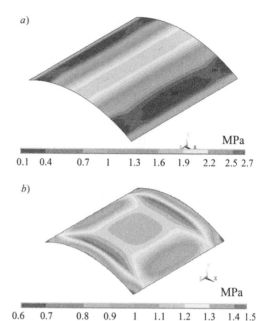

MPa

0.1 0.4   0.7   1   1.3   1.6   1.9   2.2 2.5 2.7

b)

MPa

0.6   0.7   0.8   0.9   1   1.1   1.2   1.3   1.4 1.5

Figure 3. Field stress intensity.

a function of the unit force, taking into account small displacements, and get the coefficient of the structure's own stability value: 15.391.

We apply a concentrated force of 15.391 N to the Central node. the force dependence on displacement (Figure 4) is close to linear. Here and further in the graphs, the red curve W_c corresponds to the node in the center of the structure$(x = a/2, y = b/2)$, and the blue curve W_4 corresponds to the node in the quarter of the structure$(x = a/4, y = b/4)$.

The maximum vertical movement of the middle node is 0.0006 m. Figure 5 shows the view of the deformed shell at time points 0.99 and 1.00. Judging by the image, the shell is beginning to collapse.

a)

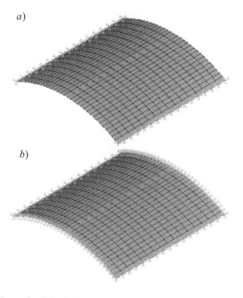

b)

Figure 2. Calculation model of the shell structure: var. 1 and 2, hinged support on two faces; b) var. 3, hinged support on all sides.

Table 2. Movement of nodes in the center and a quarter of their own weight.

Var	The deflection from its own weight, m		Max. pressure, MPa
	in the middle	at a quarter	
1	$0.12 \cdot 10^{-4}$	$1.79 \cdot 10^{-6}$	2.7283
2	$0.12 \cdot 10^{-4}$	$1.79 \cdot 10^{-6}$	2,7283
3	$1.89 \cdot 10^{-4}$	$88.94 \cdot 10^{-6}$	1,4505

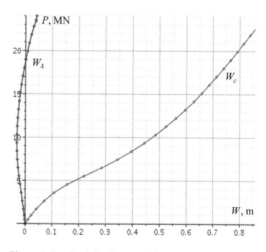

Figure 4. Load – deflection graph for option 1 construction.

215

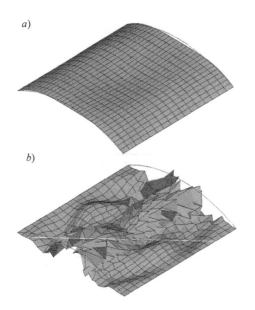

a)

b)

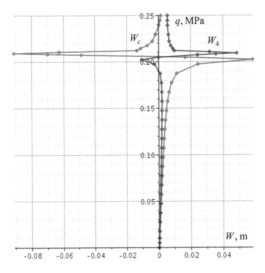

Figure 5. Scheme of shell deformation at time points 0.99 and 1.00.

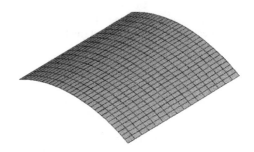

Figure 7. Scheme of deformation of the shell of option 2 at time 1 (0.25 MPa).

We apply a load of 96.490 PA (as for option 2, the load is evenly distributed). Figure 8 shows the load-deflection relationship graph, and Figure 9 shows the result of applying a load of 0.8 MPa. There is a significant deflection of the Central part of the shell, formed after the loss of stability.

Table 3 shows the values of stability loss loadsq$_{cr}$, deflection values, and stress intensity detected at the

Next, consider the process of deforming the structure under the action of a uniformly distributed load (option 2). For it, the coefficient of eigenvalue of stability is 15.391. Figure 6 shows the "load-deflection" relationship graph, and Figure 7 shows the result of applying a load of 0.25 MPa. As can be seen from Fig. 7, the deformation of the structure is almost not noticeable.

Next, consider the construction of option 3, when the shell is fixed pivotally-motionless along the entire contour. The coefficient of the eigenvalue of stability of the structure will be equal to 96.490.

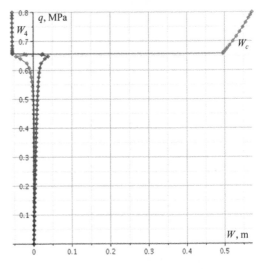

Figure 8. Load-deflection graph for option 3 design.

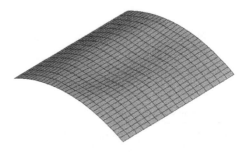

Figure 9. Scheme of shell deformation 3 at a load of 0.8 MPa.

Figure 6. Load-deflection graph for option 2 construction.

216

maximum calculated load. The corresponding stress intensity fields are shown in Figure 10.as can be seen from the data obtained, for the shell of option 3, the

maximum stress intensity values are significantly over-estimated, which indicates irreversible deformations of the material. Thus, the shell of option 3 simultaneously loses stability and strength.

Table 3. The results of the calculations.

Var.	q_{cr}, MPa	Max. design load	Deflection, m		Max. pressure, MPa
			W_c	W_4	
1	–	24 MN	0.8935	0.0415	74.5
2	0.2025	0.25 MPa	0.0014	0.0055	81.42
3	0.6562	0.8 MPa	0.5725	0.0555	1892.0

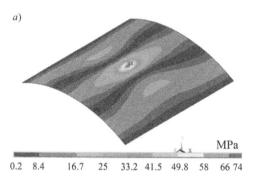
a)

MPa

0.2 8.4   16.7   25   33.2 41.5   49.8   58   66 74

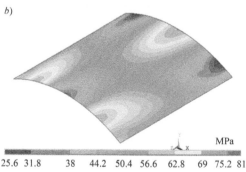
b)

MPa

25.6 31.8   38   44.2 50.4 56.6   62.8   69 75.2 81

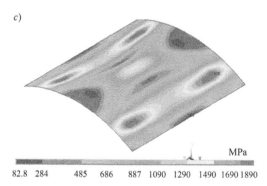
c)

MPa

82.8 284   485   686   887 1090 1290 1490 1690 1890

Figure 10. Stress intensity fields for shells of variants 1, 2 and 3, respectively.

5 CONCLUSIONS

According to the results obtained, the following conclusions can be drawn:
• the proposed method for studying the stability and stress-strain state can be used to analyze the process of deformation of cylindrical shell structures under the influence of various loads;
• the ANSYS Mechanical APDL 19.2 software package allows you to get the necessary data for building load-deflection dependencies, however, there are difficulties in obtaining values in the critical area;
• when calculating models of variants 2 and 3 it was possible to detect a loss of stability at 0.2025 and 0.6562 MPa respectively;
• graphs of the deflection dependence on the load are obtained for all the considered variants of structures. the process of de-formation is studied taking into account the geometric non-linearity and proper weight of shell structures;
• the results of calculations can be used for further comparison with the results obtained by analytical and semi-analytical methods.

ACKNOWLEDGMENTS

The research was supported by RSF (project No. 18-19-00474).

REFERENCES

Ashok, R.B., Srinivasa, C.V., Suresh, Y.J. & Prema Kumar, W.P. 2015. Buckling Behaviour of Cylindrical Panels. *Nonlinear Engineering*, 4(2): 67–75.
Bekkiev, M.Y., Chepurnenko, A.S. & Yazyev, B.M. 2018. Finite element modeling of creep of three-layered shallow shells. *IOP Conference Series: Materials Science and Engineering*, 463: 022014.
Błachut, J. 2016. Buckling of externally pressurized steel toriconical shells. *International Journal of Pressure Vessels and Piping*, 144: 25–34.
Calladine, C.R. 2018. Shell buckling, without 'imperfections'. Advances in Structural Engineering, 21(16): 2393–2403.
Chen, B., Zhong, P., Cheng, W., Chen, X. & Yang, Q. 2017. Correlation and Combination Factors of Wind Forces on Cylindrical Roof Structures. *International Journal of Structural Stability and Dynamics*: 1750104.
Feng, K. & Xu, J. 2017. Buckling Analysis of Composite Cylindrical Shell Panels by Using Legendre Polynomials Hierarchical Finite-Strip Method. *Journal of Engineering Mechanics*, 143(4): 04016121.
Iskhakov, I. & Ribakov, Y. 2016. *Design principles and analysis of thin concrete shells, domes and folders.*

Boca Raton, Florida: CRC Press, Taylor & Francis Group.

Levchenya, A.M., Kolesnik, E.V. & Smirnov, E.M. 2019. Disturbing effects of a cylinder-form macro-roughness on the turbulent free-convection boundary layer: Large Eddy Simulation. *Journal of Physics: Conference Series*, 1400: 077031.

Li, D., Qing, G. & Liu, Y. 2013. A layerwise/solid-element method for the composite stiffened laminated cylindrical shell structures. *Composite Structures*, 98: 215–227.

Sankar, A., El-Borgi, S., Ganapathi, M. & Ramajeyathilagam, K. 2016. Parametric instability of thick doubly curved CNT reinforced composite sandwich panels under in-plane periodic loads using higher-order shear deformation theory. *Journal of Vibration and Control*.

Sengupta, J., Ghosh, A. & Chakravorty, D. 2015. Progressive Failure Analysis of Laminated Composite Cylindrical Shell Roofs. *Journal of Failure Analysis and Prevention*, 15(3): 390–400.

Sonat, C., Topkaya, C. & Rotter, J.M. 2015. Buckling of cylindrical metal shells on discretely supported ring beams. *Thin-Walled Structures*, 93: 22–35.

Trushin, S. 2016. Numerical algorithm for solving of non-linear problems of structural mechanics based on the continuation method in combination with the dynamic relaxation method V. Andreev, ed. *MATEC Web of Conferences*, 86: 01006.

Trushin, S. & Goryachkin, D. 2016. Numerical Evaluation of Stress-Strain State of Bending Plates Based on Various Models. *Procedia Engineering*, 153: 781–784.

Zeybek, Ö., Topkaya, C. & Rotter, J.M. 2017. Requirements for intermediate ring stiffeners placed below the ideal location on discretely supported shells. *Thin-Walled Structures*, 115: 21–33.

Reconstruction and Restoration of Architectural Heritage – Sementsov et al (eds)
© 2020 Taylor & Francis Group, London, ISBN 978-0-367-65357-6

System of automated quality control for operating materials of construction equipment

R.N. Safiullin, V.A. Treyal & R.E. Baruzdin
Saint Petersburg State University of Architecture and Civil Engineering, Saint Petersburg, Russia

ABSTRACT: This article presents the results of studying the problem of improving the efficiency of energy conversion in construction equipment power units. Improving the efficiency of power units in construction is ensured by reducing friction losses in all operating conditions when using one of the most important operating materials to prolong their service life — engine oil. The completed experimental and theoretical studies in this area show that the developed technical solutions allow for the substantial reduction of friction losses and the improvement of economic and environmental parameters as well as reliability in all operating conditions of power units. Consequently, oil quality control ensures optimized oil viscosity, reduced hydraulic losses at inlet and outlet, and may promote a significant reduction of mechanical losses.

1 INTRODUCTION

The quality characteristics of mining machines greatly depend on the combination of properties of a power unit that supplies energy required to move equipment. Modern conditions of using power units are characterized by a significantly expanded range of requirements applied to them. In addition to functional requirements related to energy supply, the following has become extremely important: improving their readiness and faultless operation in various operation conditions of construction equipment as well as adhering to stringent requirements with regard to total resource saving and addressing the issues related to the complicating environmental situation (Safiullin 2011). Therefore, it is required to switch from individual improvements of power units to a comprehensive solution regarding the efficiency of such units as parts of a specific system.

The process of ICE energy conversion plays a central role in forming the entire combination of power unit properties, from its energy indicators to reliability and resource intensity characteristics. Hence, the problem of improving the energy conversion efficiency considered based on a systemic approach is the most important reserve to enhance the qualitative characteristics of construction equipment. Along with that, the issue of reducing emissions of exhaust gases during the operation of power units of construction equipment within a range of operating conditions is rather important. In this manner, the improvement of environmental and efficiency indicators of diesel construction machines is of special interest at this stage of development with simultaneous improvement of operating qualities of materials used in them.

2 MAIN PART

Friction losses in mechanisms are the primary component of mechanical losses, which account for 60–72% of the total amount of N_m. In addition to that, 70–72% of power N_{fr} are spent to overcome the friction of the piston and piston rings and 20–23% are spent to overcome the friction in the crankshaft bearings. Up to 60–65% of piston group friction losses are related to overcoming the friction of the piston rings. The specific weight of pump losses is evaluated as 15–30% of N_m, and the power spent for driving auxiliary units (fuel liquid and oil pumps, electrical tools) is 12–17%.

In the case of the normal operation of piston engine mechanisms and systems, specific hydrodynamic lubrication conditions are maintained in joints so that the friction pairs are reliably divided by a lubricant layer 5–15 μm thick. When the piston is reversing its stroke and its speed relative to the cylinder changes the sign going over zero, the oil film may get thinner between the piston rings and the cylinder down to tenth points of μm, which corresponds to boundary lubrication conditions (Safiullin & Morozov 2019b).

Friction between parts in hydrodynamic conditions is caused by inter-molecular interaction in the lubricant layer (engine oil). This interaction characteristic for real (Newton) liquids is defined by their viscosity (Safiullin & Kerimov 2017). The following indicators are used for the quantitative description of viscosity: μ — dynamic coefficient of viscosity measured in N-s/m²; v — kinematic coefficient of viscosity.

$$v = \frac{\mu}{\rho} \, \text{m}^2/s \left(\text{cm}^2/s\right) \qquad (1)$$

where ρ is the liquid density, kg/m^3.

According to the rheological Newton's equation, the friction force in hydrodynamic conditions of lubrication is as follows:

$$\rho_{fr} = \mu F \frac{W_0}{\delta_{oil}} \qquad (2)$$

The above fundamental dependency shows that the friction force is defined by viscosity of lubricant μ, area F of contact between mating parts, relative speed of their movement W_0 and oil layer thickness δ_{oil} equal to the clearance in friction pairs. The study results show that the value of clearance of δ_{oil} depends on the part loading, speed of their mutual movement and oil viscosity, which is, for a piston ring, approximately expressed by the following equation:

$$\delta_{oil} = \frac{S_P \mu}{P}, m \qquad (3)$$

where S_P is the piston speed, m/s; P is the component of force acting on the piston ring, perpendicular to the cylinder bore surface, N.

Taking into account the last expression and a number of conversions, the piston ring friction force equals:

$$K_t = A_{CC} Dh_{kj} (\mu n)^{0.5} p^{n1} \qquad (4)$$

where A_{CC} is the constant coefficient; Dh_{kj} is the cylinder diameter and piston ring height, m; n is the crankshaft rotation speed, min^{-1}; p is the average specific loading on the piston ring, N/m^2; $n1$ is the exponent.

For a set of k piston rings, the friction force is as follows:

$$P_{cc} = \sum_{j=1}^{k} \rho_{cc_j} \qquad (5)$$

For approximate evaluation of mechanical losses, empirical equations can be used, such as:

$$P_M = A + BS_P \qquad (6)$$

where A, B are the constant coefficients; S_P is the average piston speed, m/s.

These expressions allow us to evaluate the factors causing friction losses as the primary component of mechanical losses. Friction losses in mechanisms are defined by the dimensions of their parts, loading, speed conditions and viscosity of oil (Safiullin et al. 2019a).

To reduce friction losses, it is very important to decrease the size of the contact surface and reduce oil viscosity provided it does not cause an increase in the specific loading and reduction of the oil film bearing capacity (Safiullin et al. 2019b).

These methods are widely used in modern power units (Safiullin et al. 2018). The height of the piston rings is reduced, their number is decreased, the height and diameter of the piston guiding part and the diameter and length of the crankshaft bearings are reduced. There is also a trend to decrease engine oil viscosity, which is 8 cm2/s for gasoline engines and 10 cm2/s for diesel engines at 373 K.

Friction losses are also reduced by a limitation or even decrease of the crankshaft rotation speed, which is used in modern engine engineering (Kerimov et al. 2017). It should be noted that mechanical losses mainly depend on compliance (during manufacturing and assembly) with specifications for the roughness of surfaces of friction pairs, nature of their mating, deviations in macrogeometry, and quality of engine oil in construction equipment.

However, it is extremely difficult to determine the potential properties of oil during the operation of power units of construction equipment. Existing methods of quality control for engine oils are time-consuming and require laboratories to have special equipment. To ensure the reliability of construction equipment, it is necessary to develop in-process methods for assessing the quality of oils directly in the engine, which are intended to determine the generalized quality criterion for further influence on the operated vehicle via the electronic control system (ECS) in order to increase its service life and operational reliability (Safiullin & Morozov 2019a).

The in-process engine oil control method for construction equipment consists of the automation of oil quality assessment, adjustment of power unit operation within all ranges of operating conditions in case of using low-quality oil by optimizing the ECS using feedback with the oil quality control system, and protection of parts and components of the power unit against premature wear, which directly characterize the energy conversion efficiency of the construction equipment power unit (Di et al. 2016). Optimal control of the defined indicators by the developed system represents an extremely complicated technical task. To address this issue, a simulation system was developed to assess oil quality in vehicles (utility model patent No. RU194054U1 dd. 07.06.2019) that will expand capabilities of oil quality measurement methods in all operating conditions (Safiullin & Morozov 2019, Safiullin 2017). All the above components of this technical solution are given in Figure 1.

The developed engine oil quality control system for vehicles comprises a crankshaft rpm sensor, a mass fuel flow meter sensor, a gas pressure sensor in the cylinder, a throttle position sensor, a knock

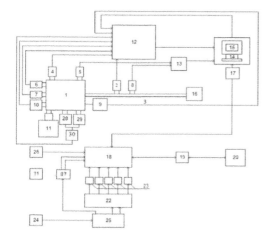

Figure 1. Overall diagram of the simulated engine oil quality control system for mining equipment: 1 — tested engine; 2 — crankshaft rpm sensor; 3 — camshaft sensor; 4 — gas pressure sensor in the cylinder; 5 — throttle position sensor; 6 —knock sensor; 7 — crankshaft position sensor detecting angle marks; 8 — oxygen concentration sensor; 9 — mass air flow meter sensor; 10 — gas analyzer of polluting emissions in combustion products; 11 — engine control unit; 12 — electronic control unit; 13 — analog-digital converter; 14 — personal computer with a monitor; 15 — loading device; 16 — control unit; 17 — electronic control unit; 18 — communication interface; 19 — ignition key simulator; 20 — generator-simulator; 21 — switch; 22 — mode setting unit; 23 — operation control device; 24 — interface between the engine control unit and the electronic control unit; 25 — interface between the electronic control unit and the operation control device; 26 — oil quality control sensor; 27 — fuel temperature sensor; 28 — electronic unit for evaluation of fuel sensor results; 29 — engine oil temperature sensor; 30 — engine oil quality control sensor; 31 — electronic unit for the evaluation of the results.

sensor, a crankshaft position sensor detecting angle marks, an oxygen concentration sensor, a mass air flow meter sensor, and a gas analyzer of polluting emissions in combustion products, installed on the tested engine, an electronic control unit for the tested engine, an analog-digital converter, a personal computer with a monitor, a model of the engine mockup electronic control unit, its communication interface with the personal computer and monitor, an ignition key simulator, a generator-simulator of signals of the above sensors, a switch for these signals, and a mode setting unit; it is also additionally equipped with an engine oil quality assessment sensor (capacitance measurement element), an engine oil temperature sensor, and an electronic unit to evaluate the measurement results of these sensors.

3 CONCLUSION

The completed experimental and theoretical studies in this area show that the developed technical solutions allow for the substantial reduction of friction losses and the improvement of economic and environmental parameters as well as reliability in all operating conditions of power units of construction equipment. Consequently, oil quality control ensures optimized oil viscosity, reduced hydraulic losses at inlet and outlet, and may promote a significant reduction of mechanical losses. An important feature of the oil quality assessment system is that it provides for in-process assessment in order to optimize power unit control using the system feedback, thereby ensuring an increased service life of construction equipment (Kozlov et al. 2009).

REFERENCES

Di, X. et al. 2016. Second best toll pricing within the framework of bounded rationality. *Transportation Research Part B: Methodological* 83: 74–90.

Kerimov et al. 2017. Evaluation of functional efficiency of automated traffic enforcement systems. *Transportation Research Procedia* 20: 288–294.

Kozlov, L.N. et al. 2009. On the conceptual approaches to forming and development of intellectual transport systems in Russia. *Transport of Russian Federation* 3–4 (-22–23): 30–35.

Safiullin, R.N. 2011. *Theoretical basis of the efficiency of energy conversion in ICEs and its enhancement methods. Monograph.* Saint Petersburg: Saint Petersburg State University of Architecture and Civil Engineering.

Safiullin, R.N. & Kerimov, M.A. 2017. *Intelligent onboard transportation systems for automobile transport.* Moscow, Berlin: Direkt-Media.

Safiullin, R.N. & Morozov, E.V. 2019a. Theoretical basics for improving the technology of technical systems operation intended for monitoring the movement of motor vehicles at various levels. *Bulletin of Civil Engineers* 2 (73): 153–160.

Safiullin, R.N. & Morozov, E.V. 2019b. *Vehicle motor oil state simulation system. Patent No. RU194054U1.*

Safiullin, R.N. et al. 2017. *Automated data control system on the technical condition of the vehicle internal combustion engine. Patent No. RU174174U1.*

Safiullin, R. et al. 2018. A model for justification of the number of traffic enforcement facilities in the region. *Transportation Research Procedia* 36: 493–499.

Safiullin, R.N. et al. 2019a. The concept of development of monitoring systems and management of intelligent technical complexes. *Journal of Mining Institute* 237: 322–330.

Safiullin, R.N. et al. 2019b. *Traffic control automation systems for automobile transport. Monograph.* Saint Petersburg: Lan.

Environmental risk analysis in construction under uncertainty

E. Smirnova
Saint Petersburg State University of Architecture and Civil Engineering, Saint Petersburg, Russia

ABSTRACT: The article analyzes the issue of determining environmental risk for the purposes of environmental safety. When using risk assessment methods, we should understand that the concept of risk implies the probability of a negative event multiplied by the amount of damage. In this regard, the issue of risk and uncertainty is considered. The author believes that the essence of assessing any risk is not to completely avoid or decrease it but to calculate it and determine whether it is acceptable or unacceptable and whether the damage is controllable or uncontrollable. All this determines the safety (acceptability) of the risk and the amount of losses that should be accepted when carrying out construction works. A matrix of distributions of environmentally extreme conditions gives an idea of quantitative characteristics of possible ecosystem "failures" as a result of anthropogenic interference.

1 INTRODUCTION

Investigation of risk and its prevention methods is an extremely important task. According to GOST R 58771-2019 (Risk management. Risk assessment technologies), risk is a consequence of some uncertainty, and risk assessment technologies help to understand it and take reasoned decisions and actions. Broadly speaking, uncertainty may cover both various types of its manifestation and the absence of any knowledge of it, unpredictability as such, against which the human mind can do nothing. That is why risk assessment technologies are relevant. They help to make a decision considering the uncertainty and choose the best option for a construction project. A number of researchers believe that "risk" is the probability of an event and its negative consequences (Medvedeva 2016). It should be noted that risk is compared with a danger that can harm both business and the environment. Risk can be attributable both to natural and anthropogenic factors, and it depends on situation complexity. In one case, it can be related to minor consequences, requiring decisions following the established rules and procedures. In another case, in conditions of high uncertainty and lack of experience, it results in a search for non-traditional technologies for its assessment. In this regard, its anthropocentric nature is highlighted, i.e. it is considered only in the "nature–man" aspect. This is hardly surprising: the quantitative measure of risk is defined as the product of hazard being studied and the amount of damage done (Kolesnikova & Novikov 2019). Various kinds of risks cause both financial and human losses. Regarding the concept of "environmental risk", its anthropocentric characteristic is obvious: risk is interpreted as probable losses due to negative anthropogenic impact on the environment, as a probability initiating such anthropogenic changes (intentional or accidental; gradual or catastrophic) in nature and society, which change the structure and functioning of natural and technical systems being a major factor in ensuring environmental safety. On the other hand, we should not forget that the likelihood of accidents and other risk events is related to special environmental conditions (biological, hydrological, geological, climatic, seismic, etc.) that "program" unpredictable negative consequences: natural, natural and man-made, man-made disasters. At its core, environmental risk is a natural and man-made phenomenon. Thus, according to the governing technical standards, the frequency and consequences of a certain hazard should be distinguished in risk; according to some researchers, the understanding of risk is related to the analysis of the uncertainty of expected events, which most often result in some damage (Borch 2016). The purpose of the study is to analyze the concept of environmental risk within the environmental balance of the "construction–man–environment" system.

2 MATERIALS AND METHODS

The analysis of the "environmental risk" concept is based on the identification of system processes. To create a conceptual model, the functional approach is applied, which provides the basis for the synthesis of system components. This approach makes it possible to model each key component of the "construction–man–environment" system and integrate them into a single complex according to the structure of the conceptual model (Smirnova & Larionov 2020, Smirnova & Larionova 2020). Speaking about environmental risk, it should be noted that the risk analysis scheme involves the following steps: 1) risk factors should be established; 2) then, sources and recipients of risk should be identified; 3) risk assessment methods should be selected according to the requirements and criteria of anthropogenic impact on the environment; 4) using the chosen methodology, further development of events should be predicted in the form of scenarios that correspond to a specific environmental regime from the classification of parameters of anthropogenic impact by the degree of increase of degradation processes on the ecosystem; 5) finally, recommendations should be given on minimizing risk and preventing risk situations, and also, in order to prevent a spontaneous increase in the anthropogenic level, monitoring of the ecosystem state should be established in the range of man-made impacts, which may still be dangerous from the point of view of riskology.

If the totality of methods makes it possible by assessing the risk to identify hazards, make recommendations for managing and reducing the threat of the consequences of extreme processes and phenomena dangerous to human life and the entire ecosystem, it can be used as a risk analysis tool (Smirnova 2020).

3 RESULTS

By assessing the process of erecting construction facilities, the factors of environmental risk (internal and external, objective and subjective) can be reduced to three large groups: natural, man-made, and social. Further, the risks can be structured as follows: natural and environmental, technical and environmental, risks of sustainable man-made and catastrophic impacts, social and environmental, environmental and regulatory, environmental and political, environmental and economic.

Normally, the multidimensional nature of the impact on environmental components should be considered in construction. The same factor can have both a negative (increasing the risk) and a positive (reducing the risk) impact on the ecosystem. Thus, toughening of environmental requirements may increase the risk of construction due to expenses for the purchase of scientific and technological innovations that reduce production costs, however, changes in the regional environmental situation may significantly reduce the risks of competition if competitors fail to master the new technology. It is important that the magnitude of risk related to possible damage through the probability of event occurrence remained within the standardized acceptable limits. Otherwise, in case of long-term negative impacts on the environment, the acceptable limits (of direct or indirect actions) may be exceeded, which will result in significant imbalances causing degradation of the environmental balance

towards uncertain states. For instance, the impact of acts of God and regional climate on the production process may result in an unpredictable risk.

It makes sense to say a few words about the environmental response to the man-made challenge. Any industrial impact causes an ecosystem response in three forms: adaptive, characterized by local imbalance and a shift in the environmental balance; self-recovering, i.e. returning to the initial equilibrium state, and partially recovering (or non-recoverable due to an irreversible shift of the "industrial facility – nature" system towards instability and uncertainty). Specifics of construction is such that only a strictly regulated and accident-free operation mode of industrial facilities can ensure the environmental balance. Within the environmental standards, the level of impact of man-made factors (e.g. explosive plants) on the ecosystem corresponds to the adaptive capabilities of a particular natural system in the region. The higher the reserves of natural recovery and the more effective the methods of artificial recovery of biocoenoses are, the faster the "industrial facility – nature" system recovers its initial equilibrium state. Due to the fact that the absolute losses incurred during the construction of a facility by causing negative environmental impacts necessarily serve as a starting point to calculate the subsequent risks of environmental damage, the idealized equilibrium state is no longer a question. However, there is a concern about whether nature (after a possible monotonous or abrupt shift in the ecosystem balance) is able to return to the initial level of safety and whether there are enough environmental reserves to withstand the increase in the total man-made impact.

The majority of modern risk assessment techniques are based on the concept of acceptable risk. This points to the fact that risk is a relative concept, i.e. it is considered in comparison with a certain (acceptable) risk. The axiological (evaluative) aspect of risk in the case of an impact on the ecosystem serves as a necessary analytical basis for determining the social value of human life in the ecosystem. It is clear that, in comparison with financial or professional risk, the environmental one affects the value of human life in a completely different way: if environmental losses have become irreversible and negative, then this shift in ecological equilibrium poses both dangers on a local scale and a widespread threat to human life. For example, sulfur dioxide, which enters the atmosphere from burning gas and oil, when copper is smelted at plants, and as a result of the widespread distribution of sulfate aerosols, this points to a clear threat to human health. It is not for nothing that the methods of artificial recovery of the environment are becoming more relevant due to the increased man-made load. Any risk can be evaluated in comparison with some acceptable ones only with consideration for the regulated boundaries of man-made changes. In this case, maintaining the equilibrium state of an ecosystem prone to harmful effects is the goal of risk assessment.

This provision can be mathematically expressed in the following system:

$$\begin{cases} S_{min} = f\left[(Q_{used})_{min}; (Q_{remain})_{max}; t\right] \\ Q_{total}(t) = const \\ Q_{acceptable}\left(t < t_i\left[Q_{acceptable}\right]\right) \end{cases} \quad (1)$$

where S_{min} = the value of the potential displacement of the ecosystem equilibrium,

Q_{used} = the amount of natural sources used, Q_{remain} = remaining natural reserve, Q_{total} = total ecosystem reserve $Q_{acceptable}$ = acceptable level of man-made changes in the environment.

Next, a model of sudden and gradual environmental failures can be built. If extreme conditions affect only a single indicator — e_i — within a single object of nature (e.g. $e_{atmosphere}$, $e_{lithosphere}$, etc.), then the environmental failure in relation to the entire natural and technical system will be partial. If an extreme situation covers the entire object or a group of natural objects, then such a failure will be complete. The quantitative value of both partial and complete failures will result in the fact that the value of man-made changes in a particular species will exceed the limit of environmental tolerance, i.e. for a partial failure, the following is true:

$$\Delta e_i \leq \Delta e_i \begin{cases} e_i > e_{top_i} \\ e_i > e_{bottom_i} \end{cases} \quad (2)$$

where $i = 1, 2, \dots$ n.

For a complete failure:

$$\Delta e_j \leq \Delta e_j \begin{cases} e_j > e_{top_j} \\ e_j > e_{bottom_j} \end{cases} \quad (3)$$

where $j = 1, 2, \dots$ 6.

In equations (2) and (3), i and j are the indices of unit parameters of an individual object of nature (i = 1, 2, ... n) and nature as a whole (j = 1, 2, ... 6). We construct a standard model for distributions of environmentally extreme conditions in nature using the binary matrix (Table 1). As Table 1 shows, the safety review of natural and technical systems includes a risk assessment procedure. The main goal of environmental risk assessment is related to the identification of hazards, analysis of the intensity of occurrence of risk situations with the calculation of possible damage. Exceeding the specified environmental tolerance [Δe] under the influence of man-made changes affects entire ecosystem. The cumulative effect of environmental "failure" depends on the relationship vector of anthropogenic changes: as they accumulate in the ecosystem, the probability of sudden and gradual "failures" and extreme situations increases. The direction of the vector (gray cells) indicates an unacceptable risk and the need to take measures to reduce it.

As already noted, the technology of risk assessment depends on the complexity of the situation. Event Tree Analysis (ETA), which visually represents the possible development of an accident and emergency, serves as the main tool to identify hazards for ecosystems. This is a method of accounting, evaluating, and studying possible "failures" and ranking measures to improve security. GOST R IEC 62502-2014 (Risk management. Event tree analysis) gives an example of an "event tree" for fire (2014). This example draws attention to the fact that the designation of the event is accompanied by an indicator of the phenomenon likelihood. Since the probability of a case takes values from zero to unity — 0 <P (A) <1, the ignition frequency due to fire isolation is 1.0e-7 per year, i.e. 0.0000001 (1 case per million). It is calculated by multiplying the frequency of the initiating event (1.0e-4 per year, i.e. 0.0001) by the probability of inability to detect a fire (1.0e-3 per year, i.e. 0.001). As a result of hazard identification, the

Table 1. Matrix of distributions of environmentally extreme conditions.

Object of nature	Environmentally extreme situations by objects of nature					
	A	G	L	Fl	Fn	H
Atmosphere (A)	$\Delta e_A >$ $[\Delta e]_I$	$e_G A$	$e_L A$	$e_{Fl} A$	$e_{Fn} A$	$e_H A$
Hydrosphere (G)	$e_A G$	$\Delta e_G >$ $[\Delta e]_{II}$	$e_L G$	$e_{Fl} G$	$e_{Fn} G$	$e_H G$
Lithosphere (L)	$e_A L$	$e_G L$	$\Delta e_L >$ $[\Delta e]_{III}$	$e_{Fl} L$	$e_{Fn} L$	$e_H L$
Flora (Fl)	$e_A Fl$	$e_G Fl$	$e_L Fl$	$\Delta e_{Fl} >$ $[\Delta e]_{IV}$	$e_{Fn} Fl$	$e_H Fl$
Fauna (Fn)	$e_A Fn$	$e_G Fn$	$e_L Fn$	$e_{Fl} Fn$	$\Delta e_{Fn} >$ $[\Delta e]_V$	$e_H Fn$
Human (H)	$e_A H$	$e_G H$	$e_L H$	$e_{Fl} H$	$e_{Fn} H$	$\Delta e_H >$ $[\Delta e]_{VI}$

order of events that could cause an accident, as well as the probabilities of their occurrence, are provided.

Any other method suitable for calculating the probability of success (Yes) or failure (No) can be combined with the "event tree" method. Some authors believe that an integrated approach that complements the "event tree" helps to substantiate the decision scenario in detail (Leach 2014). The first stage of this approach is devoted to the identification of critical risks. The Fuzzy Analytic Hierarchy Process (Fuzzy AHP) and Fuzzy Failure Mode and Effect Analysis (FMEA) are used to identify potential risks. The Risk Criticality Number (RCN) is then determined. At the modeling stage, using Monte Carlo Simulation (MCS), a matrix is built to assess the risks related to the intended actions, the variability and duration of which was previously calculated. Then, the total project time is calculated and a critical chain of events is identified. As a result, it becomes possible to obtain information on the degree of compensation for deficiencies due to risk situations in uncertain circumstances. In other words, since decisions are made with insufficient information based on random parameters and the situation development process as well as the inevitability of alternativeness, it is important to identify the probability of an event, the intensity of its impact on the ecosystem, and the possibility of detection/control (Salah 2015). Obviously, the model is determined by the algorithm, with which it was created. Mathematical modeling enables us to meet the established requirements, avoid unforeseen errors, and optimize system performance. At the stage of minimizing risks, the overall probability of the implementation of a specific project scenario according to the critical chain model increases (Abdelgawad & Fayek 2012). Here is the thing: even at the time of making a decision, it allows us to assess general risks and then take specific preventive measures (to calculate temporary, intermediate reserves of the "non-critical chain") against the environmental hazard of an object, based on the decision-making method from beginning to end and from end to beginning. The main aim is to help reduce the risk of damage to the environment.

Here it is appropriate to return to the definition of environmental risk again. The analysis of the first stages of risk assessment allows us to see one of its key aspects in the quantitative characteristics. Environmental risk is calculated as the product of the probability of an accident at a facility and losses incurred by the environment directly as a result of a man-made incident; the damage is expressed in the monetary form of negative man-made impact and its consequences on the ecosystem.

Risk indicates an event that may or may not occur. But if it does, there are three possible outcomes: negative (loss, damage), zero (no loss, no gain), positive (gain, profit). It is clear that if risk is constantly avoided, then it is impossible to make a decision on ensuring the safety of the environment against man-made impact. The so-called "risk-free"

zone preceding the "tree of events" is, in fact, only a certain assumption, initiating the further process of winning or losing. The background risk is inherent in the entire natural and technical system since the "risk-free" state is also determined with respect to risk, i.e. danger and damage are inherent in any relation to the environment; they are potentially present to an uncertain degree. Therefore, a number of authors compare the quantitative characteristics of risk with uncertainty (Gollier 2018, Jaeger et al. 2013, Knight 2013). According to Knight, uncertainty refers to many possible outcomes disclosing nothing as to their likelihood (2013). Risk refers to a situation with a finite number of outcomes, the probability frequency values of which are known. Risks can be insured. Uncertainty cannot be insured. Nevertheless, rapid technological progress is associated precisely with uncertainty. Other authors believe that the essence of technological progress reduces to the transition to an environment with a lower value of uncertainty. The goal of managerial control is to minimize it, level out the effect of uncertainty on the natural and technical system, and completely overcome it (Sutton et al. 2013).

Thus, it turns out that risk is an activity carried out under uncertainty and alternativeness to choose a course of action and associated with the probability of least damage. According to the assertions of Gaponenko and Pankrukhin (2001), management control generally aims to overcome all uncertainties. Being able to avoid undue risk is an ideal of a person in charge of his/her decisions. For such a person, the legitimate amount of risk when achieving the goal is that the level of losses does not exceed the result achieved. However, the magnitude of risk related to the expected damage and the likelihood that the damage will actually occur cannot be legitimate every time: stochasticity cannot be excluded. In the best case, the risk can be predicted in mathematical modeling in the form of systems of differential equations that are well developed (say, according to the law of large numbers, a random variable gets the values of the mean mathematical expectation and serves to estimate the probability of large deviations, i.e. with a large number of random values it is reliably known that their arithmetic mean as a random variable differs infinitely little from the non-random (normal) value of its mean mathematical expectation; in other words, the action of a combination of random factors gives a result that is almost independent of chance (Kremer 2010). Here we have the well-known approximation of calculations of probable quantities to a constant.

But the following problem arises: if a risky decision does not have a factor of randomness, random parameters, a random chain of events, then what about alternativeness and stochasticity? If the criteria for a risky decision are based on the achievement of the established task, distinguished by their consistency with regulations, follow modern scientific and technical requirements, and aimed against the risk of losses and risk consequences, this kind of desire not to lose will turn risk into directives handed down. The actual devaluation of risk will occur. In the latter case, the effectiveness and competitiveness of construction are out of the question. Obviously, in a risk situation, the correlation of probable losses and the significance of the final result and gain will prevail (Grote 2009, Kreider 2012).

According to Kuzmin (2012), overcoming uncertainty leads to stability. He emphasizes that mitigation measures — the creation of reserves, smoothing and protective measures such as regulation, structuring, standardization, unification, integration, co-opting, choice of field of activity, and control — all together makes it possible to achieve greater certainty and thereby manage uncertainty (Kuzmin 2012). It is important to find the quantum of managerial impact on uncertainty. As Pavlov (2006) concludes, each process is characterized by a certain measure of uncertainty, intrinsic to it, which may or may not correspond to the degree of uncertainty of the conditions, in which it occurs.

4 CONCLUSIONS

Thus, in order to ensure modern requirements for the environmental safety of construction, riskology provides for the following assessment steps: after identifying the danger to the ecosystem (this refers to exceeding the MAC (maximum allowable concentrations) and MAE (maximum allowable emissions) standards), the stage of risk assessment in qualitative and quantitative aspects should be proceeded with. Qualitative analysis solves the problem of determining whether the risk is acceptable or unacceptable, and calculates the risk value in specific conditions. When analyzing numerous definitions of risk, the following key definitions can be called the main features of a risk situation. In it:

1) there should be alternative solutions since we do not have an exact knowledge of the consequences of man-made interference in nature and there is no exact correspondence between the conceived and implemented due to fallibility (it should be noted that the risk manager has to make a choice: a) more reliable and acceptable, but less success, or b) expected with less confidence and with doubt, but greater gain; it is obvious that greater success as a kind of provocation to achieve an impossible goal indicates an unwillingness to completely eliminate uncertainty from practice);
2) the stochastic nature of event or action is implemented determining one of the possible outcomes of a decision;
3) after a decision is made, the development of a risk situation should lead to losses or additional benefits;
4) when calculating risk on the basis of data processing, interest is shown in the future big benefit, the

success of which is unlikely and extremely uncertain, and, on the contrary, in the case of a small but guaranteed success, the individual does not have the urge to take risky decisions.

Three types of uncertainty can be distinguished.

– lack of information as ignorance of all the reasons that could affect the actions of a company; the numerical value of probability (distribution according to the law of indifference based on expectations) regarding the outcome of decision-making is unknown;
– unpredictable, objective occasionality, deviation due to the inherent unknowableness of things (accidental equipment failure, inaccuracy, errors in project forecasts and calculations, market fluctuations, force majeure circumstances, political and economic instability, etc.). People do not just feel that they don't know why the coin will fall "heads" or "tails" — they firmly know that such a reason does not exist, and only under such a condition can they have any kind of confidence in making a probabilistic statement. The whole science of probability, as a branch of mathematics, is based, according to Knight (2013), on the dogmatic assumption that the elementary alternative outcomes are equally probable which means objective non-determinism. However, through the classification and grouping of cases, the objective randomness can be reduced to objective quantitatively defined probability.
– the highest form of unmeasurable uncertainty. It cannot be eliminated, however, it is the uncertainty, which is the source of additional income in a risky situation. In relation to a single and unique case, measurable risk and immeasurable uncertainty are identical.

REFERENCES

Abdelgawad, M. & Fayek, A.R. 2012. Comprehensive hybrid framework for risk analysis in the construction industry using combined failure mode and effect analysis, fault trees, event trees, and fuzzy logic. *Journal of Construction Engineering and Management* 138 (5): 642–651.

Borch, K.H. 2016. *The Economics of uncertainty.* Princeton, NJ: Princeton University Press.

Gaponenko, A.L. & Pankrukhin, A.P. (eds) 2001. *General and special management.* Moscow: Russian Academy of Public Administration.

Gollier, C. 2018. *The economics of risk and uncertainty.* Cheltenham; Northampton, MA: Edward Elgar Publishing.

Grote, G. 2009. *Management of uncertainty: Theory and application in the design of systems and organizations.* London: Springer.

Jaeger, C.C. et al. 2013. *Risk, uncertainty and rational action.* London: Routledge.

Knight, F.H. 2013. *Risk, uncertainty and profit.* Miami, FL: HardPress Publishing.

Kolesnikova, L.A. & Novikov, A.S. 2019. Methodical approach to environmental risk assessment to achieve sustainable development of an industrial enterprise. *Ugol' – Russian Coal Journal* 6: 98–101.

Kreider, W.S. IV. 2012. Systemic uncertainty: an examination of its causes and repercussions. *Undergraduate Economic Review* 9 (1): 5.

Kremer, N.Sh. 2010. *Probability theory and mathematical statistics*. Moscow: Unity.

Kuzmin, Ye.A. 2012. Measures of uncertainty preventive management in business systems. *Vestnik NSUEM* 4: 269–284.

Leach, L.P. 2014. *Critical chain project management*. Boston, London: Artech House.

Medvedeva, S.A. 2016. Environmental risk. General concepts and assessment methods. *Technosphere Safety. XXI Century* 1(1): 67–81.

Pavlov, K.V. 2006. "Black hole" in economy and extreme level of uncertainty in manufacturing processes and economic environment. *Proceedings of Donetsk State National University. Series: Economics* 30: 4–14.

Salah, A. 2015. *Fuzzy set-based risk management for construction projects. PhD Thesis in Building Engineering*. Montreal: Concordia University.

Smirnova, E. & Larionov, A. 2020. Justification of environmental safety criteria in the context of sustainable development of the construction sector. *E3S Web of Conferences* 157: 06011.

Smirnova, E. & Larionova, Y. 2020. Problem of environmental safety during construction (analysis of construction impact on environment). *E3S Web of Conferences* 164: 07006.

Smirnova, E., 2020. *Problems of ensuring environmental safety in relation to toxic "Krasny Bor" dump site* (in press).

Sutton, A. et al. 2013. The strategic choice approach to managing uncertainty. In L. Wilkin & A. Sutton (eds), *The management of uncertainty: approaches, methods, and applications*: 14–52. Dordrecht: Springer.

Reconstruction and Restoration of Architectural Heritage – Sementsov et al (eds)
© 2020 Taylor & Francis Group, London, ISBN 978-0-367-65357-6

Parameters of heat-shielding in prerevolutionary residential buildings

V.M Ulyasheva, A.Y. Martyanova & G.A. Ryabev
Saint Petersburg State University of Architecture and Civil Engineering, Saint Petersburg, Russia

ABSTRACT: The article presents the results of laboratory studies of bricks heat conductivity factor for pre-revolutionary buildings constructed based on the method of continuous heat flux creation. The numerical methods with Ansys software system was used to determine R-value of outer walls of the so called "tenement buildings" of Saint Petersburg. The analysis of field studies of the heat-shielding characteristics of eleven facilities of the above category was carried out, depending on the characteristics of the space-planning decisions, volume-to-size, glazing-to-wall ratio. Specific heat-shielding characteristics of the studied buildings were obtained. The specific heat consumption for heating and ventilation has been summarized taking into account the actual schemes of heating systems and coolant parameters. The correspondence of buildings heat consumption level to current energy saving classes D and E has been revealed. The obtained data can be used during development of heating systems overhaul projects.

1 INTRODUCTION

A significant number of papers of local and foreign authors (Cabeza et al. 2018, Claude et al. 2019, De Carli, 2013, Dmitriyev et al. 2008, Makrodimitri et al. 2010, Murgul 2014, 2015, Murgul & Pukhkal 2015, Park 1991, Penića et al. 2015, Pukhkal et al. 2015, Razakov et al. 2019, Roberts 2019, Wood et al. 2010), including those on historical buildings of Saint Petersburg (Murgul 2014, Razakov et al. 2019), are devoted to the study of problems arising from restoration and modernization of historic buildings. The calculation methods with the examples of the use given in some studies (Atmaca & Gedik 2019, Bhattaa et al. 2019, Denga et al. 2019, Harmati et al. 2015, Kirimtat & Krejcar 2018, Sari et al. 2019) require serious attention.

In many European countries, including Russia, the requirements for thermal insulation of buildings have increased significantly, taking into account the current trend of energy usage reduction. However, for historic buildings of cultural and architectural merit, the desire to ensure energy efficiency can result in the loss of unique character not only of sep-arate buildings (Murgul 2014), but also of historical areas of cities.

The unique architectural appearance of cities is largely determined by not only unique, but residen-tial buildings. As mentioned in the paper by Claude et al. (2019), residential buildings of medieval con-struction make up a significant proportion of the French building stock. Herewith, to preserve the architectural merit of the external facade, the only way to reduce energy usage and increase their ther-mal comfort is internal insulation. However, many researchers (Murgul & Pukhkal V. 2015, Pukhkal et al. 2015) note that internal insulation can have an adverse effect on the heat and mass transfer charac-teristics of the wall, causing internal moisture con-densation and mold formation. To exclude these effects, for example, the used of extruded polystyr-ene foam panels is proposed. Several options for the use of advanced materials for internal insulation of external walls of "tenement houses" are presented in the paper by Razakov et al. (2019). Careful selection of construction materials based on the assessment of tactile heat and thermal behavior of construction materials can significantly reduce energy consump-tion in buildings (Bhattaa et al. 2019).

One of the current trends of historic buildings modernization is the creation of atria, which is espe-cially important for the central districts of Saint Petersburg, where traditionally residential building were created with formation of a courtyard (Murgul 2015, Penića et al. 2015). The use of atria allows solving problems of improving natural lighting and solar irradiation, as well as helps to reduce heat loss through claddings.

The paper by Sari et al. (2019) presents the results of studying the process of accumulation of thermal energy by cement composite gypsum, the use of which can contribute to energy saving in building envelopes.

Taking into account that cladding thermal insula-tion has limitations in terms of architecture, the main focus of studies in creating comfortable conditions in rooms is the improvement of technical solutions for microclimate maintenance systems. First of all it relates to the heating systems. Mandatory require-ment for heating arrangement in popular buildings has determined the special trend in studying heat and mass transfer processes, heating and ventilation sys-tems functioning, the use of renewable energy sources, taking into account the specific features of these buildings (De Carli 2013, Makrodimitri et al. 2010, Wood et al. 2010).

The analysis of buildings heat consumption performed by Wood et al. (2010) allows identifying the main historical differences in this aspect:

- use of heating habitable premises only, and not the entire building;
- use of thermal accumulation of structure blocks;
- the use of traditional energy saving methods (shutters, interior doors);
- other technology of buildings construction from permeable materials, porosity allowed to keep many buildings in good condition;
- advanced household heating systems are often installed with timers that actuate the heating for a while in the morning and then again in the evening, which is suitable for a new building with a low thermal mass.

In connection with the above differences in the modernization of historic buildings, the following problems arise:

- the use of materials such as solid cement mortars and plaster, plastic paints, synthetic waterproofing coatings, etc., which reduce the air permeability of the building and adversely affect the durability of historic buildings cladding. In this case, building thermal performance is likely to be reduced, and adequate compensation methods may be required;
- energy bulk may be lost due to cracks and gaps in structures that increase the volume of infiltrating air;
- locally controlled heating systems are inefficient for a conventional building with a relatively small amount of insulation and high thermal mass;
- features of current life associated with a more intense flow of moisture and a simultaneous decrease in the intensity of rooms ventilation.

To ensure old buildings durability and the necessary state of their microclimate, it is suggested:

- restore its initial hygrothermal characteristics by using membranes to protect the structural elements against excessive moisture;
- to search for alternative methods of evaporation of moisture in the envelope thickness;
- install "secondary" glazing that does not change the windows appearance, but reduces infiltrating air volume from 2.5 ÷ 3 to 0.7 h-1 to ensure natural air exchange;
- use shutters with the installation of "secondary" glazing;
- provide more economical change in the heating mode in accordance with the natural time of the building thermal response;
- use the most advanced and flexible heating control systems, for example, using highly efficient condensing boilers, which are relatively

harmoniously controlled with conventional buildings design;
- use modern combined compact energy systems with renewable energy sources (solar cells, wind turbines, geothermal and air heat pumps);
- taking into account a variety of historic buildings, the actual trend is full-scale studies of microclimate state and buildings heat consumption.

In many cases, the main challenge in studying the thermal condition of historic buildings is the absence of reliable data on the thermal performance of enclosing structures. Modern computer technologies allow for simulating various physical processes. For example, the Delphin software product was used in the paper by Claude et al. (2019) to simulate the hydrothermal characteristics of facades with different internal layers of insulation. The results make it possible to predict the state of building structures including mold formation and spread depending on some morphological characteristics of the city and the hygrothermal characteristics of the structures, to select the most efficient insulation material. However, as noted in the paper by Wood et al. (2010), the results of numerical calculations do not always properly correlate even in studies of modern buildings due to significant differences in actual hydrothermal characteristics from reference ones. In addition, the program algorithms do not highly take into account the dynamic interaction of heat and mass flows in enclosing structures, especially permeable materials with high thermal mass, as well as the influence of adjacent structures. An attempt to numerically study non-steady heat and mass transfer processes was undertaken in the paper by Denga et al. (2019). It has been established that the high thermal mass of the external walls is to be interact with the room air in order to avoid overheating and overcooling under appropriate cooling and heating conditions.

Modern programs for dynamic energy modeling provide an opportunity to analyze the energy performance of buildings (Harmati et al. 2015). However, in this case, the first problem is obtaining reliable information about the thermal performance of building structures. One of the promising trends in this area remains the well-known method of infrared thermography (Kirimtat & Krejcar 2018), which, taking into account the development of advanced measuring instruments, can detect not only thermal disturbances and air leaks, but even moisture variations on the external surface of building structures.

The above analysis shows that historic buildings, including the so-called "tenement buildings", represent a special category of buildings with porous structures that require studying both the thermal performance of materials using current methods and tools, and energy indicators of the microclimate systems.

The Housing Committee of the Saint Petersburg Administration adopted the Regional Program for overhaul of apartment buildings for the period until

2035 and includes the work for heating systems overhaul in 11.5 thous. buildings while 32% of the buildings from the address list are pre-revolutionary buildings.

Despite the fact that the current requirements for thermal protection of buildings do not apply to this category of buildings, however, the publications of local and foreign researchers recommend to meet the following requirements in terms of improving energy efficiency:

– element requirements, i.e. bring the specified heat transfer resistance of separate enclosing structures to a value not less than standardized values;
– comprehensive requirement is to provide the standardized value of the specific heat-shielding characteristic of the building;
– sanitary requirement, according to which the temperature on the internal surfaces of the enclosing structures shall mot be below the minimum permissible values.

For buildings specifying the unique architectural look and, consequently, having cultural value, the greatest difficulty is heat insulation outside facades. However, insulation of the internal surface of external walls as was shown above is also difficult, and this work is not included in the overhaul program.

Thus, for this buildings category, an increase in energy efficiency can follow the path of finding energysaving solutions in building utilities, in particular, heating systems, provided that the rated temperature of indoor air is ensured. The heating system overhaul project specifies a value of heat load transferred to the building to enter into the contract with heating Supply Company.

It is known that to design a heating system, besides the space-planning data of the object, it is necessary to know thermal performance of enclosing structures. These values to be determined as a result of full-scale and laboratory tests are given in the regulatory documents and documentation of material manufacturers. However, for pre-revolutionary buildings there is no information about the thermal performance of the materials of enclosing structures, which specifies the relevance of studies to determine the actual thermal conductivity of bricks for the above buildings.

The paper by Ryabev (2019) presents the results of full-scale, laboratory and numerical experiments to determine the thermal conductivity factor of bricks and the heat transfer resistance of brick masonry. Materials for studies are obtained from the restored building (tenement buildings) built in 1908. The value of the thermal conductivity factor of ceramic brick in a dry state according to the test results with an average density of 1956 kg/m3 is 0.57 W/(m·K). For the current similar structure of ceramic brick in a dry state with a density of 1800 kg/m3, the thermal conductivity factor is 0.55 W/(m·K). Accordingly, the factors of heat transfer resistance change. According to the technology of tenement

houses construction, the external brick walls thickness varies in a building height - this is 3.5; 3.0 and 2.5 bricks. The final data obtained as a result of mathematical modeling of the heat exchange process in brickwork using the ANSYS software system are given in Table 1.

In comparison with the current requirements for thermal protection of residential buildings, then the standardized value of heat transfer resistance factor in the climatic conditions of Saint Petersburg is 1.26 m^2 K/W. That is the current requirements for thermal protection can be provided only for floors with a wall thickness of 3.5 bricks.

To assess a building energy consumption, specific characteristics of heat energy consumption are used, W/(m^3·K) depending on the purpose, area and a number of floors in the facility. Based on a comparison with the standard values, these characteristics determine not only the energy saving class, but also the choice of design solutions for the building utilities.

This paper covers the performed analysis of the actual heat consumption of pre-revolutionary residential buildings. The construction performance of buildings is given in Table 2. The facilities under study are buildings of various geometry (extended and tower buildings), different number of floors (from 2 to 6), with different composition of floors, buildings located in different districts of the historical part of Saint Petersburg (Figure 1). The results of full-scale studies performed during 2017-2018 are used in the calculations.

For connection to the heat supply system, a heat point is used located either directly in the heated building or in the neighboring one. Connection to the urban heat network is carried out either through an uncontrolled electro hydraulic elevator, or with the help of a mixing unit and a pump. As a rule, there is no weather-dependent control, which does not provide effective control of coolant parameters.

Almost all heat points are equipped with heat meters. However, when several houses are connected to a single header in the heat point (coupled houses), individual accounting of the consumed heat energy is not performed at several addresses. The design coolant parameters- 95/70 °C.

The facades glazing rate has a significant negative effect on the amount of transmission loss; the results of this characteristic study are shown in Figure 2.

Table 1. Heat transfer resistance factor of brick walls with various thickness.

Brickwork thickness	Heat transfer resistance factor, m^2·K/W
2.5 bricks	1.03
3 bricks	1.19
3.5 bricks	1.35

Table 2. Construction indicators of buildings.

Address	Number of floors	Year of construction
35, Savushkina st., lit. A	2	1916
58-60, 13th Line V.O., lit. B	2	1917
90. Fontanka River naberezh-naya, bld. 3, lit. C	3	1806
29, Politekhnicheskaya st., bld. 1, lit. F	4	1901
5, Aleksandara Nevskogo, lit. B	4	1917
6, Guseva st., lit. A	4	1917
22, Chaikovskogo st., lit. A	5	1828
1/1, Birzhevaya line, lit. G	5	1881
18, Moscovsky prospekt, lit. A	5	1884
72, 9th Line V.O., lit. B	5	1917
14, Tverskaya, lit. B	6	1908

Address	Heated volume,m^3	Volume-to-size ratio
35, Savushkina st., lit. A	1723.80	0.55
58-60, 13th Line V.O., lit. B	3275.00	0.52
90. Fontanka River naberezh-naya, bld. 3, lit. C	11,405.32	0.47
29, Politekhnicheskaya st., bld. 1, lit. F	19,997.88	0.36
5, Aleksandara Nevskogo, lit. B	1427.64	0.60
6, Guseva st., lit. A	3463.46	0.42
22, Chaikovskogo st., lit. A	24,974.68	0.36
1/1, Birzhevaya line, lit. G	11,382.39	0.32
18, Moscovsky prospekt, lit. A	18,934.81	0.35
72, 9th Line V.O., lit. B	3908.97	0.41
14, Tverskaya, lit. B	5369.28	0.35

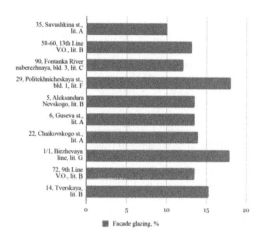

Figure 2. Facades glazing factor.

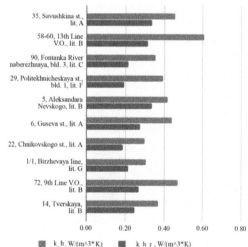

Figure 3. Comparison of the specific heat-shielding characteristics of buildings.

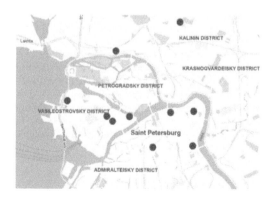

Figure 1. Location of the studied buildings.

Comparison of the design values of the specific heat-shielding characteristics of the buildings under study (k_h) and those required (k_h_r) for similar modern buildings is presented in Figure 3. The deviation is from 36 to 103%. The main reason is the inconsistency of wall structures with modern conditions for thermal protection conditions.

Even though this characteristic takes into account the buildings geometry, its value does not depend on a number and a height of floors and, thus, does not indicate an increased consumption of thermal energy by individual rooms.

An indicator that comprehensively takes into account the efficiency of building heat consumption is the specific characteristic of the heat energy rate of building heating and ventilation during the heating period, which is determined according to the formula:

$$q_h^r = [k_h + k_v - (k_{hh} + k_r) \cdot v \cdot \xi] \cdot (1 - \xi) \cdot \beta_h \quad (1)$$

where k_h is a specific heat-shielding characteristic of the building, W/(m^3·K); k_v - a specific ventilation characteristic of the building takes into account an outside air amount flowing the building through ventilation and infiltration, provision of air flow, air penetration characteristics of the building envelope elements; k_{hh} is a specific characteristic of household heat generation; k_r is a specific characteristic of internal heat gains from solar radiation; v - heat gains reduction factor due to thermal inertia of enclosing structures; ξ - efficiency factor of automatic control of heat supply in heating systems takes into account the type and equipment of heat supply systems and indoor heating; β_h - a factor taking into account the additional heat consumption of the heat supply system, depending on the heat supply system performance and the building geometry (Figure 4).

The results of the characteristics calculation are given in Figure 5.

$$v = 0.7 + 0.000025 \, (HSDD - 1000) \qquad (2)$$

where $HSDD$ is a heating season degree-day, $°$C·day.

According to the results of studying buildings of former tenement buildings carried out during the

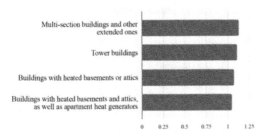

Figure 4. Factor values.β_h.

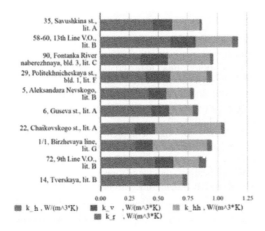

Figure 5. Specific building characteristics: heat-shielding k_h, ventilation k_v, heat generation k_hh, heat gains from solar radiation k_r.

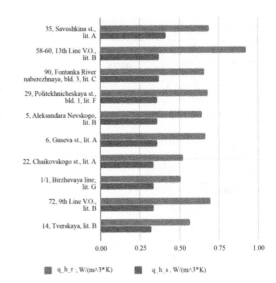

Figure 6. Comparison of calculated g_h_r and standard g_h_s values of specific rate of buildings heating and ventilation.

implementation of the overhaul program, it became clear that these buildings are almost universally equipped with single-pipe riser heat supply systems with the upper position of flow main and the lower position - return main. This type of riser arrangement does not allow constructing a system of another type, since this requires the possibility of laying new pipelines in all apartments of the building, which in practice cannot be implemented without re-housing of residents during the overhaul. Improvement of the system is achieved by installing control devices on radiators and risers. In this case, the factor value for newly designed systems is .

The results of the calculation for specific rates of heating and ventilation of tenement buildings in comparison with the regulatory data for modern buildings are shown in Figure 6. Design specific rates of heating and ventilation of tenement houses are from 0.50 to 0.92 W/(m3·K). Taking into account the modern indicator of the rated specific consumption of 0.32–0.41 W/(m3·K) for heating and ventilation, these buildings could be classified as energy saving classes D and E, requiring restoration with appropriate economic feasibility.

2 CONCLUSIONS

1) tenement buildings are a group of objects with different architectural, planning solutions, indicators of enclosing structures;
2) the values of the specific heat-shielding performance of buildings are 0.3–0.45 W/(m³·K) that does not meet the up-to-date requirements due to the low heat-shielding performance of separate envelope elements;

3) design specific consumption for heating and ventilation of tenement buildings are from 0.50 to 0.92 $W/(m^3 \cdot K)$ that significantly exceeds the current parameter of the standard value of specific consumption for heating and ventilation of 0.32–0.41 $W/(m^3 \cdot K)$;

4) the buildings can be classified as energy saving classes D and E - the lowest ones that require buildings restoration with economic feasibility.

REFERENCES

Atmaca, A.B. & Gedik, G.Z. 2019. Evaluation of mosques in terms of thermal comfort and energy consumption in a temperate-humid climate. *Energy and Buildings* 195: 195–204.

Bhatta, S.R. et al. 2019. Quantifying the sensation of temperature: A new method for evaluating the thermal behaviour of building materials. *Energy and Buildings* 195: 26–32.

Cabeza, L.F. et al. 2018. Integration of renewable technologies in historical and heritage buildings: A review. *Energy and Buildings* 177: 96–111.

Claude, S. et al. 2019. Evaluating retrofit options in a historical city center: Relevance of bio-based insulation and the need to consider complex urban form in decision-making. *Energy and Buildings* 182: 196–204.

De Carli, M. 2013. *Technologies and examples of heating and cooling historical buildings.* Seminar: Historical Architecture & Energy Efficiency - Proposals and Solutions, Bruxelles, 25 September, 2013.

Denga, J. et al. 2019. Effectiveness of the thermal mass of external walls on residential buildings for part-time part-space heating and cooling using the state-space method. *Energy and Buildings* 190: 155–171.

Dmitriyev, A.N. et al. 2008. *Energy saving in buildings reconstructed.* Moscow: ASV.

Harmati, N. et al. 2015. Energy consumption modelling via heat balance method for energy performance of a building. *Procedia Engineering* 117: 786–794.

Kirimtat, A. & Krejcar, O. 2018. A review of infrared thermography for the investigation of building envelopes: Advances and prospects. *Energy and Buildings* 176: 390–406.

Makrodimitri, M. et al. 2010. Heating historic structures. A review of heating systems in historic church buildings and implications related to conservation and comfort. The case of four historic churches in Cambridge. Accessed June 15, 2019. http://www.an-patrimoine-echanges.org/IMG/pdf/session_s2e_-_magdalini_makro dimitri_-_universite_de_cambridge_angleterre.pdf

Murgul, V. 2014. Features of energy efficient upgrade of historic buildings (illustrated with the example of Saint-Petersburg). *Journal of Applied Engineering Science* 12 (1): 268.

Murgul, V. 2015. Reconstruction of the courtyard spaces of the historical buildings of Saint-Petersburg with creation of atriums. *Procedia Engineering* 117: 808–818.

Murgul, V. & Pukhkal, V. 2015. Saving the architectural appearance of the historical buildings due to heat insulation of their external walls. *Procedia Engineering* 117: 891–899.

Park, S.C. 1991. Heating, ventilating, and cooling historic buildings—problems and recommended approaches. Accessed June 15, 2019. http://www.oldhouseweb.com/how-to-advice/heating-ventilating-and-cooling-historic buildings.shtml.

Penića, M. et al. 2015. Revitalization of historic buildings as an approach to preserve cultural and historical heritage. *Procedia Engineering* 117: 883–890.

Pukhkal, V. et al. 2015. Studying humidity conditions in the design of building envelopes of "passive house" (in the case of Serbia). *Procedia Engineering* 117: 859–864.

Razakov, M.A et al. 2019. Results of energy-saving measures in commercial apartments of Saint Petersburg. *Plumbing, Heating, Ventilation* 2: 79–81.

Roberts, B. 2019. 19th-century radiators and heating systems. Accessed June 15, 2019. https://www.building conservation.com/articles/heating-systems/heating-sys tems.htm.

Ryabev, G.A. 2019. Determination of the actual coefficient of brickwork thermal conductivity of pre-revolutionary buildings. *Bulletin of Civil Engineers* 1 (72): 132–137.

Sarı, A. et al. 2019. Preparation, characterization, thermal energy storage properties and temperature control performance of form-stabilized sepiolite based composite phase change materials. *Energy and Buildings* 188–189: 111–119.

Wood, C. et al. 2010. *Energy efficiency and historic buildings. Application of part L of the Building Regulations to historic and traditionally constructed buildings.*

Reconstruction and Restoration of Architectural Heritage – Sementsov et al (eds)
© *2020 Taylor & Francis Group, London, ISBN 978-0-367-65357-6*

Assessment of quality of adhesive joints of amber mosaic

I.I. Verkhovskaia
Saint Petersburg State University of Architecture and Civil Engineering, Saint Petersburg, Russia

ABSTRACT: A solution to the problem of evaluating the quality of adhesive joints of mosaic products made of amber (succinite) is proposed. Method. To study the strength characteristics of adhesive joints, the method of cutting adhesive joints was used. Zwick/Roell Z100 equipment was used for the research. The subject of the study was samples of amber (Kaliningrad, Russian Federation) as a decorative material; ceramics, marble, aviation plywood, carbon fiber as a base and polyvinyl acetate, epoxy adhesives, five variants of wax-vanifol mastic as adhesives. Main results. The data of strength characteristics of adhesive joints in artistic mosaic products are presented in comparison. It is established that in the manufacture and restoration of mosaic products as adhesives, the use of reversible wax-cauliflower adhesives is a priority, and aviation plywood is the most acceptable base. Practical significance. The proposed method may be of interest for research into the processes of manufacturing and restoration of artistic mosaic products.

1 INTRODUCTION

Mosaic is one of the most ancient techniques in decorative and applied art. Currently, mosaic technology is widely used on an industrial scale in interior and exterior solutions.

In the manufacture of mosaic products, various technologies are used and a variety of materials are used, the choice of which largely depends on the goals and final operational purpose. However, the key purpose of using mosaic products is a decorative purpose, as a means of achieving artistic expression in simple and complex forms of different areas of art objects [1-2].

Depending on the functional purpose of mosaic products, there are various methods, techniques and technologies for their production, while the choice of materials also depends on the final goals [3-8]. For example, depending on the location and spatial shape of mosaic products, the choice of materials, the method of fixing the decorative layer and the technique of the set is determined, and the aesthetic goal determines the choice of the type of mosaic technique (Figure 1).

As noted earlier, mosaic technology is more often used in interior solutions, for example, when finishing surfaces of various shapes and areas with decorative facing materials, such as stones and gems. In decorative and applied art, mosaic products can be created as independent artistic solutions, or be decorative elements. Examples are the decoration of the Amber room, the mosaic panel "Industry of socialism" of the Museum of the Central research and exploration Museum named after academician F. N. Chernyshev (Saint Petersburg, Russia), unique stone vases of the Hermitage interiors, as well as a variety of decorative and applied interior items: caskets, candlesticks, cigarette cases, writing materials, etc.

Adhesive mosaic products (hereinafter referred to as CMI) are structures that allow you to implement a variety of individual projects of any complexity and have a number of advantages and advantages: ease of construction, high aesthetic properties, as well as high strength, which ensures safety in interior solutions. Sometimes, in the design plans, mosaic products are artistic design of functional structures and can be implemented, for example, in modern interiors when finishing various items with different technical and aesthetic purposes, including bent structures.

An adhesive mosaic structure is a complete set of various parts with a certain mutual location, which consists of parts of certain parameters connected to each other by an adhesive layer.

In real conditions, CMI is a combination of layers of various possible materials, which can be divided into three groups: decorative surface, adhesive layer and base. Glue joints (hereinafter referred to as CS) have a number of advantages and allow you to connect a variety of materials, and in some cases be the only practical acceptable method. In the CS, the stresses are more evenly distributed, and technical holes that weaken the bonded elements are excluded, thus ensuring the strength of the structure.

However, in the process of gluing materials, various defects inevitably arise related to both the adhesive material (non-adhesive, porosity, low adhesion, under-curing of the glue, etc.) and decorative materials and bases, which leads to a decrease in the strength of the entire structure and leads to the formation of defects during operation [9].

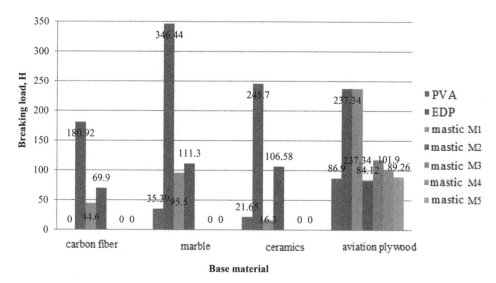

Figure 1. Strength of the art mosaic adhesive joint.

It is important to note that the CMI can be a monolithic CS, in which the elements are glued with continuous seams throughout their entire length, and composite CS, in which there are gaps between the elements (non-adhesive).

Modern methods and means of non-destructive testing (optical, acoustic, radiation, radio-wave, etc.) allow detecting various defects, but at present there are practically no engineering methods for calculating the effect of defects on the operation of the CMI. At the same time, there is no analysis of the influence of the choice of glued materials for mosaic products on the quality of adhesive connections in the literature. There is not enough data on the adhesives used in the mosaic technique of decorative and applied art.

2 MATERIALS AND METHODS

Fixing the decorative layer of the mosaic product with the base is carried out by means of an adhesive substance, the choice of which is determined by

Table 1. Systematization of art mosaic: types, materials and techniques.

Mosaic	Antique		Roman		Florentine	Russian
Technique			Venetian	Roman	Florentine	Russian
Location	Exterior			Interior		
Architecture (form, space, composition)	volumetric	twodimensional				volumetric
Hardness of decorative material (Mohs scale)	solid			soft		
Method of fastening	cement composition, concrete + reinforcement	cement composition		cement composition + adhesive substance	adhesive substance + base	

a number of factors. The mechanical properties of adhesives are evaluated by the minimum strength, guaranteed indicators given in the technical specifications or GOST of the corresponding brands for technical products. But when choosing an adhesive, you should take into account not only their properties and features, but also the physical, chemical and other influences on the materials to be joined. In the manufacture of decorative CMI, it is important to take into account such a factor as the reversibility of the adhesive joint [10-11], since during restoration work, this will preserve the integrity of the materials of the decorative layer and the base.

The study presents adhesives that are used in the manufacture of mosaic products in the decorative and applied arts, and are divided into reversible, when removing which will preserve the integrity of the materials, and irreversible, in which the dismantling of adhesive connections will cause the destruction of the base materials and the decorative layer.

The materials of the adhesive substance are presented in four variants: epoxy glue (EDP), polyvinyl acetate glue (PVA), wax-cauliflower mastic of five variants with different compositions (M1, M2, M3, M4, M5) and a variable content of wax, rosin, as well as dammara, shellac and paraffin. Aviation plywood of the BS-1 brand (GOST 102-75), APG 300x200 ceramic tile (GOST 6787-200), monochrome homogeneous marble, carbon fiber of the CMU-11 brand (Table 1) were studied as the base material. Homogeneous transparent Baltic amber of the Palmniken Deposit was chosen as a decorative material.

The choice of materials for testing is determined by the conditions of the experiment.

The study of mechanical properties of adhesive joints of mosaic products was carried out by the cut-off method. The purpose of testing samples is to quantify the bond strength of the decorative layer and the base with different versions of adhesive compositions, namely, to determine the adhesion strength of the materials being glued under uniaxial loading, in which the load is evenly distributed over the entire longitudinal section of the sample. The maximum load is set for each variant of the base and adhesive combination.

The quality of adhesive joints of mosaic products was evaluated on specially prepared samples, which are a duplex system and consist of plates of various base materials with a size of 8x12x6-12 mm, adhesive substance and decorative material with a size of 8x12x6 mm. The shape and size of the samples meet the requirements of the test method.

80 samples were made, which represent four groups of base plates with four combinations of adhesive substance, five copies in each group. However, preliminary studies have shown different mechanical characteristics of adhesive compounds with different compositions of wax-cauliflower mastics, so an additional study of three variants of wax-cauliflower mastics was conducted.

In the production of samples of ceramic, marble and Baltic amber tiles of specified sizes and rectangular shapes, the Master Max MCM-2400 electric cutting machine was used, and for obtaining samples from carbon fiber and plywood, the Corvette 53 universal grinding machine with a universal spindle and a Stripping circle was used. For the purity of the experiment, the samples preserved the parallelism of the processed surfaces for free movement in parallel guides of the test facility, which allow to obtain a simultaneous cross-section over the entire cross-section area.

Surface preparation of samples before gluing, the method of applying the glue, the number of layers, the flow rate, the modes of gluing are determined by the technical documentation intended for the appropriate name, brand and purpose of the adhesive substance.

Studies of the mechanical properties of adhesive joints were carried out by the cut test method using a universal testing machine Zwick/Roell Z100.

The Zwick/Roell Z100 universal testing machine is designed for performing mechanical quasi-static tests for uniaxial tension, creep, compression, 3x and 4x point bending. The test complex has a maximum compression force of 100 kN, force sensors of 1 kN and 100 kN with a unit price of 0.0008 N and 0.008 N, a speed of 0.001-600 mm/min, equipped with hydraulic grippers with a maximum clamping pressure of 500 bar, an electronic control unit TestControl II and software TestXpert III.

The tests were performed at a temperature of 23°C (±1) and a relative humidity of 50% (±5). The samples were placed in the grips of the universal testing machine so that the longitudinal axis of the sample

Table 1. The properties of the basis materials mosaic.

Properties	Unit.	Material			
		carbon fiber	marble	aviation plywood	ceramics
density, ρ	кg/m^3	1.3	2.73	0.8	3.6
compression strength, σ_c	MPa	1450	400	470	140
tensile strength, σ_{st}	MPa	2060	1050	735	985
transverse strength, σ_{st}	MPa	1500	18.0	650	1100
hardness	MOHS	-	3-4	-	5-6
absorption of water	%	0.2	1	8	3.5

coincided with the axis of application of the load, while each substrate was fixed and held in the grips. The test sample was loaded at a set speed of 13 mm/min until its destruction and the highest load reached was recorded. Then the samples were subjected to visual inspection to determine the type of destruction. The tests were performed under the same conditions, by the same operator. The obtained results were analyzed for compliance with the criteria of repeatability (exceeding the standard deviation by no more than 2.5 times) and reproducibility (exceeding the arithmetic mean of individual values by no more than 20%).

3 RESULTS AND DISCUSSION

Processing of the obtained data allowed us to obtain the results of the strength of the adhesive joint under shear P (H), which is expressed by the value of the shear strength σ (Pa) and is calculated by the formula:

$$\sigma = P/F$$

where P is the breaking load, N;
F is the bonding area, m², calculated to an accuracy of 0.000001 m² using the formula:

$$F = lb,$$

where l is the overlap length, m;
b - overlap width, m.

Then the arithmetic mean of the breaking load or breaking stress is calculated.

4 CONCLUSIONS

The results of the study allowed us to obtain data that the greatest strength of the adhesive connections of the decorative material with the base was shown by samples with epoxy resin. However, during the test in 25% of cases, there was a partial destruction of amber, which characterizes the irreversibility of the adhesive connection and, thus, is unacceptable in the manufacture and restoration of mosaic products.

It is necessary to note the strength characteristics of wax rosin mastic, the characteristics of which depend on the quality content and % ratio of substances in the composition (Figure 3).

Thus, adhesives used in the manufacture of mosaic products can be represented as reversible, the removal of which will preserve the integrity of the materials, and non-reversible, in which the dismantling of adhesive joints will cause the destruction of the base materials and the decorative layer.

The results obtained allowed us to conclude that the strength characteristics of the adhesive connection of mosaic products with various types of adhesives are most optimal with the basis of aviation plywood. At the same time, the obtained data on the strength characteristics of the M1 and M3 wax-canifol bridges allow us to present them as universal and reversible adhesives for the manufacture and restoration of mosaic products with any combination of base materials.

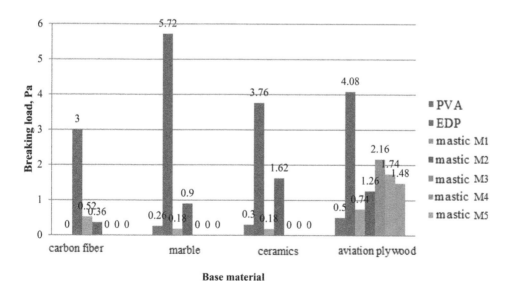

Figure 2. Tensile strength of the art mosaic adhesive joint.

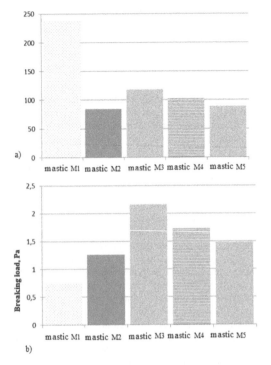

a)

b)

Figure 3. Strength of the art mosaic adhesive joint of aviation plywood with different wax-cauliflower mastics.

REFERENCES

Andreescu-Treadgold, I., & Treadgold, W. 1997. *Procopius and the imperial panels of S. vitale. Art Bulletin*, 79(4): 708–723. doi:10.1080/00043079.1997.10786808

Izzo, F., Arizzi, A., Cappelletti, P., Cultrone, G., De Bonis, A., Germinario, C., ... Langella, A. 2016. The art of building in the roman period (89 B.C. - 79 A.D.): Mortars, plasters and mosaic floors from ancient stabiae

(naples, italy). *Construction and Building Materials*, 117: 129–143. doi:10.1016/j.conbuildmat.2016.04.101

James, L. 2006. Byzantine glass mosaic tesserae: Some material considerations. *Byzantine and Modern Greek Studies*, 30(1): 29–47. doi:10.1179/030701306X96582

Silvestri, A., Tonietto, S., & Molin, G. (2011). The palaeo-christian glass mosaic of st. prosdocimus (padova, italy): Archaeometric characterisation of 'gold' tesserae. *Journal of Archaeological Science*, 38(12), 3402–3414. doi:10.1016/j.jas.2011.07.027

Silvestri, A., Tonietto, S., Molin, G., & Guerriero, P. (2012). The palaeo-christian glass mosaic of st. prosdocimus (padova, italy): Archaeometric characterisation of tesserae with antimony- or phosphorus-based opacifiers. *Journal of Archaeological Science*, 39(7): 2177–2190. doi:10.1016/j.jas.2012.03.012

Thunø, E. (2015). The apse mosaic in early medieval rome: Time, network, and repetition. doi:10.1017/CBO9781107707078

Bescher, E., Pique, F., Table, Doctor, & McKenzie, J. D. (2000). Long-term protection of the Domesday mosaic in Prague. *Journal of Sol-gel science and technology*, 19 (1-3): 215–218. doi:10.1023/A: 1008784221259

Croveri, P., Apartments Fragalà, I., & Ciliberto, E. (2010). Analysis of glass tesserae from the mosaic "Villa del Casale" near Piazza Armerina (Enna, Italy). Chemical composition, conservation status and production technology. *Applied physics a: materials Science and processing*, 100 (3): 927–935. doi:10.1007/s00339-010-5670-8

Potapov A.I., Makhov V.E. Experimental Ultrasonic Study of the Elastic Modulus of Glass Fiber Plastics in Constructions. *Russian Journal of Nondestructive Testing*, 2018, Vol. 54, No. 1, pp. 1–16.

Trostyanskaya E. B., Tomashevich, Sorokina E. V. 1960. Glue compositions for duplication, Issues of restoration and conservation of works of fine art, Moscow: Academy of Arts of the USSR.

Ya. I. Verkhovskaya, L. T. Zhukova, E. N. Petrov, 2008. On the issue of control and evaluation of the quality of adhesive compounds in the manufacture of artistic mosaics, Conference proceedings of *14s International scientific and practical conference Tomsk*: 471–473.

Reconstruction and Restoration of Architectural Heritage – Sementsov et al (eds)
© 2020 Taylor & Francis Group, London, ISBN 978-0-367-65357-6

Analysis of methods of nondestructive testing of heat pipelines

D. Zakharova & A. Potapov
Saint Petersburg Mining University, Saint Petersburg, Russia

ABSTRACT: Modern control of the condition of heat pipelines is based on hydraulic tests. These tests are accompanied by an increase in pressure in the pipe. The appearance of metal ruptures and fluid leaks indicates the need to replace the pipeline. However, often an increase in pressure in the pipeline can lead to the appearance and development of internal defects. Nevertheless, often the pipe can be recognized as suitable for exploitation while the increase in pressure in the pipeline can lead to the appearance and development of internal defects. The article describes the methods of Nondestructive testing of pipelines of heat networks in operation. The analytical review of thermal, magnetic, radiation and acoustic methods is carried out. These methods allow to testing the equipment without its decommissioning, and do not lead to destructive consequences.

1 INTRODUCTION

The length of Russian heating networks is very large. For many years, there has been a problem of pipeline failure. At the beginning of the century, there was a sharp deterioration of the situation in the heat power industry. According to the Department of Energy of the Russian Federation, the average percentage of pipeline wear was 60%, among which 15% required immediate replacement. This is due to the fact that most of the pipelines were built more than 30 years ago and are already in disrepair. The average age of pipes in different cities is 30.8 years. For 2014, the sources of the centralized heat supply has worn out more than on 70%, and the leakage of heat carrier is 18-20%, while the normal value of the leak is 0.25% of the volume of heat carrier in the system (Davydov & Abdullin, 1998). Of all the devices of heat networks, pipelines are the most damaged. Exploitative damage to pipelines includes mechanical damage, wear out and corrosion. Table 1 shows some of the defects that occur during the maintenance of the pipeline.

According to the degree of danger of defects of pipelines are divided into three categories:

"A" - defects and damage to the main load-bearing structures, pipes, representing a direct threat to their destruction;

"B" - defects and damage to the pipe, do not present an immediate danger of the destruction of load-bearing structures, but is able to continue to cause damage to other components and units or in the development of damage - go to category "A»;

"C" - defects and damages of local character that at the subsequent development cannot influence the main bearing designs of pipes (Sukhorukov 1992).

Defects and damages of pipes of category "B" and separate damages with insignificant development of category "B" can be liquidated according to the technical documentation developed by design divisions of the organizations, which operate object of control. Defects and damages of category "A" and damages of category "B", capable at fast development to pass to category "A", shall be eliminated only according to the technical documentation designed by the specialized organization having permission (license) of Federal Agency for construction and housing and communal services of Russia on this type of activity and examination of industrial safety of the technical documentation (project) approved by Federal mining and industrial supervision of Russia.

Every year, at the end of the heating season, pipelines of heat networks are subjected to high-pressure tests. These tests often lead to failures in performance, because high pressure leads to the development of internal defects, leaks, etc.

The actual problem is the diagnostics of the pipelines without devastating consequences. Nondestructive testing methods are becoming more and more widespread in the field of control and diagnostics of thermal networks. Next, the testing methods that are used to testing the pipelines of heat networks during operation will be considered.

2 DISCUSSION OF METHODS

Visual-optical method. Visual-optical method is based on the inspection of the object of control before the main diagnosis. As a rule, the inspection is carried out to the naked eye or with the help of optical devices. Also, the inspection is accompanied by the measurement of the size and shape of the pipeline and the detected surface defects with the use of devices such as ruler, tape measure, caliper, probes, templates and others. At Visual-optical

Table 1. Some of the defects of the pipeline.

Mechanical damage	Corrosion
- fatigue crack	- solid corrosion
- scratches	- local corrosion
- dents	(pitting and corrosion spots)
- erosion damage to	- linear corrosion
internal surfaces	- intergranular corrosion

inspection of pipelines of thermal networks, the thermal insulation of a pipe is removed, and the temperature of a surface of a pipe decreases to temperature not exceeding 40°C. The controlled section of the pipe is inspected for the presence or absence of corrosion, cracks, delaminations, undercuts, compliance with the size of the weld and the geometric dimensions of the pipe (Kanevsky & Salnikova 2007). This simple method does not require expensive complex material instrument base and special skills of specialists. But this method does not allow to fully assessing the condition of the pipeline, it has a low sensitivity, so it must be used in conjunction with other methods.

Radiation method (X-ray method). This type of testing has not yet been widely used in the diagnosis of pipelines of heat networks. More often it is used in the field of oil and gas industry than in heat power engineering. Typically, this method is used to inspect pipe welds. Figure 1 shows an example of radiation monitoring of a pipeline with a diameter of 80 to 800 mm (Kanevsky & Salnikova 2007).

Ionizing radiation (R) is passed through the test object; this radiation passes through the pipe wall and is registered by the detector (D). A special film is used as a detector. The presence of defects can be judged by the background on the developed film. Dark spots indicate a decrease in the thickness of the metal in the weld, and the presence of light areas on the film indicates a thickening of the weld. The depth of the defect is determined by the degree of change in hue and clarity of the print on the film - the clearer the image of the spot on the film, the

closer the defect is to it. The sensitivity of the detector can be determined by the formula:

$$D = \lg\left(\frac{F_0}{F}\right) \qquad (1)$$

Here D is the optical density of blackening;
F_0/F is the opacity of the film;
F_0, F - the intensity of the light flux emitted and transmitted through the film.

Thus, the more transparent the image, the greater the blackening density value tends to zero. Decoding of the film requires high qualification of a specialist. The film is a document that can be transported and stored. However, the film thickness is quite thin, so it can easily be damaged. During the decoding of the image on the film should not be scratches, stains, stripes to obtain reliable results. Radiation monitoring is more effective in the diagnosis of large thicknesses, it is not suitable for the control of angular welds, it does not allow to detect small defects and to judge the danger of the defect by the concentration of mechanical stresses (Potapov & Kondrat'ev 2014). In addition, radiation monitoring must comply with safety regulations and protection against ionizing radiation. Magnetic methods.

Magnetic methods. Magnetic methods of nondestructive testing are used to detect surface and subsurface defects of pipelines. The application of this testing method requires removing the insulation coating from the pipeline and preparing the surface for inspection. In the process of magnetization of the object of control, a magnetic field is formed in it (Potapov et al. 2014a). Magnetic flows are formed in the object of control under the influence of a magnetic field. In the absence of defects, they will pass through the product. And if there are discontinuities in the object of control, magnetic poles are formed along their walls, then part of the magnetic flux will tend to go around the defect and come to the surface of the metal. Then local stray fields will be formed over the defects, as shown on Figure 2.

Thus, the places of the highest concentration of power lines can be determined in various ways, for example, using a magnetic powder containing ferromagnetic particles such as Fe_2O_3 (brown), Fe_3O_4

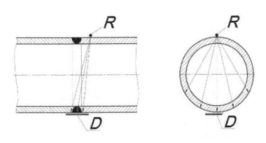

Figure 1. Radiation monitoring of a pipeline with a diameter of 80 to 800 mm.

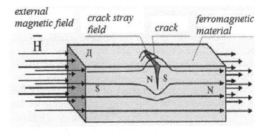

Figure 2. Stray field above the defect.

(black). The type of defect can be determined from the powder drawings on the surface of the test object. Unacceptable is the appearance of powder drawings on the sites working on a break and bend, the number of defects on the area of 100x100 mm should not exceed two defects, and the defect length is allowed no more than 3 mm. Magnetic stray fields above the defect can be detected by Hall sensors (Potapov et al. 2014b). Hall sensor is a thin semiconductor plate. Such converters are based on the occurrence of EMF Hall (E), the value of which is associated with the induction (B) of the magnetic field following dependence:

$$E = -\beta B \qquad (2)$$

The coefficient β is determined by the formula:

$$\beta = \frac{R_{H^\mathsf{I}}}{h}, \qquad (3)$$

where R_H - Hall constant, Ohm·mm/T;
I - current, A;
h - plate thickness, mm.

Thus, in the presence of defects in the object of control, Hall sensors record the magnetic induction of the scattering fields above them. According to the data, it is possible not only to conclude that there are defects, but also to determine the various magnetic characteristics of the object under testing (Chernyaev et al. 1997).

In addition to the above methods of magnetic inspection of pipelines, the main ones are fluxgate technique and tape. In addition to the above methods of magnetic inspection of pipelines, the main ones are fluxgate method and a method using magnetic tape (magnetographic method). The magnetographic method, for example, is effective for the testing of angle welds of pipelines. All these methods are used to determine the surface and subsurface defects of products made of ferromagnetic materials at a depth of 15 mm. During the inspection, the product should be magnetized in two directions or use a combined magnetization, since it is easier to detect defects that propagate perpendicular to the direction of the magnetic flux lines (Potapov et al. 2014b). After the inspection, it is necessary to carry out the demagnetization, because the residual magnetized can lead to unintended consequences and bad results the next time control.

Thermal imaging method. The thermal testing method is used in industries in which the heterogeneity of the thermal field can provide information about the technical condition of the objects of control. The method of thermal testing allows to detect such types of internal and external defects as: deviations of physical parameters, cracks, porosity, voids, casting defects, foreign inclusions, local overheating. It does not require contact with the object of control; it is possible to carry out remote diagnostics of pipelines running underground (Potapov et al. 2019).

For diagnostics of pipelines of thermal networks by a thermal method, the thermal imager is used. This is a device for monitoring the distribution of temperature on the surface of the object. Its principle of operation is to receive infrared radiation from the test surface. By the nature of the image is judged on the quantitative and qualitative characteristics and determine the location of the detected defects. Figure 4 shows the results of temperature testing over the heating main surface.

In red color on the screen of the thermal imager the sites of object of control with the damaged thermal insulation are displayed (Potapov et al. 2019). These are places of increased heat loss, indicating the presence of a defect in the pipeline. According to the data, it is possible to plot the temperature distribution over the surface of the object of control and to determine the zones of urgent and planned repair. However, the thermal method is very affected by environmental conditions. High humidity, temperature, precipitation and solar radiation have a great on the accuracy of the control results.

Acoustic methods

There are active and passive acoustic methods. When monitoring by active methods, the ultrasonic signal is introduced into the product by special piezoelectric transducers. The ultrasonic method is based on the registration of the reflection of elastic vibrations from the interface of media. If there is a defect in the test object, part of the ultrasonic signal will be reflected from it and return to the receiving transducer. Then, in the time interval between the probe and the bottom pulse, a signal reflected from the defect will appear. The attenuation of the signal passing through the object of control is

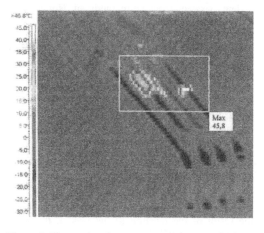

Figure 4. The results of temperature testing over the heating main surface.

determined by the formula of the acoustic path. Formula for the acoustic path is recorded:

$$\frac{U_d}{K_u U_g} = \frac{S_s S_d e^{-2\delta h}}{\lambda^2 h^2} \qquad (4)$$

where U_d - voltage amplitude of the first transmitted signal;

K_u - the ratio of the electric double conversion;

S_d, S_s - the area of the emitter and the area of the defect;

δ - attenuation;

h - distance to the defect;

λ - wavelength.

Thus, the larger the defect area and the smaller the distance to it, the greater the amplitude of the received signal. The attenuation coefficient (δ) is the sum of the absorption and scattering coefficients. The decrease in the amplitude of particle oscillations and the intensity of ultrasound due to absorption is exponential:

$$I = I_0 \cdot e^{-2\delta nh}, \qquad (5)$$

I, I_0 – the intensity of the ultrasonic wave at a depth of h and introduced into the object of control;

δ_n – absorption coefficient.

The absorption coefficient can be determined from the attenuation of the signal amplitude when it is re-reflected from the boundaries of the object of control. Having found the approximating exponential function by the method of least squares, it is possible to obtain the value of the absorption coefficient of ultrasound in the object of control. Thus, it is possible to trace the dependence of the absorption coefficient on the degree of corrosion of the object of control, which will determine the corroded sections of pipes (Belov & Podlevskikh 1997).

The passive acoustic methods of Nondestructive testing include the method of acoustic emission. Here the source of the signal is the object of control; the sensors are mounted on the object of control, while the full volume of the product is monitored. Control occurs when the object is loaded. Before conducting acoustic emission testing, it is necessary to assess the level of acoustic noise, to establish their source and, if possible, to eliminate it (Potapov et al. 2018a). The coordinates of the defect are determined by the difference in the time of arrival of acoustic signals from the defect to the sensors of the system. It is convenient to present the results of defect localization on pipelines and other linear objects in the form of a histogram. Figure 5 (A, B) shows the result of the location of the leak in the pipeline sensors installed at a distance of 50 meters from each other.

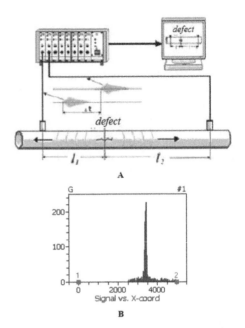

Figure 5. Location of the leak in the pipeline sensors installed.

During acoustic emission testing there should be no rain, hail or snow, because it creates interfering noise signals. Wind speed during the testing should be no more than 12 m/s. Gusts of wind should not be more than 1 time per minute (Potapov et al. 2018b).

Acoustic emission is caused by the growth and development of various types of defects such as cracks, stratifications, corrosion and others. This is due to the fact that the deformation of the material of the object of control there are elastic vibrations, which register the sensors installed on the surface of the object (Nosov 2017). Thus, it is possible even to predict the emergence of new defects.

The method of acoustic emission differs from other methods of Nondestructive testing in that it allows to detect developing defects, that is, the most dangerous. This method is effective in the testing of hard-to-reach places, because it allows you to control the full volume of the product. It allows you to determine the degree of danger of the defect and does not require special preparation of the object of control for diagnosis.

3 CONCLUSION

As a result of the analysis of methods of Nondestructive testing of pipelines of thermal networks, it was concluded that the acoustic method has a number of advantages over all others.

First, this method allows to testing not only steel pipelines, but also, for example, polyethylene pipes, in contrast to magnetic methods, in which it is possible to diagnose only ferromagnetic materials. Even when using the method of magnetic memory of metal, which is based on the detection of places of stress concentration, it is necessary to carry out additional ultrasound diagnostics to obtain more information about the existing defects.

Secondly, acoustic methods of Nondestructive testing provides a greater depth of penetration of ultrasonic rays and allows you to testing the pipelines of large thicknesses, detect internal defects, determine the thickness of the product, while magnetic methods reveal only surface and subsurface defects at shallow depths, and visually measuring methods reveal only external defects.

Third, during the diagnosis by acoustic methods of Nondestructive testing does not require additional protection from radiation, as in the x-ray method. The method of ionizing radiation testing is used to testing objects of large thickness. He cannot testing the angle welds, to determine the presence of discontinuities in the products of small thickness and to detect small defects. With the help of acoustic methods, it is possible to testing the entire volume of material of different thicknesses (depending on the method of testing); also, it can control the welds in full

Fourth, unlike thermal Nondestructive testing methods, acoustic methods are independent of environmental conditions such as humidity and air temperature, wind, precipitation, etc.

Fifth, according to the data obtained as a result of the acoustic testing, we can determine the type of defect, the location of the defect, the depth of its occurrence, the degree of danger and the spread of defects. Some of the methods listed above cannot allow you to do this.

Among other things, even during the testing by other methods and hydraulic tests of pipelines, it is often recommended to carry out additional ultrasonic acoustic testing.

REFERENCES

Belov, V.M. & Podlevskikh, M.N. 1997. Methods and equipment for acoustic emission diagnostics of trunk pipelines. *Occupational Safety in Industry* 11: 36–38.

Chernyaev, V.D. et al. 1997. System reliability of pipeline transportation of hydrocarbons. Moscow: Nedra.

Davydov, S.N. & Abdullin, I.G. 1998. *Equipment and methods of corrosion tests.* Ufa: Ufa State Petroleum Technological University.

Kanevsky, I.N. & Salnikova, E.N. 2007. *Nondestructive testing methods.* Vladivostok: Far Eastern State Technical University.

Nosov, V.V. 2017. Control of inhomogeneous materials strength by method of acoustic emission. *Journal of Mining Institute* 226: 469–479.

Potapov, A.I. et al.2014a. *Electromagnetic and magnetic methods of nondestructive testing of materials and products. Vol. 1. Electromagnetic and magnetic methods to control the thickness of coatings and walls of products.* Saint Petersburg: Nestor-Istoriya.

Potapov, A.I. et al. 2014b. *Electromagnetic and magnetic methods of nondestructive testing of materials and products. Vol. 2. Electromagnetic and magnetic methods of flaw detection and control of material properties.* Saint Petersburg: Nestor-Istoriya.

Potapov, A.I. et al. 2018a. *Nondestructive physical methods and means of control of the natural environment, materials and products. Vol. 13. Physical basis, methodological, instrumentation and metrological support of acoustic emission methods of Nondestructive testing of materials and products.* Saint Petersburg: Polytekhnika-Print.

Potapov, A.I. et al. 2018b. *Nondestructive physical methods and means of control of the natural environment, materials and products. Vol. 14. Application of acoustic emission methods of Nondestructive testing of materials and products.* Saint Petersburg: Polytekhnika-Print.

Potapov, A.I. et al. 2019. *Nondestructive physical methods and means of control of natural environment, materials and products. Vol. 15. Physical basis, methodological, instrumentation and metrological support of thermal methods of nondestructive testing of materials and products.* Saint Petersburg: Polytekhnika-Print.

Potapov, A., Pavlov, I., & Verkhovskaia, I. 2019. Nondestructive monitoring and technical evaluation conditions of the monument alexander III. *Architecture and Engineering*, 4(1): 38–46. doi:10.23968/2500-0055-2019-4-1-38-46

Potapov, I.A. & Kondrat'ev, V.A. 2014. Remote monitoring of pipelines using telecommunications technology. *Journal of Mining Institute* 209: 138–143.

Sukhorukov, V.V. (ed.) 1992. *Nondestructive testing.* Moscow: Vysshaya Shkola.

Technological basis for use of composite materials when reinforcing wooden rafters in heritage buildings of Saint Petersburg

D. Zhivotov, Yu.I. Tilinin & V.V. Latuta
Saint Petersburg State University of Architecture and Civil Engineering, Saint Petersburg, Russia

ABSTRACT: Preservation of cultural heritage sites in the process of restoration and construction works during overhaul and reconstruction of facilities in the historical center of Saint Petersburg depends largely not only on architectural but also on technological solutions adopted in method statements and flow diagrams for individual works. One of the factors for the preservation of culturally significant elements of historic buildings is the high-quality repair of roofs using new building technologies and materials to reinforce wooden rafters. In addition to the traditional methods of reinforcing rafters, the authors propose to analyze the use of composite profiles and rods for the manufacture of reinforcing plates and ties. The results of experiments conducted to assess the deformation of composite materials are considered. Fiberglass profiles to reinforce rafters are proposed.

1 INTRODUCTION

The territory of the historical center of Saint Petersburg — the central part of the city delineated by Admiralteysky, Vasileostrovsky, Petrogradsky, and Tsentralny administrative districts — is characterized not only by the presence of architecturally valuable urban development facilities but also by non-compliance of many facilities with operation requirements due to the physical wear of structural elements and facility obsolescence in general (Government of Saint Petersburg 2016). Poor roof performance is one of the main reasons behind the depreciation of the architecturally valuable facades of cultural heritage objects. Therefore, an urgent need arises to develop efficient technologies restoring the performance of wooden structural elements in roofs of historic buildings.

Repair and restoration of roofs and facades represent works aimed to preserve a cultural heritage site and performed by a company having a corresponding license. It is prohibited to conduct construction works without a permit from heritage site protection agencies and in the absence of examined and approved design documentation. During repair and restoration, changes in the architecture of a building are not allowed. It is also prohibited to make temporary openings in walls, construct apertures, or erect partition walls on floors and in attics.

This article addresses the selection and application of technologies for the restoration of wooden rafters by means of their replacement, reinforcement, or use of prostheses. Due to a wide selection of new building materials and methods of their application to repair and reinforce wooden rafters, the issue of choosing a rational technology (especially when

restoring rafters in architecturally, historically and artistically valuable buildings) is becoming more urgent.

2 METHODS

Many researchers have been studying the technical-and-economic, organizational, and engineering aspects of heritage site restoration. For instance, the technical-and-economic and organizational aspects of reconstruction, as well as repair with elements of restoration, are addressed in papers by such Russian researchers as professor Asaul A.N. (2003, 2005), professor Rybnov Ye.N. (Asaul 2003), professor Drozdova I.V. (Drozdova et al. 2016). The engineering aspect of reconstruction and restoration is considered in papers by professor Kazakov Yu.N. (Asaul 2003), professor Yudina A.F. (Yudina 2016, 2017, Yudina & Rozantseva 2012). Professor Mangushev R.A. (Mangushev & Osokin 2014, Mangushev et al. 2018) and professor Verstov V.V. (Gaydo et al. 2014) deal with issues related to the reinforcement and reconstruction of bases and foundations.

When it comes to the reconstruction of cultural and historic heritage sites, it should be noted that such operations are carried out without changes in the structural volume (and especially the protected part of a heritage site) but with the obligatory repair of rafters, lathing, and roof.

As the operation practice shows, the roof is one of the most wearable building elements (Sokol & Golovina 2014).

Improving the technology for the restoration of attic roof elements is one of the crucial tasks in preserving cultural heritage sites. Figure 1 shows restored roof elements of a fire station located in the historical part of Saint Petersburg. The building has cultural heritage elements of regional significance.

Rafters, purlins, and posts are usually made of pine logs 200–300 mm thick at a rafter spacing of 1.2–2.1 m. The standard service life of wooden rafters is 50 years (Afanasyev & Matveyev 2008).

Typical defects in wooden load-bearing attic roof structures are the following:

– unacceptable deflection of rafters;
– wood shrinkage, occurrence of gaps between joints;
– wood decay in areas of roof leakage (usually, in places where rafters and wall plates join);
– fungus defects in wood in case of moisture more than 23% and inadequate natural roof ventilation (Zhivotov & Tilinin 2020).

When restoring the performance of wooden rafters in dilapidated roofs, rotten lower ends of the rafters are cut off and the rafters are extended. The main methods of rafter extension imply the use of wooden plates attached to the rafters with steel threaded studs 12–16 mm thick (Figure 2).

When reinforcing a rafter in a span, two wooden boards 40–50 mm thick are installed in the damaged area. They are attached to the steady parts of the rafter with steel threaded studs and washers for nuts (Mavlyuberdinov & Almeev 2017).

In case there is no high-quality wood available in the construction market, steel rod prostheses can be used for rafter extension. The cut end of the rafter is supported by the lower part of the rod prosthesis, which prevents it from moving (Figure 3).

Some roof elements are subjected not only to climatic effects (wind, snow, rain) but also to the action of condensate in case of inadequate roof ventilation, which results in wood decay due to fungi.

Fungi defects severely affect rafter performance (Figure 4). Their significance shall be considered

Figure 1. Restoration of wooden load-bearing attic roof elements using standard plates attached with steel studs.

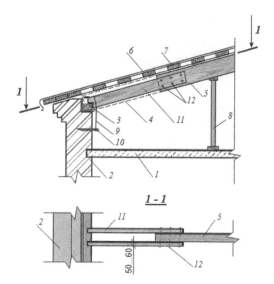

Figure 2. Rafter extension using wooden plates: 1 — roofing; 2 — wall; 3 — wall plate; 4 — extended part of the rafter; 5 — rafter; 6 — roof coating; 7 — lathing; 8 — temporary post; 9 — wire twisting; 10 — spike; 11 — wooden plate; 12 — studs (Mavlyuberdinov & Almeev 2017).

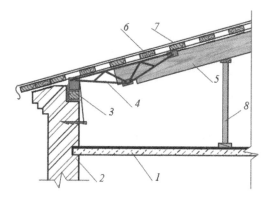

Figure 3. Rafter extension using a metal prosthesis: 1 — attic floor; 2 — wall; 3 — wall plate; 4 — metal prosthesis; 5 — rafter; 6 — roof coating; 7 — lathing; 8 — temporary post (Zhivotov & Tilinin 2020).

during rafter restoration (Biryukov et al. 2017, Fyodorov 2003).

Wood damaged by fungi maintains its strength properties for a long time and can be used after treatment with special preservatives.

3 RESULTS

The authors analyzed the physical wear of rafters damaged by fungi. After the statistical analysis of expert survey results, they obtained a relationship (Figure 5) between the rafter wear accumulation

Figure 4. Rafters with fungi defects in operational condition.

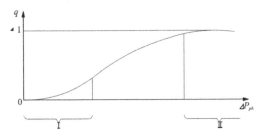

Figure 5. Increase in the rafter wear (q) depending on the area damaged by fungi (P_{ph}). I — initial (operational condition; damaged area in fractions $\Delta P_{ph} \leq 0.3$) stage of physical wear $q \leq 0.27$; II — final (at $\Delta P_{ph} \leq 0.8$) (emergency) stage of physical wear $q \leq 0.9$; intermediate (overhaul) stage at a damaged area in fractions of the whole in the range of 0.3% $< \Delta P_{ph} < 0.8$ and wear in the range of $0.27 < q < 0.9$.

q (as a decimal quantity) and the rafter area damaged by fungi ΔP_{ph} (in fractions of the whole).

The graph shows that if we do not prevent fungus defects from developing at the initial stage, the damage will grow intensively. That is why when a fungus is detected, it is required to perform mycological studies of wood and treat the surface with wood preservatives.

The authors believe that it is reasonable to reinforce a wooden rafter treated with preservatives with polymer composite profiles and ties that are rust-proof and fungi-resistant (in contrast to wooden plates and steel rod prostheses).

Table 1 describes the stress–strain performance of glass-fiber-filled polymer samples tested by the authors.

Figure 6 shows relative tensile strain in a sample (in %) with increasing tensile load (N), and Figure 7 shows absolute tensile strain in mm.

The samples were manufactured by 3D printing using a digital model and Total GF-30 glass-fiber-filled thermoplastic polyurethane (TPU) (Figure 8) (BCG, Tilinin et al. 2018).

As can be seen from the name (GF-30), the matrix is 30% glass-fiber. TPU is soluble only in hot tetrahydrofurane (choosing from more or less available materials). It is suggested to use the study results in the reinforcement of wooden rafters. In contrast to steel, the composite material is light and characterized by some peculiar features, e.g. absence of plasticity (though, in terms of this property, it is similar to wood). Glass-fiber-reinforced plastic profiles are currently available in the construction market.

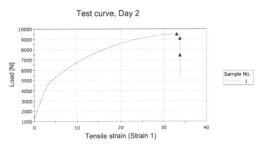

Figure 6. Tensile strain (in %) in a sample manufactured by 3D printing using Total GF-30 glass-fiber-filled composite depending on tensile load (N).

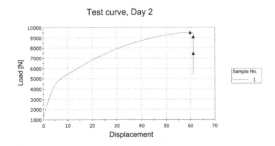

Figure 7. Tensile strain (in mm) in a sample manufactured by 3D printing using Total GF-30 glass-fiber-filled composite depending on tensile load (N).

Table 1. Test results of samples manufactured by layer-by-layer build-up using Total GF-30 glass-fiber-filled composite.

No.	Max load	Ulti mate strength-	Modulus	Relative tensile strain-	Color	Diameter
	kN	MPa	MPa	%	mm	
1	9.53	46.95	370.51	33.81	White	10.1 × 20.1

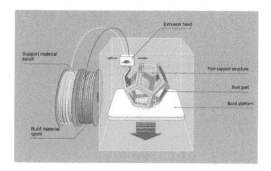

Figure 8. Flow diagram for the process of glass-fiber sample 3D printing, http://3d.globatek.ru/3d_printing_technologies/fdm/(accessed October 19, 2019).

In the authors' opinion, the following glass-fiber-reinforced composite profiles are suitable for rafter reinforcement:

1. glass-fiber-reinforced composite plates with standard dimensions (6000 × 400 × 6 mm);
2. L-bars (75 × 75 × 6; 50 × 50 × 5; 40 × 40 × 4; 25 × 25 × 3 mm);
3. U-sections (200 × 60 × 8; 150 × 50 × 6; 100 × 40 × 5; 45 × 20 × 3 mm);
4. H-sections (200 × 100 × 10; 150 × 100 × 6 mm);
5. Deformed bars with a diameter of 4, 6, 8, 10, 12 mm.

The following polymers are used as the glass-fiber matrix:

1. synthetic resins (phenol, epoxy, polyester) resistant to heating (thermosetting);
2. polymers that are plastic when heated (polyamides, polyethylene, polystyrene).

Glass-fiber-reinforced plastic profiles based on an acrylic binding polymer are characterized by increased fire-resistance.

4 DISCUSSION

After the analysis of traditional technologies for the extension and reinforcement of rafters with wooden plates and steel rod prostheses, the authors have concluded that it is reasonable to extend and reinforce rafters that are slightly damaged by fungi with glass-fiber-reinforced plastic profiles after rafter treatment with wood preservatives. The main arguments in favor of glass fiber as a material to reinforce wooden rafters are that it is rust-proof and fungi-resistant. Besides, it has required strength and strain properties experimentally verified by the authors.

REFERENCES

Afanasyev, A.A. & Matveyev, Ye.P. 2008. *Reconstruction of residential buildings. Part I. Technologies for the restoration of the operational reliability of residential buildings.* Moscow: OOO TsPP.

Asaul, A.N. (ed). 2003. *Issues of economic evaluation regarding the reconstruction of facilities in the historical center of Saint Petersburg.* Saint Petersburg: Saint Petersburg State University of Architecture and Civil Engineering.

Asaul, A.N. (ed). 2005. *Real estate reconstruction and restoration.* Saint Petersburg: Humanistica.

BCG. Five keys to unlocking digital transformation in engineering & construction. A global industry council report. https://media-publications.bcg.com/Oracle-Acco nex-BCG-Unlock- Digital-Transformation-E-C.pdf

Biryukov, D.V. et al. 2017. Protection of structures from humidification as a priority direction of buildings and facilities preservation. *Bulletin of Civil Engineers* 1 (60): 127–134.

Drozdova, I.V. et al. 2016. Technical evaluation as a tool to determine the feasibility of overhaul and reconstruction of the historical residential development in Saint Petersburg. In *8th International Conference on Contemporary Problems of Architecture and Construction,* Yerevan, 26–28 November, 2016.

Fyodorov, V.V. 2003. *Reconstruction and restoration of buildings.* Moscow: INFRA-M.

Gaydo, A.N. et al. 2014. *Pit sheeting technologies under conditions of water development and water areas.* Saint Petersburg: Lan Publishing House.

Government of Saint Petersburg 2016. List of cultural heritage sites in the territory of Saint Petersburg. Committee for the State Preservation of Historical and Cultural Monuments. Accessed May 04, 2016, https://kgiop.gov. spb.ru/uchet/list_objects/.

Mangushev, R.A. & Osokin, A.I. 2014. Arrangement of the underground space when reconstructing administrative building. *Zhilishchnoe Stroitel'stvo* 9: 3–10.

Mangushev, R.A. et al. 2018. *Construction and reconstruction of bases and foundations on weak and structurally unstable soils.* Saint Petersburg: Lan Publishing House.

Mavlyuberdinov, A.R. & Almeev, I.M. 2017. Technological features of roofing overhaul of apartment houses, built in 30-50 years of the last century. *News of the Kazan State University of Architecture and Engineering* 1 (39): 257–263.

Sokol, Yu.V. & Golovina, S.G. 2014. Current issues of urban development reconstruction. In *Challenging Issues of Modern Construction; III International Congress of Young Scientists,* Saint Petersburg, 9–11 April 2014. Saint Petersburg: Saint Petersburg State University of Architecture and Civil Engineering.

Tilinin, Yu.I. et al. 2018. Construction of buildings with the use of 3 D print. *Colloquium-journal* (Part 2) 12 (23): 6–8.

Yudina, A.F. 2016. *Reconstruction and technical restoration of buildings and structures.* 4th edition. Moscow: Academia Publishing Center.

Yudina, A.F. 2017. Modern technologies applied at reconstruction of buildings and structures. *Bulletin of Civil Engineers* 3 (62): 117–123.

Yudina, A.F. & Rozantseva, N.V. 2012. Reconstruction of pitched roofs of civil buildings. *Bulletin of Civil Engineers* 6 (35): 92–95.

Zhivotov, D. & Tilinin, Y. 2020. Experimental studies of the strength of nodal joints of geodesic domes made of wood and fiberglass made on a 3D printer for the Arctic and Northern territories. In Tuuli Mirola (eds), *Becoming greener — digitization in my work. International Week 10.–14.2.2020. The Publication Series of LAB University of Applied Sciences,* part 2: 57–65. Lappeenranta: LAB University of Applied Sciences.

Dust and environment: Saint Petersburg, Russian Federation, case study

A. Ziv
Saint Petersburg Electrotechnical University "LETI", Saint Petersburg, Russia
Voeikov Main Geophysical Observatory, Saint Petersburg, Russia

E.A. Solov'eva
Saint Petersburg State University of Architecture and Civil Engineering, Saint Petersburg, Russia

ABSTRACT: Saint Petersburg residents often consider the city as rather dusty. Corresponding nuisances andaesthetic damage are extremely negatively perceived. Current paper analyses the state of particulate matter (PM) pollution in Saint Petersburg based on five-year data on dust concentrations. To revel the possible contribution of auto transport to dust pollution we use also nitrogen dioxide concentration data at the same monitoring stations. It was obtained that dust levels in residential areas are generally higher than in the city center. Approximate analysis permits to suggest that monitoring stations register PM concentrations of the size up to 70-100 μm with about 30-40 % of PM10. Probably the main contribution comes from fugitive dust from the surface, which creates the impression of a dusty environment.

1 INTRODUCTION

Dust in the air has an evident negative impact on the state of the urban environment. We concentrate here on the case of, Saint Petersburg, Russian Federation, but the same problem is common for many cities in Country. We analyze total suspended particle (TSP) concentrations monitoring data using also "dust" as synonym that's why say "dust in the air" bearing in mind also other components of environment, e.g. pavements, houses, historical and architectural heritage, etc. Particles of size less than 10 μm have well known health impact and it is common to evaluate the environment form the point of their content in the air. At the same time, coarse PM fractions and fugitive dust in particular make nuisances that are more sensible for citizens. Confirmation of this is that residents of the metropolis increasingly complain about the inconvenience that is associated with dust.

The subjective assessment by residents of perceived environmental qualities is an equally important indicator for determining, predicting and making decisions regarding environmental quality (Khan et al. 2015).

We conducted a brief survey of young residents of Saint Petersburg in order to determine how much dust pollution and air pollution affect the feeling of comfort/discomfort in an urban environment. We developed a questionnaire that included five questions, two of which were general, i.e. dust in them was considered as one of the factors of environmental risks. The other three questions dealt directly with dust: how often they pay attention to it, what are the negative manifestations of dust and where it is most often encountered. The survey involved students and graduate students of Saint Petersburg State University of Architecture and Civil Engineering. The results are as follows.

Air quality is in first place among the factors characterizing the ecological well-being in the urban environment. The most important organoleptic indicators of air pollution, respondents called odors, dust pollution, smog, insufficient oxygen saturation of the air, lack of freshness (factors of anthropogenic nature). On a three-point scale of significance, dust received a rating of 2.5.

Most of all, the dust bothers respondents in enclosed spaces - houses and apartments, commercial public institutions, and in transport (2.6). It is perceived not only visually, but also tactilely (touch). The subjective dust content on the city streets is also high (2.4). Above all, it is near roads and in the city center. Below - in the parks.

The respondents showed the most activity when answering the question regarding negative manifestations of dust. By analogy with Chris Anderson's "The Long Tail", a large number and variety of responses suggest that dust is not just an indicator of air purity, but an important environmental quality parameter. The presence of dust causes a lot of trouble and is extremely negatively perceived by people.

Three blocks of the mentioned inconveniences from dust can be distinguished, adverse effects, nuisances/aesthetic damage, visibility reduction with percent ratios 37.5, 50, 12.5, correspondingly

Adverse effects on physical well-being: allergies, eye pain, cough and runny nose, shortness of breath, skin problems, etc.;

Nuisances/Aesthetic damage: dirty windows, dust on window sills and furniture surfaces, dirty cars, dust on tree leaves, pets, clothes and shoes, house facades covered with dust and their unaesthetic appearance. The latter is especially critical for the historical part of the city;

Visibility reduction: this is less often, possibly because the city has few wide open spaces and because of the climate, the air is rarely transparent.

In the rest of the paper we consider TSP monitoring data for 2014-2018 as seems above mentioned nuisances are mainly relate to PM fraction more than 10 μm. PM2.5, PM10 data are also available at least as annual mean values.

2 MONITORING DATA AND SIMPLE STATISTICS

There are three sets of monitoring stations with PM measurements in Saint Petersburg. The first monitoring network conducts total suspended particle (TSP) measurements three times daily. Second and third measure PM2.5 and PM10 continuously but none of these three sets is coincide.

We consider mainly TSP data together with NO_2 concentration data at same location (Databank "Atmospheric pollution"). A joint analysis of the atmospheric content of dust and nitrogen dioxide (or NOx) is quite common (Norman et al. 2016), although mainly in the combination of PM2. 5 - PM10 – NO_2. Mention here that nowadays most of publications concern PM 2.5, PM10 air pollution that makes our study little bit more complicated. The main reason for combined use of both data is significant contribution of particles from auto transport as a result of exhaust and non-exhaust emissions. The last ones are especially valuable in North Countries where studded tires are used in winter and roads are treated with anti-icing compounds. As mentioned in Denby et al. (2016) NOx concentrations represent somewhat tracer. From this point, joint analysis permits to understand at least very approximate the origination of high dust concentrations and/or part of dust, which can be attributed to the auto transport. It even more essential due to the lack of any information on the spectrum of the measured particles and their composition in our data.

Data set consists of 5 years (2014- 2018) measurements at 9 stations inside Saint Petersburg. Their numbers correspond to ones in all Russia data base: 1,2,4,6,7,8,10,12,27. Hereafter each station will be identified as S"number",S1 - S27. At most of stations observations were made three times daily (07:00, 13:00, 19:00) 6 days a week, Mon - Sat. There is one station (St.1) with additional time 01:00. Both dust and NO_2 concentrations data available from any station but NO concentrations from St.7 and St.8 only. The map with stations (created by Google Maps application) is shown in Figure 1.

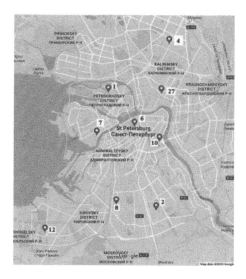

Figure 1. Location of the monitoring stations.

Stations 2, 4, 8, 12 are in residential areas far from city center. Station 27 is located not exactly in such area but close to it. The surroundings nearby are very similar - roadways with medium traffic intensity are fairly close, as well as lawns, small parks, and numerous footpaths. It follows from Table 1 however, that concentrations at stations 2, 4, 12, 27 being almost the same for NO2 differ significantly for dust from those at 8-th one. Stations 1, 6, 7, 10 are inside the city center and again splits into 1,7, 10 and 6.

These features can be partly explained by the assumption that significant contribution to TSP level gives dust from nearby surfaces, trees, houses, etc., which in its turn come from more far places as a result of some kind dust migration or saltation. Dust from construction, industry and roads consist in particular from PM fractions, which deposited not far from the sources. Then however, these particles may spread by wind, cars, pedestrians, etc. Another essential contribution is a lot of places with open soil (as there is now vegetation at least half a year) and non-paved small roads between houses. S8 is most dusty though located in residential to similar to S(2,4,12,27) close surroundings. Probably difference is in of highways and roads with very intensive traffic especially of heavy-duty vehicles in 1-3 kilometers West and South. There are also big industry regions including cargo harbor in 4-5 kilometers West. Taking into account, that prevailing wind directions in Saint Petersburg are from South to West these roads are might be the additional source of dust in station vicinity due to above mentioned dust migration. In addition to that, there is a rather big park in about 300 m North with large open ground surfaces and unpaved paths

One can see from Table 1 that dust concentrations at stations 1, 7, and 10 are rather low though annual

Table 1. Main statistics of observed data 2014 – 2018, µg/m³.

Station #	1	2	4	6	7	8	10	12	27
	Dust								
mean	90	128	132	158	51	243	67	140	130
median	24	58	69	92	16	169	18	73	66
max	500	500	500	500	400	900	400	500	500
StDev (σ)	91	114	105	120	60	170	79	115	109
V= σ/mean	1.01	0.89	0.80	0.76	1.18	0.70	1.17	0.82	0.83
	NO2								
mean	62	49	39	87	57	32	57	46	52
median	58	40	33	80	54	30	50	40	50
max	260	246	210	330	170	120	240	200	246
StDev	44	40	34	59	32	22	45	38	41
V Mean	0.71	0.82	0.87	0.68	0.56	0.70	0.80	0.83	0.79
(NO2)/Mean(Dust)	0.69	0.38	0.30	0.55	1.12	0.13	0.85	0.33	0.40

NO_2 concentrations are moderate high and more than National air quality standard (AQS) for annual mean concentrations equals 40 g/m3. If we consider again the hypothesis of dust migration, we find out that both 1 and 7 stations are protected from it by rivers and also surrounded by mostly paved roads where freight traffic is prohibited. Station 10 is nearby the river as well and there is a rather high fence between station and medium sized park. The rest of the area consists of houses and paved roads. Note in an addition to this description that main traffic is from the North to South-East with less frequency then South and West directions. It follows from Table 1 that mean concentration at this station is higher than AQS for annual mean TSP equals 150 mg/m3. Second "dusty" station S6 located in a typical urban center with dense buildings and paved roads. However, two small parks are not far away. Continuing the theme of long-range transference mention that large part of city center with intensive traffic on South-West from this station. Mean TSP concentration at S6 is also higher than AQS. Almost the same with mean NO_2 concentration - the highest in the city. It differs this station from S8 with lowest NO_2 concentration. Mean TSP concentrations at the rest of stations are inside AQS though remember Introduction Saint Petersburg is coincided as dusty by citizens. Opposite for nitrogen dioxide mean values are inside AQS (40 mkm/m3) at only two stations S8 (most dusty) and S4.

It looks worth to compare our data with those published in Eeftens et al. (2012). Last paper shows results for PM2.5, PM10, NO_2 for cities and territories in Europe. First two concentrations vary from 5 to 30 µg/m3 and from 15 to 50 correspondingly being lower in North countries. Authors characterize territories also by the median of ratios of annual mean NO_2 to annual mean PM10, which varies in range 0.32 – 1.25 being in 70% cases between 0.95 and 1.25. Data of Saint-Petersburg Committee for Nature Use, Environmental Protection, and Ecological Safety (Environmental Web-Portal of Saint Petersburg 2020) for 2014-2017

give the range of PM2.5 concentrations 5 – 18 µg/m3 and the range of PM10 concentrations – 5 – 32 µg/m3. These data suggest that Saint-Petersburg is one of "clean" North city. Median that calculated for the same data (NO_2 from the same stations) equals 2.875 in 2017. Maximum value in Eeftens et al. (2012) is 1.82 (London/Oxford). Last row in Table 1 show the ratios of NO2 to dust concentrations and median equals 0.4. Although this value is in the abovementioned range, it correspond to TSP but not PM10.

PM10 concentrations at stations considered in this work may be evaluated using approximate ratio PM10/TSP= 0.55 for annual mean concentrations (Dockery & Pope 1994). At least this value utilized in risk assessment in Russian Federation (Rakhmanin & Onishchenko 2002). Plots in (Koch & Rector, 1987) give very close value as an upper bound for annual means. Later extensive research by Cohen et al. (2004) show that this ratio varies significant but 0.49 is acceptable. Air quality monitoring data over the territory of Belorussia (Prosviryakova & Shevchuk 2018) give little higher value about 0.58. Bearing in mind this coefficient of 0.55, we obtain for our data the median value 0.73.

3   TIME VARIATION

As follows from the table of elementary statistics, the average values of concentrations of both dust and nitrogen dioxide varies significantly over the city. This, apparently, is explained by the variability in the territory of Saint Petersburg of the urban background, and local features of the location of the stations. It is of interest to consider the similarities in time variation of concentrations despite different particular levels. As hourly data at different stations are practically uncorrelated we consider series of monthly average values, in five years, 60 values in total, and monthly averages for 5 years, 12 values. In both cases, we also add corresponding variation of precipitation indicator obtained as follows. We calculate the probability Ppr of occurrence of precipitation in the each month as the ratio of the number of meteorological observations with any precipitation to the total number of observations.

It is mentioned in a number of publications (De la Paz et al. 2015, Molina et al. 2010, Shen et al. 2016) that precipitation in general reduces dust pollution, so finally precipitation factor is set to Ppr-0.3. The 0.3 degree was chosen for better representation in Figure 2 and better correlation as well. Figure 2 shows 60 month variations of city averaged values for dust and NO_2 scaled by corresponding 5 years averaging (123 μkg/m3 for dust and 55 μkg/m3 for NO_2.

Year-month values are coded by integer number equals to (Year-2014)*100+Month, OX axis tick marks.

Curves in the plot look rather similar especially from the middle of 2016 (206 as OX tick mark). One can mention the evident shift in concentrations' levels since that time. As long as there is no explanation for this. It is possible to conclude from Figure 2 that relatively high concentrations both NO_2 and dust correspond to Spring-Summer period, which is the additional evident to the fact that significant role in dust pollution plays dust resuspended from the surface. In publications on PM pollution in North countries where studded tires are used (Denby et al. 2016, Johansson et al. 2007, Norman et al. 2016) highest concentrations near the roads are usually observed during the wintertime. However, this mainly concerns particles with less than 10 μm diameter.

There are good correlations of these 60-months series at several stations. Table 2 shows values higher than 0.5 for dust and NO_2 separately. Beside that correlations 0.51 - .61 between dust and NO2 were obtained for pairs S1-S6, S6-S4, S6-S6, S12-S4, S12-S6.

It should be noted that in some cases, relatively high correlations in the table correspond to stations with very different concentration levels. So, NO_2 concentrations at the 8th station are quite low, the

Table 2. Correlations between 60-months series higher than 0.5.

Dust	S_02	S_04	S_06	S_07
S_01	0.69		0.60	
S_04			0.55	
S_06				
S_07				
S_08				
S_10				
S_12				

Dust	S_08	S_10	S_12	S_27
S_01	0.53	0.53	0.67	
S_04			0.53	0.54
S_06	0.58	0.64	0.60	
S_07		0.63		
S_08		0.64	0.71	
S_10			0.68	0.67
S_12				0.71

NO2	S_02	S_04	S_06	S_07
S_01	0.55	0.56	0.68	
S_02		0.62		0.57
S_04			0.74	0.69
S_06				
S_07				
S_08				
S_10				
S_12				

NO2	S_08	S_10	S_12	S_27
S_01		0.51		
S_02		0.61	0.63	0.59
S_04	0.76	0.74	0.75	0.69
S_06	0.57	0.74	0.53	
S_07	0.72	0.70	0.85	0.66
S_08		0.64	0.66	0.53
S_10			0.78	0.70
S_12				0.66

average is 32 μg/m³, and at 4-7th - 39, 87, 57 μg/m³ (see table 1). In general, it turns out that the largest number of correlations corresponds to stations 10, 12, and 27 though locations of these stations are rather different, 10 is almost in the city center, 12 is in South-West residential area, 27 is on the peripheral of city center. Perhaps due to difference in their locations these stations reflect most of the variability in the concentrations of the all stations. It mostly concerns NO2 and partly dust pollution. Mean nitrogen dioxide values are nearly the same at these stations 57, 46, 52 μg/m³, mean dust levels at 12 and 27 station are very close 140 and 130 μg/m3 but at S10 this value is 67 μg/m³.

Joint factor analysis of 60-months series for all stations and NO2 and dust shows that first 3 components explain 70% of dispersion, 41% for the 1-st factor. Correlations higher than 0.7 with

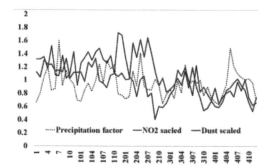

Figure 2. 60-month variation of city averaged scaled values for dust and $_{NO2}$ together with precipitation factor.

251

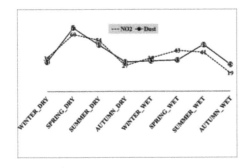

Figure 3. Variation of the sums of ranks over stations for the each season-precipitation class.

this 1-st factor correspond S4, S6, S10 for NO$_2$ and S6 for dust.

Then average values for dust, NO2 and the each station were calculated over 8 classes in accordance with year season and occurrence of any precipitation in day of observation, winter-dry, winter-wet, spring-dry, etc. One can attribute rank for the each of 8 classes for the current station (1 for the lowest value). Figure 3 represents the curves for sums of ranks over stations for the each class. The similarity of these plots for NO$_2$ and dust confirms the correspondence of both concentrations. One can distinct "spring-dry" class among others, though slight reduction in concentrations due to wetness there is for other classes as well.

As it is indicated in publications Spring is rather dusty season. It explained partly by the resuspending of sediments after Winter. It is not quite clear why NO$_2$ concentrations rise also. It might be due to features of meteorological regime together with increase in number of vehicles in comparison to Winter. City is normally specially cleaned in the end of Spring thus during Summer dust pollution decreases. Unfortunately, due to the lack of any information on particle size distribution these considerations are only qualitative

4 DUST CONCENTRATIONS AND WIND DIRECTION

The usual way to locate influenced sources is to consider the dependence of pollution on wind direction and wind speed. Concentration roses are the suitable instrument for this (Grigoratos et al. 2014, Kozawa et al. 2012). However, bearing in mind the limited data we have considered similar relationships dividing wind speed and direction (WS-WD) into 16 gradations – 8 WD and 2 WS less than 3 m/s and more than. Obtained concentration roses show rather small variability of concentrations with these gradations. Table 4 comprises some characteristics of WS-WD distributions for both dust and NO2. "Span/ Mean" indicates the ratio of the difference of maximum and minimum values over gradations to the mean value for the current station. Span can be

considered also as the maximum difference from uniform distribution correspondent to minimum value or the maximum value of relative rose obtained after subtraction of minimum.

It follows from Table 4 that relative variation in values for different WD and WS are almost the same in case of dust and NO2, higher for moderate wind speeds, St.7 is exclusion. Correlations of concentration roses are quite high for St. 1 and St.2 at moderate WS. In other cases, correlations are hardly to take into account. Comparison of Tables 1 and 4 gives that variations of dust concentrations over different wind sectors are mostly inside 50 – 80 µg/m^3.

Examining of the stations locations gives the conclusion that relative dust concentration roses are oriented mainly on nearby traffic, which permits to suggest that indicated in Table 4 variability is mainly connected with light PM. Low dust concentrations at stations with good dust-NO2 correlations supports this suggestion. Moreover the percentage of light PM (if this supposition is true) in good agreement with PM10/TSP ratio already discussed.

As in case of our research dust concentration data comprise all the diversity of PM we assume that great part of coarse particles arrive from nearby territory and randomly depend on wind direction. At least we can see e.g. from Grigoratos et al. (2014) that urban background station located at city periphery indicates city center as dust source for PM10-PM2.5 more than city center station. Note also that following Fairweather et al. (1965), the visible unpleasant dustiness is mainly due to particles with diameters greater than 40 µm. In the well-known

Table 3. Span values of concentration roses relative to average concentrations at the same station and correlations between roses.

	St_1	St_2	St_4	St_6	St_7
Dust					
Span/Mean WS ≤ 3	0.31	0.31	0.24	0.18	0.23
Span/Mean WS > 3	0.17	0.21	0.20	0.14	0.46
NO2					
Span/Mean WS ≤ 3	0.31	0.28	0.31	0.35	0.14
Span/Mean WS > 3	0.22	0.32	0.32	0.22	0.11
Correlations					
Dust – NO2 WS ≤ 3	0.78	-0.06	-0.21	0.11	0.39
Dust – NO2 WS > 3	0.18	0.45	-0.04	0.34	0.21

	St_8	St_10	St_12	St_27
Dust				
Span/Mean WS ≤ 3	0.28	0.50	0.19	0.39
Span/Mean WS > 3	0.27	0.38	0.14	0.20
NO2				
Span/Mean WS ≤ 3	0.19	0.26	0.24	0.26
Span/Mean WS > 3	0.19	0.21	0.20	0.20
Correlations				
Dust – NO2 WS ≤ 3	0.28	0.71	0.48	0.43
Dust – NO2 WS > 3	0.49	0.29	0.15	0.38

EPA method regarding the factors of the emission of resuspended dust (EPA 2006), the main parameter is the surface sediment on the road, which refers to particles with a size of up to 70 microns. Thus, it possibly follows from above considerations, that the range of particle sizes, which is recorded at the measurement stations, refers to an interval up to 70-100 μm.

5 CONCLUSION

This study considered the state of TSP air pollution in Saint Petersburg, Russian Federation. The main reason for this brief analysis is the perception of city environment by citizens as dusty. Dust concentrations levels and their 5-year variability studied not only alone but also in comparison with simultaneous nitrogen dioxide concentrations.

This joint analysis show definite relation of these two pollutants. It is shown in particular that seasonal maxima for both pollutants is Spring with dry weather. Five-year variability of monthly mean dust and NO_2 concentrations reveal their general correspondence and the correspondence with precipitation index specially introduced for this study.

Spatial distribution of dust levels indicates in general lower values for city center with mostly paved roads and rare spots of open ground and unpaved paths.

The dependence of dust and NO_2 concentrations on wind speed and direction (WS, WD) is not definite. Maxima relative differences of concentration roses from uniform very similar for dust and NO_2 and are in range $0.11 - 0.55$ with median value equals 0.27 for wind speeds less than 3 m/s and 0.21 for the rest of wind speeds. Correlations of concentration roses for TSP and NO_2 are negligible for all stations except Station 1 and Station 10 with values 0.78 and 0.71 in case of moderate wind speeds. Both stations are with rather low dust levels and located in the city center. The deviations from uniform roses are mostly directed on nearest roadways, which probably indicated that they are mostly due to the direct and indirect contribution from auto-transport.

The performed study permits to suggest that the main contribution to the TSP pollution comes from dust resuspended from the surface. It seems that monitoring stations register particles to 70-100 microns and coarse dust accounts for about 70%.

Note as a conclusion that despite fulfilment of TSP annual air quality standards in most place in the city Saint Petersburg appears as rather dusty which is not facilitate its attractiveness.

REFERENCES

Cohen, A. et al. 2004. Urban air pollution. In M. Ezzati, et al. (eds), *Comparative quantification of health risks: global and regional burden of disease attributable to selected major risk factors*: 1353–1433. Geneva: World Health Organization.

Databank "Atmospheric pollution". "RSBD Pollution of the atmosphere". Reg. No. 2012620233.

De la Paz, D. et al. 2015 Implementation of road dust resuspension in air quality simulations of particulate matter in Madrid (Spain). *Frontiers in Environmental Science* 3: 72.

Denby, B.R. et al. 2016. Road salt emissions: A comparison of measurements and modelling using the NORTRIP road dust emission model. *Atmospheric Environment* 141: 508–522.

Dockery, D.W. & Pope, C.A. 1994. Acute respiratory effects of particulate air pollution. *Annual Review of Public Health*, 15: 107–132.

Eeftens, M. et al. 2012. Spatial variation of $PM_{2.5}$, PM_{10}, $PM_{2.5}$ absorbance and PM_{coarse} concentrations between and within 20 European study areas and the relationship with NO_2 — Results of the ESCAPE project. *Atmospheric Environment* 62: 303–317.

Environmental Web-Portal of Saint Petersburg 2020. Reports on the environmental conditions in Saint Petersburg for 2014–2018. http://www.infoeco.ru/index.php?id=982.

EPA 2006. Air emissions factors and quantification. http://www.epa.gov/ttn/chief/ap42/ch13/index.html

Fairweather, J.H. et al. 1965. Particle size distribution of settled dust. *Journal of the Air Pollution Control Association* 15 (8): 345–347.

Grigoratos, T. et al. 2014. Chemical composition and mass closure of ambient coarse particles at traffic and urban-background sites in Thessaloniki, Greece. *Environmental Science and Pollution Research* 21 (12): 7708–7722.

Johansson, C. et al. 2007. Spatial & temporal variations of PM_{10} and particle number concentrations in urban air. *Environmental Monitoring and Assessment* 127 (1–3): 477–487.

Khan, M.F. et al. 2015. Quality of urban environment: a critical review of approaches and methodologies. *Current Urban Studies* 3 (04): 368–384.

Koch, R.C. & Rector, H.E. 1987. Network design and optimum site exposure criteria for particulate matter.

Kozawa, K.H. et al. 2012. Ultrafine particle size distributions near freeways: Effects of differing wind directions on exposure. *Atmospheric Environment* 63: 250–260.

Molina, L.T. et al. 2010. An overview of the MILAGRO 2006 Campaign: Mexico City emissions and their transport and transformation. *Atmospheric Chemistry and Physics* 10 (18): 8697–8760.

Norman, M. et al. 2016. Modelling road dust emission abatement measures using the NORTRIP model: Vehicle speed and studded tyre reduction. *Atmospheric Environment* 134: 96–108.

Prosviryakova, I.A. & Shevchuk, L.M. 2018. Hygienic assessment of PM10 and PM2.5 contents in the atmosphere and population health risk in zones influenced by emissions from stationary sources located at industrial enterprises. *Health Risk Analysis* 2: 14–22.

Rakhmanin, Yu.A. & Onishchenko G.G. (eds) 2002. Fundamentals of risk assessment for public health when exposed to chemicals polluting the environment. Moscow.

Shen, Z. et al. 2016. A comparative study of the grain-size distribution of surface dust and stormwater runoff quality on typical urban roads and roofs in Beijing, China. *Environmental Science and Pollution Research* 23 (3): 2693–2704.

Problems of training architects and restorers

Reconstruction and Restoration of Architectural Heritage – Sementsov et al (eds)
© 2020 Taylor & Francis Group, London, ISBN 978-0-367-65357-6

Peculiarities of forming architecture of universal youth centers in Republic of Crimea

Z.S. Nagaeva & D.S. Mosyakin
V.I. Vernadsky Crimean Federal University, Simferopol, Republic of Crimea, Russia

ABSTRACT: The article discusses the main directions of forming the architecture of universal youth centers and recommendations for designers.

1 INTRODUCTION

Young people are an important part of modern society. They are the people who will take over most of the processes that take place in the state and our public life in the future. For the quality and full use of their resources in the future, young people need to be provided with a place where they can meet all their needs for communication, interaction with each other and providing information.

An important component of the organization of a young persons' time is their leisure time. However, now the most popular is not just leisure of an entertaining nature, where young people could get distracted from pressing problems and just stay in the company of their own kind, discussing their problems, interests and just relaxing, but educational leisure with a large share of education (Barkhin 1975). Only specialized institutions – universal youth centers (hereinafter referred to as UYC) – can provide the full range of needs and aspirations of young people. Specialized leisure institutions began to appear in the early twentieth century in Europe and later in the USSR. However, the directions of development of leisure centers in Europe and in Russia were different. In European countries, the centers gathered many functions, becoming multifunctional centers of attraction. Domestic leisure centers have gone the other way, monofunctional centers (sports centers, theater clubs, etc.) have been greatly developed. Multifunctional institutions – homes and palaces of pioneers, began to appear only in the 60s.

These days, UYC are just beginning to develop and replace the stagnant monofunctional complexes. Similar centers are being established in Russia on the Black sea coast, in a number of major cities, such as Moscow, Saint Petersburg, Izhevsk, Samara, and others. The design of the UYC is complicated, since there is no necessary methodological base for designers and architects.

2 MATERIALS AND METHODS

The purpose of the study is to develop recommendations for architects and designers, which they can use to guide them when creating a smart design center. And also to set the main directions for the development of architecture of youth complexes in the Republic of Crimea.

The research methodology is based on complex and systematic approaches, which include:

– analysis of historical and modern domestic and foreign experience in the design, construction and operation of UYC;
– theoretical modeling and drawing up of typological schemes of UYC;
– conducting a sociological survey using the questionnaire method;
– study and comprehensive analysis of functional-planning, volume-spatial solutions for the design of UYC in the Russian Federation and in particular in the Republic of Crimea. Peculiarities of forming the architecture of universal youth centers in the Republic of Crimea.

3 MAIN PART

Universal youth centers are a new type of multifunctional complexes that are now actively developing and beginning to be introduced into the culture of young people. The emergence of UYC is associated with changing needs of young people, with the expansion of their resources, the development of tourism and public leisure in the country as a whole (Bezrodny 1970). The beginning of mass introduction of multifunctional complexes for young people has a number of advantages over existing centers and leisure facilities:

1. A number of space management tools can be used for most of the necessary conditions (in

addition to urban planning, climate and social conditions, which are also necessary);

2. Creating a closed zone with its own architectural solutions and developing its own functional saturation in them, which may not be applicable anywhere else.

UYC create a favorable environment for various types of interaction with the consumer of functions (in this case, mainly youth). Youth centers can be perceived as mini-cities, where there is such an important shift in our time from monofunctionality to multi-functionality and universality. An important component of the UYC is the development of not only the leisure component of the center, but also the educational and communication parts. The main goal of such a complex is to create a zone of free communication, where young people interact with each other, with the old generation (continuity of experience) and with their environment (both urban and natural).

The development of UYC is a necessary condition for the development and improvement of social diversity in our time. This, in turn, affects the functional saturation of the youth center, which in modern conditions is still insufficient. Hence, we can conclude that the development of social diversity and functional saturation makes the UYC not only a place of gathering and interaction of young people, but also of all age and social groups.

In the domestic practice of designing complexes of this kind, a number of typical models of cultural and leisure centers (clubs, houses of culture, palaces of culture, cultural centers) have been formed over the course of the XX century, focused mainly on club and entertainment functions. However, relying on sociological research and the study of the activities of modern multifunctional complexes, they do not justify the reality of the new social life of the XXI century and the needs of various groups of the population.

As the processes in society change, the architecture must also change. Architecture is a living science, it reflects all the changes, problems and interactions in society that occur in a certain period of time. This period is characterized by the increasing development of multifunctional structures for various purposes and with a different set of functions: business, entertainment, cultural and leisure, educational, health, recreation, historical, etc. It is worth noting that an important common feature of these institutions is their complexity and universality.

Generalization of the practical experience of designing and building such universal housing complexes in the world allows us to suggest promising directions for the development of UYC as a new type of public complexes. Some aspects of their formation have already been touched upon in domestic research, in particular, in the development of All-Russian research Institute of technical aesthetics, All-Russian research Institute of theory and history

of architecture and urban planning, Central scientific-research Institute of typical and experimental design of entertainment buildings and sports facilities, Central scientific-research Institute of typical and experimental design of urban planning, Moscow Institute of Architecture (Bylinkin 1985).

The approach to the organization of complexes is changing, and socio-cultural research and analysis of the needs of the main consumers of the functions of the UYC are being widely introduced. The relationship between the social function and the commercial direction is economically advantageous, since buildings with a social orientation have a longer payback period and such institutions are less popular than complexes with a commercial orientation in their composition.

In the Russian Federation, such integration is very poorly developed and, as a result, there are difficulties in the spatial organization and interrelation of the components of the complex.

Thus, the topic of proper architectural and spatial organization of University youth centers is very relevant, and it requires the development of General recommendations for the design of youth centers, since the development of society is a continuous process and at this stage it requires a response in the architecture. Leisure is constantly being transformed, and here it is through architecture that it is possible to reflect the needs for the functioning and interconnection of parts of the UYC, and in the future, to reflect these connections in the design standards, as a separate normative act concerning the UYC.

In the conditions of constant development of the city in social, economic, demographic and territorial aspects, additional needs of the population arise for the possibility of self-realization in various spheres and areas of life (Vergunov 1982).

4 RESULTS AND DISCUSSION

Universal youth centers solve many urban problems, as they are a point of attraction for city residents. Such complexes become a place of a large concentration of people and resources, and often require non-standard urban planning solutions. The very territory of the UYC can be located both on the open territory and in the conditions of cramped urban development. Different locations and different operating conditions affect the architectural appearance and type of the youth center. The architecture of youth complexes is original and unique, which makes it a place of attraction for creative people and creativity in General. But in whatever urban development conditions the UYC, like any other complex, is a synthesis of architecture and space.

The design of youth centers is a phenomenon of time, reflecting the needs and beliefs of the time in which they function. This is reflected in the architecture, planning and spatial organization of the complex.

Constant changes in the society require important characteristics of the UYC: transformation, adaptation. Adaptation of youth centers lies in the fact that the complex itself, its design and spatial organization of the internal environment is changeable. There are no permanent, clearly delineated spaces, each space is interconnected with each other and flows from one room to another. This important feature of the modern youth center makes it possible to use it in any conditions, to repurpose it depending on different functions and different purposes. The result of adapting the space and layout to the functions is transformation is another important quality of the modern youth complex. Transformable architecture makes it possible to change the configuration, appearance, and purpose of premises within the complex using design solutions and architectural interactions. Such opportunities are an additional impetus to the development of the architecture of youth centers in the direction of modularity and universality. The modular architecture has great possibilities in terms of interaction in different conditions and variability. When non-consumer goes and visits the youth center, but the center can be located in a convenient place for visitors, even in the conditions of historical development and on the territory of an object that is significant from a cultural point of view.

Architects strive to create comfortable conditions for the functioning of youth centers that meet the requirements of the time. UYC belong to multifunctional buildings and gradually, with the development of functions and interactions within the complex, they begin to grow into a system of youth centers. It is worth noting that in the Crimea, such a transformation is predictable and necessary. Crimean cities have a number of disparate functions within the urban planning system that need to be met to meet the needs of young people, and only a system of youth centers can cope with this task. This direction of development is promising and requires further scientific justification.

The typology of universal youth centers is the next stage in the development of a system of youth centers (Gelfond 2003).

The standard functional structure of the first pre-Suga institutions has been significantly expanded and widened. They have become large complexes. UYC in the modern world, and in our country as well, have become extremely popular in recent years.

In the modern world, education is growing and the demands of consumers of services of multifunctional complexes are increasing, so the centers are developing and improving, this is an important process of architecture response.

If we consider the placement of leisure centers in the city structure, they are practically absent in the historical environment of the city. This is one of the differences from other types of multi-functional structures. If they are located in the city, then only in recreational green areas. If outside the city, then in the suburban recreation area (Korob'ina 1986). However, due to their versatility, mobility and adaptation, UYC can integrate into the historical environment and interact closely with it. This interaction additionally attracts young people to the protection of the historical environment, interests them and attracts them. Of course, this interaction is also reflected in the architecture, because historical buildings are very valuable and any interference with them is prohibited. This is another challenge for architects, because both biopositive technologies and non-standard architectural solutions (transformability of premises, introduction of lightweight structures, quickdeployed and temporary structures that do not violate the integrity of historical buildings) should be used as much as possible. Such architectural solutions, combined with the development of youth interest, make the topic of cultural heritage protection the main one in the functional saturation of such complexes.

The main task of the UYC is to create a universal synthesized environment. It is not so much the value of each object, which is an element of this peculiar city formation, that is important here, as the whole unity of the constituent elements of the UYC as a whole. The object itself must be integrated into the existing situation, landscape, or urban planning node. In the UYC, the space flows into each other and works like a maze – involving the visitor with its uniqueness, and does not let go until the end of the route (Gorokhov 2003). This can be achieved through architecture, developing small architectural forms, combining small spaces with growing into contrasting spaces of halls and atriums. This contrast develops the viewer's interest and sense of expectation, revealing new details.

UYC are modern points of attraction and potential development with a high concentration of various functions, saturated with various buildings, zones and structures that ensure free communication of people in this rich multifunctional space, where the visitor acts simultaneously as a participant, spectator and creator.

UYC spaces can be divided into several groups:

1. UYC with historical themes dedicated to the history of countries and cities;
2. UYC with scientific and technical topics that reflect the development of science and technology or the sacred future, achievements and scientific discoveries;
3. UYC of educational subjects.

Each of the groups of spaces has its own specifics, goals, tasks, and architectural approaches to disclosure (Kravchenko 1991).

It embodies the idea of creating a simultaneously educational and entertaining environment, where education is in the format of a game or quest, which attracts the visitor with its unusual form and content.

UYC are designed not only for young people, but also for visitors of different ages, while architects and designers take into account the psychological perception of children and adults, which has its own characteristics (Cashayeva 1988). Therefore, the UYC includes buildings and structures of various sizes. UYC can be considered a synthesis of space and surrounding greenery, natural landscape, where additional factors are the color, light and design solutions.

If in ordinary parks there are separate buildings, pavilions, and attractions that "dissolve" in the natural environment, then in the shopping center they are concentrated, prevail, and dominate over the greenery and landscape (Gorokhov 2003). The surrounding landscape is often an extension of buildings, their internal space opens up to the outside and plays a connecting role between them, contributing to full recreation, walks, entertainment and various activities. This statement is very important for the Republic of Crimea, because the Crimean nature is unique and the inclusion of nature in the structure of the complex, the synthesis with it, the use of biopositive technologies in the design of complexes are important conditions for the formation of the architecture of the UYC on the territory of Crimea.

The architecture of the UYC with its artistic and imaginative characteristics emphasizes the uniqueness of this commercial complex, which is primarily a tourist attraction, it is the epicenter of functions in their various manifestations. In such complexes, art architecture becomes one of the main means of attracting visitors (Girardi 1999). The architecture of the UYC is very specific, it is designed for effect and originality. And also, in conditions where the architecture of the architectural tour emphasizes and reveals the essence of the complex, makes it a recognizable and attractive tourist object.

The diverse image of architecture in this type of complex comes out on top. Here, the image of the entire UYC is equally important, which should be memorable and impressive, as well as each element of the complex separately. Architecture is always individual and has a memorable and expressive appearance. Visitors can see a panorama of the complex, spread over a large area, which gives a complete picture of the planning structure of the shopping center. Landscaping of the territory of the complex with elements of landscaping and elements of the Park complements the ensemble of structures of various functional purposes. The diverse landscape is organized as a system of intersecting spaces in the form of a labyrinth, drawing visitors further and further into the center. All elements of improvement only emphasize this.

Architecture should not contrast with the surrounding landscape with its forms and colors, but merge with it, organically fitting into the General style and structure of the surrounding area (Zalesskaya & Mikulina 1979).

Creating a multi-functional spatial environment for recreation, entertainment and education is very important for Russia as a whole and for the Crimea in particular. The search for original planning and architectural compositions is a distinctive feature in the design of a UYC center.

In the specific conditions of the Republic of Crimea, UYC should be transformed into a volume, preserving the basic urban planning principles of functional zoning, when open functional zones, spaces and sites are created at different levels under a single surface. This also applies to the composition of the UYC from a number of separate interconnected objects, for example, the UYC, where the idea of a mini-town is implemented in its internal space.

5 CONCLUSIONS

1. The study of the evolution of types of multifunctional leisure facilities in the XX century allowed us to trace the process of formation of modern recreation centers.
2. The features of multi-functional UYC are Revealed, which are manifested in their distribution, functional zoning, compatibility, architecture and design, the method of obtaining information in the process of entertainment, as well as in the presence of a highly emphasized theme or the development of creative ideas and concepts. A distinctive feature of the UYC is its special creation, when architectural means create an atmosphere of an environment that attracts visitors, promotes emotional and psychological development of a person, activates the variety of his activities.
3. The Planning organization of the UYC is determined by the main compositional axes and nodes, and the harmonic connection with the natural landscape, the arrangement of objects in the space, and in general creates the unity of the overall composition.
4. The architecture of the UYC is a landmark, one of the most popular centers for attracting the local population and tourists, as it has a bright figurative expression, concentrates as much as possible a variety of different functions, responds promptly to the needs of the time: dynamic, open to further growth, change and improvement.
5. Recommendations have been Developed for the design of UY in the Republic of Crimea, for creating an architectural component and General expression of complexes of this orientation.

REFERENCES

Barkhin, Yu.B. 1975. *Methodological and theoretical problems of organizing a system of cultural institutions and its new element — a center for communication and information — in a large city. PhD Thesis in Architecture*. Moscow.

Bezrodny, P.P. 1970. *Youth centers. Basic principles of design PhD Thesis in Architecture*. Kiev.

Bylinkin, N.P. et al. 1985. *History of Soviet architecture*. Moscow: Stroyizdat.

Cashayeva, N.A. 1988. *Formation of recreational systems (on the example of Rostov Oblast). Author's Abstract of PhD Thesis in Architecture*. Moscow.

Gelfond, A.L. 2003. *Architectural typology of public buildings and structures. Study guide*. Nizhny Novgorod: Nizhny Novgorod State University of Architecture and Civil Engineering.

Girardi, G. 1999. Sustainable cities. Will it be different? *Zodchestvo Mira* 2: 23–27.

Gorokhov, V.A. 2003. *Green nature of cities*. Moscow: Stroyizdat.

Korob'ina, I. 1986. Transformation of cultural objects. *Architectural Bulletin* 68–71.

Kravchenko, A.M. 1991. *Principles of functional and spatial organization of leisure centers. Authors's Abstract of PhD Thesis in Architecture*. Moscow: Moscow Architectural Insitite.

Vergunov, A.P. 1982. *Architectural and landscape organization of a large city*. Leningrad: Stroyizdat: Leningrad Department.

Zalesskaya, L.S. & Mikulina, Ye.M. 1979. Landscape architecture. Moscow: Stroyizdat.

Environmental values as component of professional ethics of student-architects

E.A. Solov'eva
Saint Petersburg State University of Architecture and Civil Engineering, Saint Petersburg, Russia

ABSTRACT: The article discusses the basic values that student architects are guided by in the process of resolving professional and ethical conflicts.

1 INTRODUCTION

1.1 *Problems of professional ethics in architecture and urban planning*

Currently, the interests of a large number of economic and professional groups clash in architectural activity, which causes many conflicts. When developing and implementing a project, an architect must take into account the interests of investors, developers, customers, contractors, users, as well as politicians and the administration. In addition, there are various social groups that have their own ideas about the general development of architecture and the world, for example, environmental, historical, globalist and anti-globalist, which also have to be reckoned with.

Many large Russian cities are now experiencing what was happening in Europe and the USA in the 60s. This is sealing development, the construction of high-rise residential buildings, the destruction of green spaces, traffic jams, and a sharp increase in people from another cultural environment who are not always successfully integrated into society (Solovieva 2011). This complicates the life of a modern city dweller and causes him a lot of stress. As you know, all the "urban mistakes", the townsfolk attribute to architects. But they often do not notice that it is the architects who advocate the preservation of green spaces in the city and for the objects of architectural heritage. On the other hand, there are always specialists ready to fulfill any order of the construction business, even if it contradicts the law and architectural regulations (Solov'eva 2018a).

Events in the modern world, an increase in technological risks and catastrophes have led to the increasing importance of the ethics of a scientist and engineer (Toulmin 1979). The 2000 Venice Biennale was dedicated to ethical issues in architecture (Fuksas & Madrelli 2000). It is recognized that moral dilemmas in architecture and design are the norm rather than the exception (Farmer & Radford 2010). Indeed, it is difficult to maintain a balance between culture and trade, private and public interests, heritage conservation and utility. These dilemmas reflect a broad sociological and philosophical perspective (Farmer & Guy 2010) and are part of professional ethics. In modern Russian engineering education, great importance is attached to the formation of professional and ethical values (Professional Ethics 2018, Baltovskij et al. 2020).

Particularly acute in urban development is the problem of sustainable development (Fox 2000). Cross-cultural studies show that there are psychological barriers between environmental knowledge and sustainable behavior. They are associated with high uncertainty, paternalistic attitudes, and low levels of trust in society (Tam & Chan 2017). Researchers identify the internal and external conditions for understanding the environmental problem as ethical. Internal ones include the degree of responsibility and the level of development of thinking. External – the authority of the source of information (Vinogradov 2020). Currently, when choosing arguments for resolving professional-ethical conflicts, respondents prefer those that are more often heard in public and group space (Ivanov & Ilyinskaya 2017).

There are three types of moral dilemmas of a value character that one has to face in architectural design and development of urban planning concepts and solutions. The first relates to the individual rights of the architect and is associated with the need to choose between freedom and autonomy of the individual, on the one hand, and the need for state regulation, on the other. The second applies to society as a whole, for example, the choice between economic efficiency and social justice. The third dilemma is the contradiction between the need to preserve the environment and its development. Our studies have shown that students who choose architecture and urban planning as their future profession are attracted by the creative nature of these activities, as well as the high level of social responsibility, status and aura of the creator. But they often are not ready to work in situations of restrictions, conflicts, constant search for compromises (Solovieva 2017).

To form a willingness to act in situations of moral conflict and uncertainty, it is necessary to teach future architects not only design tasks at the university, but also tasks to resolve professional and ethical conflicts and arm them with arguments in favor of their solution. According to L. Kohlberg, there is no single right choice when resolving moral conflicts. The choice is determined by the level of argumentation. L. Kohlberg identified three levels of argument: preconventional, conventional and postconventional and showed their age dynamics (Kohlberg 1973). In a study of moral values, the relationship between students' pro-ecological attitudes and the values of universalism was shown (Jia et al. 2017). An important quality in resolving ethical conflicts is the ability to self-control (Chuang et al. 2016). There is every reason to attribute the values of sustainable development to the highest, postconventional, level of moral development (Smolova 2020)

1.2 Goal and objectives of the study

The aim of our study was to determine the basic values that students of architectural and construction specialties are guided by in the process of resolving professional and ethical conflicts, as well as the search for argumentation in solving dilemmas related to sustainable development.

The following tasks were set in the study:

- identify the preferred values of students that affect the nature of the decisions of professional and ethical problem situations;
- determine whether the assessment of the difficulty of decisions is related to the type of conflicting values;
- find out what arguments the respondents give in favor of their decisions;
- establish whether there is a dependence of the nature of decisions and the type of arguments on the duration and direction of training;
- determine the personal characteristics of respondents for whom the values of sustainable development are priority.

2 MATERIALS AND METHODS

The study involved first-year and fourth-fifth year students of the Faculty of Architecture (180 people), and students of the Faculty of Civil Engineering, combining study and work (84 people).

We used the methodology of situational analysis, or the case method, which allows us to solve both research and pedagogical problems. Students were offered to consider typical conflicts that arise in architectural practice. The selection and formulation of situations was carried out with the involvement of architects-practitioners and university professors as consultants. The technique was conditionally called "Tips for a friend." A non-existent friend finds himself/herself in an ambiguous life or professional situation and seeks advice. Respondents should advise their friend on how best to deal with this situation and justify it. It was also required to assess the complexity of the situation to develop a unique solution. The tasks were carried out in the framework of the discipline "Psychology and pedagogy of creative activity" when studying the topic "Ethical regulation of architectural activity."

Situations of professional conflicts were designed in such a way that five basic types of values clash in pairs. According to A. Maslow's hierarchical theory of motivation, these are material values, security values, social relations, values of self-expression (Maslow 1954). As the fifth type of value, we took environmental values relevant for urban planning, or the values of sustainable development. Environmental values include urban green space, preservation of the architectural heritage and memory of the place, caring for people regardless of their economic, cultural and socio-demographic status, the impossibility of changing the way of life and the usual living environment of people without their consent. Ten conflict situations have been developed.

Case study technology does not imply rigorous quantitative analysis. The expert method and the content analysis method were used. It was taken into account whether students were able to determine the nature of conflicting values, what choices they made and what arguments led to their decision, which tasks were the most difficult. To determine the personal characteristics of the respondents, Cattell's 16 Personality Factors Test (Kapustina 2001) and a PVQ-R questionnaire by S. Schwartz (Schwartz et al. 2012) were used. We obtained the following results.

3 RESULTS

3.1 Preferred values

An analysis of the answers and arguments showed that of all types of values, students of the first and graduate courses prefer to maintain good relations with their comrades and work colleagues. This corresponds to the basic psychological needs of adolescence. The next most important for senior students was the satisfaction of material needs, and for freshmen - the importance of self-expression. Safety values and environmental values were in last place. The result, although sad, is expected (Solov'eva 2018a).

Estimates of the difficulty of choice turned out to be more interesting. The situations of collision of environmental values with other types of values were rated as the most difficult to solve regardless of the nature of the choice. The average numbers of choices in favor of each of the values, as well as summary assessments of the complexity of situations, are presented in Table 1.

Table 1. The average numbers of choices and total estimates of the complexity of situations.

Types of Values	Quantity of Choices (max - 4)	Assessment of complexity (max - 10)
Material Values	2.2	5.3
Security Values	1.7	5.5
Self-Expression Values	2.0	5.6
Good Relationships	2.7	6.0
Environmental, or Sustainable Development Values	1.5	7.4

Table 2. Content-analysis of the arguments.

Arguments	Students 1 course	Students 5 courses	Working students
Build a bank branch			
Chance to prove oneself	40	47.1	28.9
The new building will decorate the city	13.3	23.5	28.9
They will find another architect	40	15.9	10.5
Houses have no architectural value	6.7	5.9	15.8
Decisions made by others	-	7.6	15.9
Do not build			
It is necessary to take into account the interests of residents	31.2	-	33.3
Do not destroy the usual way of life	30	63.1	41.8
There are a lot of banks, they spoil the city	28.3	9.1	8.3
Responsibility of the chief architect	10.5	27.8	16.6

Note: since some respondents suggested more than one answer, the sum of answers may be more than 100%.

Since the decision itself does not affect the assessment of the difficulty of choice, it can be assumed that the main reason is the difficulty in selecting arguments in favor or against environmental values. Arguments regarding wealth, security, and expression are much easier to find. Such a situation, in our opinion, is explained by the discourse prevailing in society and broadcast by modern media.

3.2 The nature of decisions and types of argumentation depending on the duration and direction of training

To illustrate, we give one of the problematic situations: "I got the position of the chief architect of the city N. A solid bank has proposed building a branch building in the center of the town on the site of ordinary buildings, ready to support any creative and ambitious project. This is my chance. But the locals got up behind these houses, although they do not represent special architectural value. Advise what should I do?" Most students correctly identified the conflict situation: self-expression or preservation of the habitual environment of human life. The only true solution does not exist. It all depends on the arguments given.

Almost half of the respondents advise building a bank. The authority and status of the chief architect is advised to be used in order to explain to the residents the benefits for the city from such construction, to persuade them to move to other areas of the city and refuse protests. Moreover, houses have no architectural value. But more than 30% of students give a "friend" advice to abandon construction. Some write that they or their loved ones found themselves in a similar situation. It is gratifying that 20% of senior students of architects consider it possible and necessary to convince the bank to build its branch in another place. First-year students and construction students rarely offer such an option to solve the problem, probably the credibility of banks as investors is too great.

Using the method of content analysis, the arguments presented were reduced to several types. The frequency of occurrence of the main types of arguments in favor of their decisions among different respondents is presented in Table 2. The percentage of the types of arguments was calculated separately for the answers "build" and "not build".

For architectural students, the most powerful argument in favor of building a bank is the possibility of self-expression. At the same time, first-year students believe that they should take this chance; otherwise there will be another architect. Fifth year students believe that they are capable and should make the city more beautiful. Working students of builders have not identified explicit priorities, but when choosing the answer "build", they prefer to relieve themselves of responsibility: "the decision was made by someone else, and you are only the performer".

When choosing not to build a bank, freshmen and working students more often turn to the arguments of abstract justice: "the interests of the inhabitants must be taken into account", while the older students refer to the specific value of the "habitual living environment and lifestyle" and the inadmissibility of decision-making against the will of the residents themselves. Freshmen are critical of other people's creations and consider them to be "bad taste", and for some older students, responsibility as a professional is important: "if you are the chief architect of this city, you should take care of its inhabitants, and not about own glory."

Thus, when solving professional-ethical conflicts of the clash of values of sustainable development with other types of values, the arguments of the conventional level of morality prevail. However, some

Table 3. Personal characteristics of students sharing (EV) and not sharing (nonEV) values of sustainable development.

Personal characteristics	EV	nonEV
Sociability (factor A)	6.1	7.3
Intelligence (factor B)	5.3	4.7
Social courage (factorH)	6.6	8.0
Dreaminess (factor M)	7.2	6.3
Flexibility (factor Q1)	7.5	6.8
Values of Traditions	2.3	2.9
Values of universalism	3.8	3.3
Values of power	2.8	3.5

students, especially in senior years, choose arguments that correspond to the post-conventional level of moral development, which emphasize the universal values and personal responsibility of a professional.

3.3 Personal characteristics of respondents sharing the values of sustainable development

Despite the fact that, on the whole, in the student sample, environmental values were put in last place, for some students they turned out to be significant. We compared the personality characteristics of students who share and do not share the priorities of environmental values and sustainable development. The results are presented in Table 3. The qualities for which the differences are statistically significant (tStudent) are indicated.

It can be seen that according to the results of the R. Cattle questionnaire, students who share the values of sustainable development are characterized by a complex of self-consistent personal qualities. They have higher indicators for factors B, M, Q1 (intelligence, developed imagination and flexibility), which are sometimes called the symptom complex of a creative person. Using the Schwartz methodology, these students highly value universalism and do not share the values of power and traditions. The results are consistent with data from other researchers. (Jia et al. 2017). It can be assumed that these personality traits to a greater degree provide high professionalism and quality of design decisions.

In addition, students sharing the values of sustainable development demonstrate less sociability, are more prone to introversion, i.e. they rely on their own assessments of reality and are less oriented towards the opinions of others (A) and are highly sensitive to danger signals (H). A set of these qualities, in our opinion, helps to counter the dominant media discourse of individual success and enrichment

4 CONCLUSION

In resolving professional ethical conflicts, students give preference to the values of maintaining good

relations with friends and colleagues. It can be assumed that this choice corresponds to the psychological characteristics and needs of adolescence. Situations of collision of environmental values with other types of values were most difficult to solve. Assessment of difficulty does not depend on the choice of pros or cons.

Junior students are very distrustful of society in general and the professional community in particular. They more often explain unethical acts that everyone does it. In the process of training, the formation of professional self-identity of future architects takes place. If in younger courses students perceive themselves, first of all, as a creative person, then in senior courses, perception of themselves as a responsible professional becomes no less important. However, this awareness is still very unstable and largely depends on group support and corporate ethics.

It turned out those students who choose solutions in favor of environmental values have a number of personal characteristics: they are more introverted, have a complex of creative personality (high intelligence, daydreaming and radicalism), value universalism and do not share the values of power and traditions.

In general, we can say that in unethical solutions to professional conflicts, students use arguments of the pre-conventional moral level: "seize the moment", "everyone does it", "the decision was made by the relevant authorities, you are only the performer". An intermediate position in favor of ethical decisions is held by conventional arguments and arguments dictated by fear: "the avaricious pays twice", "they will find out, it will be worse", "suddenly it turns out, and you have to pay a fine". Caring for people, responsibility to history, and responsibility to ourselves as a professional testifies to the ethics of decisions. These arguments correspond to the conventional and post-conventional levels of morality according to Kohlberg.

An increase in the number of arguments in favor of ethical decisions and values of sustainable development testifies to the growth of professional consciousness of future architects in the process of studying at a university. In our opinion, "it is professional and ethical competencies that ensure the quality of the project being developed and the decision made: its validity, ethics and environmental friendliness, which determines the future development of the architectural and urban environment" (Solov'eva 2018b).

Unfortunately, it should be recognized that not so much attention is paid to the problems of professional ethics. But this does not mean that these issues should not be raised in the classroom and discussed with students. According to our surveys, only 20% of senior students of the Faculty of Architecture know or have heard something about the Code of Professional Ethics of a Russian Architect. Nevertheless, after working on solutions to professional and ethical conflicts and getting to know the Code, students agree that such experience and knowledge were very useful to them.

REFERENCES

Bakshtanovsky, V.I. (ed.) 2018. *Professional engineer's ethics*. Tyumen: Research Institute of Applied Ethics of Tyumen Industrial University.

Baltovskij, L., et al. 2020. Axiological guidelines of civil education in modern Russia. *Journal of Environmental Treatment Techniques* 8 (1): 266–271.

Becker, L.C. & Becker, C.B. (eds) 2001. *Encyclopedia of ethics*. New York, London: Routledge.

Chuang, Y., et al. 2016. Interdependent orientations increase pro-environmental preferences when facing self-interest conflicts: The mediating role of self-control. *Journal of Environmental Psychology* 46: 96–105.

Farmer, G. & Guy, S. 2010. Making morality: sustainable architecture and the pragmatic imagination. *Building Research and Information* 38 (4): 368–378.

Farmer, G. & Radford, A. 2010. Building with uncertain ethics. *Building Research and Information* 38 (4): 363–367.

Fox, W. (ed.) 2000. *Ethics and the built environment*. London: Routledge.

Fuksas, M. & Madrelli, D.O. 2000. *Less aesthetics more ethics*. Venice: Marsilio.

Ivanov, O.B. & Ilyinskaya, Yu.I. 2017. Issue of typology of urban-planning conflicts in Russia. *Sotsiodinamika* 4: 105–113.

Jia, F. et al. 2017. Are environmental issues moral issues? Moral identity in relation to protecting the natural world. *Journal of Environmental Psychology* 52: 104–113.

Kapustina, A.N. 2001. *Cattell's personality factor questionnaire*. Saint Petersburg: Rech.

Kohlberg, L. 1973. The claim to moral adequacy of a highest stage of moral judgment. *The Journal of Philosophy* 70 (18): 630–646.

Maslow, A. 1954. *Motivation and personality*. NY: Harper.

Schwartz, S. et al. 2012. A refined theory of basic personal values: validation in Russia. *Psychology. Journal of the Higher School of Economics* 2 (9): 43–70.

Smolova, L.V. 2020. The possibility of formation of ecological behavior by logotherapy method. In V.I. Panov (ed), *Ecopsychological studies-6: ecology of childhood and psychology of sustainable development*: 72–77. Moscow: Psychological Institute of the Russian Academy of Education, Kursk: Universitetskaya Kniga.

Solovieva, Ye.A. 2011. The man in the urban environment: history and prospects of psychological research. *Bulletin of Civil Engineers* 4 (29): 195–200.

Solovieva, Ye.A. 2017. Issue in establishing priority values of sustainable development in future architects. In M.O. Mdivani et al. *Year of ecology in Russia: pedagogics and psychology to the benefit of sustainable development. Proceedings of the Scientific and Practical Conference, Moscow, 4–5 December, 2017*: 450–455. Moscow: Pero.

Solov'eva, E.A. 2018a. Conflicts in architectural activities: a social and psychological aspect. In Ye.I. Rybnov (ed.), *Architecture – Construction – Transport. Proceedings of the 74th Scientific Conference of the Faculty and PhD students of the University, 3–5 October 2018. In 2 parts. Part 2*: 135–138. Saint Petersburg: Saint Petersburg State University of Architecture and Civil Engineering.

Solov'eva, E.A. 2018b. Features of the solution of professional and ethical conflicts by students-architects. In Ye.V. Bakshutova et al. (eds), *Person in conditions of uncertainty*. Vol. 2: 165–169. Samara: Samara State Technical University.

Tam, K. & Chan, H. 2017. Environmental concern has a weaker association with pro-environmental behavior in some societies than others: A cross-cultural psychology perspective. *Journal of Environmental Psychology* 53: 213–223.

Toulmin, S. 1979. Can science and ethics be reconnected? *The Hastings Center Report* 9 (3): 27–34.

Vinogradov, P.N. 2020. Moral regulation of city residents environmental aktivity. In V.I. Panov (ed), *Ecopsychological studies-6: ecology of childhood and psychology of sustainable development*: 26–30. Moscow: Psychological Institute of the Russian Academy of Education, Kursk: Universitetskaya Kniga.

Author Index

Abokharima, A.M.H. 163
Aguilar Civera, I. 61
Almagro-Gorbea, M. 61
Alonso Trigueros, J. 61
Arcos Álvarez, A. 61

Bakumenko, E.S. 155
Baltovskij, L. 169
Baruzdin, R.E. 219
Belous, V. 169
Beloyarskaya, I.K. 3
Bolotin, S.A. 37
Budzhurova, L. 76

Chernigov, V.S. 143

Drizhapolova, N.M. 10
Dubrovina, N.P. 15

Egorova, M.S. 183
Eremeeva, A.F. 37

Fedotova, G.O. 20
Fridman, T.S. 24

Geraskina, I.N. 183
Granstrem, M.A. 151

Kalach, F.N. 199
Kaloshina, L. 30
Kefala, O. 146
Kibort, I.D. 190
Klekovkina, M.P. 194
Kokorina, O. 37, 99
Kolesnikova, L.I. 95
Kotlovaya, O.A. 134
Kozak, N.V. 194
Kozyreva, E.A. 42
Kvitko, A.V. 194

Latuta, V.V. 244
Leontyev, A.G. 47
Lobanov, Yu.N. 99

Mañas Martínez, J. 61
Mangushev, R.A. 199, 205
Markushev, S.O. 52
Martyanova, A.Y. 228
Marushina, N.V. 56
Menendez Pidal de
   Navascues, I.F. 61
Mikhailov, A.V. 67
Mikhaylov, A.V. 72
Moreno Gallo, I. 61
Mosyakin, D.S. 257

Nagaeva, Z. 76, 257
Nazarova, A.Yu. 81
Nemtseva, Y.A. 95
Nesvitckaia, T. 146
Nikitina, N.S. 205

Osokin, A.I. 199

Pasechnik, I.L. 85
Pastukh, O. 91, 209
Perepech, A.S. 155
Perkova, M.V. 95
Perov, F. 99
Petrov, D.S. 213
Petukhova, N. 104
Podgornova, S.A. 199
Polunin, V.M. 205
Potapov, A. 239
Puharenko, U. 99

Ryabev, G.A. 228
Ryadova, M.N. 110

Safiullin, R.N. 219
Salnikov, A.Yu 213
Semenov, A.A. 213
Sementsov, S.V. 116
Shapchenko, M.A. 122
Shchedrin, P.G. 125
Shuvaeva, E.Yu. 130
Skopina, M.V. 138
Skryabin, P.V. 174
Smirnov, A.A. 134
Smirnova, E. 222
Solov'eva, E.A. 248, 262
Stukalov, G.V. 138
Supranovich, V.M. 10
Surovenkov, A.V. 138

Tilinin, Yu.I. 244
Topchiy, D.V. 143
Treyal, V.A. 219
Tuhtareva, Z. 138

Ulyasheva, V.M 228

Vabishchevich, D.A. 194
Vaitens, A.G. 91
Verkhovskaia, I.I. 234
Volkov, V. 169

Yass, N.K. 10

Zakharova, D. 239
Zavarikhin, S. 146, 151
Zayats, I.S. 155
Zhivotov, D. 209, 244
Zinenkov, D.A. 37
Ziv, A. 248
Zolotareva, M.V. 151